About Island Press

Island Press is the only nonprofit organization in the United States whose principal purpose is the publication of books on environmental issues and natural resource management. We provide solutions-oriented information to professionals, public officials, business and community leaders, and concerned citizens who are shaping responses to environmental problems.

Since 1984, Island Press has been the leading provider of timely and practical books that take a multidisciplinary approach to critical environmental concerns. Our growing list of titles reflects our commitment to bringing the best of an expanding body of literature to the environmental community throughout North America and the world.

Support for Island Press is provided by the Agua Fund, The Geraldine R. Dodge Foundation, Doris Duke Charitable Foundation, The Ford Foundation, The William and Flora Hewlett Foundation, The Joyce Foundation, Kendeda Sustainability Fund of the Tides Foundation, The Forrest & Frances Lattner Foundation, The Henry Luce Foundation, The John D. and Catherine T. MacArthur Foundation, The Marisla Foundation, The Andrew W. Mellon Foundation, Gordon and Betty Moore Foundation, The Curtis and Edith Munson Foundation, Oak Foundation, The Overbrook Foundation, The David and Lucile Packard Foundation, Wallace Global Fund, The Winslow Foundation, and other generous donors.

The opinions expressed in this book are those of the authors and do not necessarily reflect the views of these foundations.

About Ecoagriculture Partners

Ecoagriculture Partners is an international non-profit organization dedicated to supporting rural communities to produce food and enhance their livelihoods while protecting the biological diversity of plant and animal life, and to educate policymakers, institutions, and innovators about ecoagriculture management approaches.

Ecoagriculture Partners was formally established at the World Summit for Sustainable Development in Johannesburg, South Africa in 2002. It operated through 2004 as a joint initiative of the World Conservation Union (IUCN), Forest Trends, and the World Agroforestry Centre (ICRAF), and was incorporated as a non-profit organization in January 2005.

The vision of Ecoagriculture Partners is a world with widespread agricultural landscapes that support rural livelihoods and sustainable production of crops, livestock, fish, and forests, while also conserving biodiversity, watersheds, and ecosystem services on a globally significant scale. Ecoagriculture Partners seeks to strengthen collaboration and knowledge exchange among ecoagriculture innovators working in different locations, farming systems, and sectors, and to mobilize the scaling up of successful ecoagriculture approaches by catalyzing strategic connections, dialogue, and joint action among key actors at local, national, and international levels.

About the IUCN–World Conservation Union

Founded in 1948, the World Conservation Union brings together states, government agencies, and a diverse range of nongovernmental organizations in a unique world partnership of over 1000 members in all, spread across some 140 countries.

As a union, IUCN seeks to influence, encourage, and assist societies throughout the world to conserve the integrity and diversity of nature and to ensure that any use of natural resources is equitable and ecologically sustainable.

The World Conservation Union builds on the strengths of its members, networks, and partners to enhance their capacity and to support global alliances to safeguard natural resources at local, regional, and global levels.

FARMING WITH NATURE

The Science and Practice of Ecoagriculture

FARMING WITH NATURE

The Science and Practice of Ecoagriculture

Sara J. Scherr
Jeffrey A. McNeely

Library of Congress Cataloging-in-Publication Data

Scherr, Sara J.
Farming with nature : the science and practice of ecoagriculture / Sara J. Scherr, Jeffrey A. McNeely.
p. cm.
Includes bibliographical references and index.
ISBN-13: 978-1-59726-127-2 (cloth : alk. paper)
ISBN-10: 1-59726-127-0 (cloth : alk. paper)
ISBN-13: 978-1-59726-128-9 (pbk. : alk. paper)
ISBN-10: 1-59726-128-9 (pbk. : alk. paper)
1. Sustainable agriculture. 2. Biodiversity conservation. I. McNeely, Jeffrey A. II. Title.
S494.5.S86S34 2007
630—dc22

2006100959

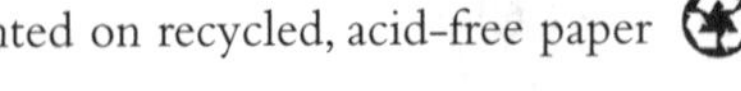

Printed on recycled, acid-free paper

Manufactured in the United States of America
10 9 8 7 6 5 4 3 2 1

This book is dedicated to the myriad people around the world who have begun to work together, across borders and across sectoral divides, to create, protect, promote, and understand ecoagriculture landscapes.

We believe that mobilizing a movement of diverse stakeholders inspired and committed to ecoagriculture and the improvement of rural livelihoods, together with preservation and restoration of ecosystem services, will build synergies and achieve globally significant benefits for food security, human health and nutrition, poverty alleviation, and environmental sustainability.

From the "Nairobi Declaration on Ecoagriculture"
Nairobi, Kenya, 1 October 2004

Contents

List of Figures, Tables, and Boxes

Figures

Tables

Boxes

Acknowledgments

This book would not have been possible without the strong support of many of our colleagues, beginning with Barbara Rose, who first brought us together as a team. Dennis Garrity of the World Agroforestry Centre (ICRAF) helped to organize and host the International Ecoagriculture Conference and Practitioners' Fair in Nairobi in September 2004, where many of the ideas in this volume were first presented and discussed. Michael Jenkins and Forest Trends hosted Ecoagriculture Partners throughout the preparation of the book.

We also sincerely thank the Gordon and Betty Moore Foundation and the World Bank Development Grants Facility, which generously supported the Ecoagriculture Partners staff involved in preparing the manuscript.

More than 60 authors—leaders in diverse fields essential to ecoagriculture—contributed to this volume. They were able to synthesize and draw lessons from literature and experience in their own fields, and make this accessible to a broad audience. Both editors are grateful for their generous input and collegiality, and for helping us understand the core advances in science and practice that are turning the concept of ecoagriculture into a reality.

Jeff particularly thanks Wendy Price for her diligence in converting many of the manuscripts into polished final products, and for helping to assure some stylistic consistency throughout the volume. Sara acknowledges the Ecoagriculture Partners staff for their invaluable input to the book: Claire Rhodes for overall strategic input, Ariela Summit for masterfully managing the final edits and formatting of a complex manuscript, and Seth Shames for editing support.

We are grateful for the valuable input of Todd Baldwin, Jessica Heise and colleagues at Island Press in producing this volume. Finally, for moral support throughout this process, Sara warmly thanks Alan Dappen and Jeff thanks Pojanan Suyaphan McNeely.

Preface

We first coined the term "ecoagriculture" back in 2001, to refer to landscapes that achieve the joint objectives of sustainable agricultural production, biodi versity and ecosystem conservation, and rural livelihoods (McNeely and Scherr 2001). At that time, we had little premonition about how the concept was going to develop. Our more complete discussion on the topic was published by Island Press (McNeely and Scherr 2003) and generated considerable interest, even controversy in some quarters. But more important, it led to significant follow-up activities, including the International Ecoagriculture Conference and Practitioners' Fair held in Nairobi in September 2004.

That meeting brought together over 200 of the world's leading innovators in ecoagriculture, including community leaders, farmers, conservationists, policymakers, researchers, technical advisers, and land-use planners. Our intention was to assess the state of ecoagriculture knowledge and practice and to develop a strategy to promote and support ecoagriculture development around the world. A conference proceedings report was published (Nairobi Declaration 2004), along with the *Nairobi Declaration*.

One outcome of the Nairobi Conference was a renewed and strengthened confidence that ecoagriculture was sufficiently mature to justify the establishment of an institutional structure to promote it further. A new nongovernmental organization known as Ecoagriculture Partners was legally incorporated in 2005 and is now engaged in activities in various parts of the world to support ecoagriculture innovators (www.ecoagriculture.org). But it was also clear that the concept of ecoagriculture needed further development, which led to the preparation of this book.

We asked many of the participants in the Nairobi conference to do additional work on their contributions, commissioned additional papers where we felt that ecoagriculture-related issues had not yet received appropriate attention, and drew on other ecoagriculture-related analyses and experiences to compile the work that you will find on the following pages.

In publishing this book, we hope to establish a baseline as of 2006 about the current science and practice of ecoagriculture, while recognizing that in a time of dynamic social, economic, ecological, and political change, the concepts will inevitably evolve. We anticipate that ecoagriculture innovators will adopt the concept of "adaptive management" put forward by Hans Herren and his colleagues in this book, learning lessons in a structured way and applying those lessons to new challenges.

Sara J. Scherr
Jeffrey A. McNeely
May 2007

References

McNeely, J.A. and S. Scherr. 2001. *Common Ground, Common Future: How Ecoagriculture Can Help Feed the World and Save Wild Biodiversity.* A Joint Publication with IUCN and Future Harvest. IUCN, Washington, DC.

McNeely, J.A. and S. Scherr. 2003. *Ecoagriculture: Strategies to Feed the World and Save Wild Biodiversity.* Island Press, Washington, DC.

Nairobi Declaration on Ecoagriculture. 2004. International Ecoagriculture Conference, Nairobi, Kenya. September 27–October 10, 2004. http://www.ecoagriculture partners.org/whatis/nairobideclaration.htm.

Chapter 1

The Challenge for Ecoagriculture

Sara J. Scherr and Jeffrey A. McNeely

Agriculture dominates land and water use like no other human enterprise, with landscapes providing critical products for human sustenance. Yet because of their predominance, agricultural landscapes must also support wild species biodiversity and ecosystem services (MA 2005). Moreover global demand for associated agricultural products is projected to rise at least 50% over the next two decades (UN Millennium Project 2005). These conflicting trends are prompting farmers and policymakers alike to identify innovative ways of reconciling agricultural production and production-dependent rural livelihoods with healthy ecosystems (Acharya 2006; Breckwoldt 1983; Jackson and Jackson 2002; McNeely and Scherr 2003). Unfortunately, the dominant national and global institutions for policy, business, conservation, agriculture, and research have been shaped largely by "mental models" that assume, and even require, segregated approaches.

During the 21st century, a continuing and growing demand for agricultural and wild products and ecosystem services will require farmers, agricultural planners, and conservationists to reconsider the relationship between production agriculture and conservation of biodiversity.

This chapter introduces a new paradigm, ecoagriculture, defined as integrated conservation–agriculture landscapes where biodiversity conservation is an explicit objective of agriculture and rural development, and the latter are explicitly considered in shaping conservation strategies. The rationale for scaled-up action to promote ecoagriculture landscapes, and the defining characteristics of this new approach, are developed further in this book.

The Current Ecological Footprint of Agriculture

Nearly a third of the world's landmass has agricultural crops or planted pastures as a dominant land use (accounting for at least 30% of total area), which has a profound ecological effect on the whole landscape. Another quarter of land is under extensive livestock grazing, and approximately 1 to 5% of food is produced in natural forests (Wood et al. 2000). The "human footprint" analysis of Sanderson et al. (2002) estimated that 80 to 90% of lands habitable by humans are affected by some form of productive activity. More than 1.1 billion people—most directly dependent on agriculture—live within the world's 25 biodiversity "hotspots," areas described by ecologists as the most threatened species-rich regions on Earth (Cincotta and Engelman 2000; Myers et al. 2002).

Both extensive lower-yield and intensive higher-yield agricultural systems have profound ecological effects. Millions of hectares of forests and natural vegetation have been cleared for agricultural use and for harvesting timber and wood fuels. Half the world's wetlands have already been converted for production (MA 2005). Overuse and mismanagement of pesticides poison water and soil, while nitrogen and phosphorus inputs and livestock wastes have become major pollutants of surface water, aquifers, and coastal wetlands and outlets. Between 1890 and 1990, the total amount of biologically available nitrogen created by human activities increased ninefold, and human activity now produces more nitrogen than all natural processes combined (MA 2005). Agrochemical nutrient pollution from the US farm belt is the principal cause of the biological "dead zone" in the Gulf of Mexico 1500 km (932 miles) away (Rabalais et al. 2002), and similar impacts are felt in the Baltic Sea and along the coasts of China and India. Water supplies and quality for major urban centers and industries are threatened by poor soil and vegetation management in agricultural systems in their watersheds.

Some introduced agricultural crops, livestock, trees, and fish have become invasive species, spreading beyond their planned range and displacing native species (Matthews and Brand 2004; Mooney et al. 2005). Additionally, there are concerns about genetically modified crop varieties potentially becoming invasive species or hybridizing with wild relatives and leading to a loss of biodiversity (Omamo and von Grebmer 2005; NRC 2002; Oksman-Caldentey and Barz 2002). On a broad scale, agriculture fragments the landscape, breaking formerly contiguous wild species populations into smaller units more vulnerable to extirpation. Farmers have generally sought to eliminate wild species from their lands, seeking to reduce the negative effects of pests, predators, and weeds. However, these practices often harm beneficial wild species like pollinators (Buchmann and Nabhan 1996), insect-eating birds, and other species that prey on agricultural pests.

The threats posed by agriculture have been a key motivator for conservationists to develop protected areas where agricultural activity is officially excluded or seriously limited. Nonetheless, the Millennium Ecosystem Assessment (MA) Hassan et al. 2005 calculated that more than 45% of 100,000 protected areas had more than 30% of their land area under crops. In light of political and economic realities, many recently designated protected areas in several African countries explicitly permit biodiversity-friendly agriculture, usually in areas considered category V or VI in the World Conservation Union (IUCN) system (IUCN 1994).

As populations and economies grow around the world, meeting increased demand for both agricultural products and ecosystem services will require that many agricultural landscapes be managed through ecoagriculture approaches.

Meeting Increased Demand for Agricultural Products in Ecologically Sensitive Areas

Human population is expected to grow from a little over 6 billion today to over 8 billion by 2030, an increase of about a third, with another 2 to 4 billion added in the subsequent 50 years (Cohen 2003). Food demand is expected to grow even faster as a result of growing urbanization and rising incomes (OECD-FAO 2005), and assuming hunger is reduced among the over 800 million people currently undernourished (UN Millennium Project 2005). More land will surely be required to grow crops, even more so if biofuels become a greater contributor to energy needs. In Africa alone, land in cereal production is expected to increase from 102.9 million ha in 1997 to 135.3 million ha in 2025 (Rosegrant et al. 2005). Global consumption of livestock products is predicted to rise from 303 million metric tons (t) in 1993 to 654 million t in 2020 (Delgado et al. 1999).

Tilman (2001) predicts that feeding a population of 9 billion using current methods would mean converting another 1 billion ha of natural habitat to agriculture, primarily in the developing world, together with a doubling or tripling of nitrogen and phosphorous inputs, a twofold increase in water consumption, and a threefold increase in pesticide use. A serious limiting factor will be water, because 70% of the freshwater used by people is already devoted to agriculture (Rosegrant et al. 2002). Scenarios prepared by the MA thus suggest that agricultural production in the future will likely have to focus more explicitly on ecologically sensitive management systems (Carpenter et al. 2005).

There are four major reasons why meeting increased demand for agricultural products will often require ecoagriculture systems (Scherr and McNeely 2007).

Most of the Increased Food Production Will Be Grown Domestically and in More "Marginal" or "Fragile" Lands

An estimated 90% of food products consumed within most countries will be produced by those same countries. Total agricultural exports increased sharply between 1961 and 2000, but exports still accounted for only about 10% of production (McCalla 2000). A reduction in developed world subsidies and growing demand from China and India could further spur export agriculture in the developing world (Runge et al. 2003). The general pattern of increased trade with most production for domestic markets seems unlikely to change over the next few decades, even though continuing globalization of agriculture will influence product mix and prices. Changes will depend not only on productivity and quality but also on shifts in relative costs for international shipping and internal overland transport. In addition the distances that need to be covered between major population centers and ports and agricultural regions populations fluctuate, and new centers emerge. Interior populations in large countries will continue to be fed mainly by local and national producers.

The declining rate of growth in agricultural yields in places like the Punjab in India, the US Midwest, and the Mekong Delta indicate that most new production may *not* come from the areas of highest current grain productivity, and some areas are already experiencing declining yields or productivity of inputs (Rosegrant et al. 2002). Although yields in these places may increase through greater input use, plant breeding, biotechnology, and improved irrigation efficiency (Runge et al. 2003), economic and environmental costs are likely to be high.

Lower-productivity lands (drylands, hillsides, rainforests) now account for more than two-thirds of total agricultural land in developing countries (Nelson et al. 1997). Because current yields are relatively low, technologies that already exist can double or even triple current yields, provided adequate investments, market developments, and attention are given to good ecosystem husbandry (UN Millennium Project 2005). Extensive grain monocultures are not likely to be environmentally sustainable in such areas, calling for more diversified land-use approaches. Though the bulk of new production will come mainly from existing croplands, the most promising areas with significant new land for agriculture are in places like the forest and savanna zones of Brazil and Mozambique. These places are also the main remaining large reservoirs of natural habitat in the world. These habitats would be seriously damaged by simplified, high-external-input production systems, but an ecoagriculture approach could both provide a means to increase food production and retain the natural value of the landscape.

Wild Products Will Continue to Be Important for Local Food Supply and Livelihoods

People in low-income developing countries and subregions will continue to rely on harvesting wild species. Wild greens, spices, and flavorings enhance local diets, and many tree fruits and root crops serve to assuage "preharvest hunger" or provide "famine foods" when the economy or crops fail. Frogs, rodents, snails, edible insects, and other small creatures have long been an important part of the rural diet in virtually all parts of the world (Paoletti 2005). Bushmeat is the principal source of animal protein in humid West Africa and other forest regions, and efforts to replace these with domestic livestock have been disappointing. Fisheries are the main animal protein source of the poor worldwide. In Africa and many parts of Asia, more than 80% of medicines still come from wild sources. Gathered wood remains the main fuel for hundreds of millions of people, while forests and savannas provide critical fodder, soil nutrients, fencing, and other inputs for farming (McNeely and Scheer 2003). Achieving security in food and livelihood will therefore require the conservation of the ecosystems providing these wild foods and other products.

Agricultural Systems Will Need to Diversify to Adapt to Climate Change

Strategic planning for agricultural development increasingly focuses on adaptation of systems to climate change, anticipating rising temperatures and more extreme weather events. The US Department of Agriculture and the International Rice Research Institute have both concluded that with each 1°C increase in temperature during the growing season, the yields of rice, wheat, and maize drop by 10% (Brown 2004; Tan and Shibasaki 2003). Cash crops such as coffee and tea, requiring cooler environments, will also be affected, forcing farmers of these crops to move higher up the hills, clearing new lands as they climb, meaning that montane forests important for biodiversity are likely to come under increasing threat. Effective responses to climate change will require use of alternative seed varieties, modified management of soils and water, and new strategies for pest management as species of wild pests, their natural predators, and their life cycles change in response to climates. Increasing landscape- and farm-scale diversity is likely to be an important response for risk reduction (Diversitas 2002).

Agricultural Sustainability Will Require Investment in Ecosystem Management

The ability to meet food needs and economic demand for agricultural products will be constrained by widespread natural resource degradation that is already either reducing supply or increasing costs of production. Up to 50% of the globe's agricultural land and 60% of ecosystem services are now affected to some degree by land or water degradation, with agricultural land use the chief cause (MA 2005; Pretty et al. 2006). Half the world's rivers are seriously depleted and polluted, and 60% of the world's 227 largest rivers have been fragmented by dams, many built to supply irrigation water. Up to 20% of irrigated land suffers from secondary salinization and waterlogging, induced by the buildup of salts in irrigation water (Wood et al. 2000). The food system will also have to confront the collapse in harvests of wild game and wild fisheries in many regions around the world due to overexploitation and habitat loss or pollution (Hassan et al. 2005). Considerable investments will be required to rehabilitate degraded resources and ecosystems upon which food supplies, particularly those of the rural poor, depend (UN Millennium Project 2005).

Meeting Increased Demand for Ecosystem Services

Many consider conservation of wild biodiversity (genes, species, and ecosystems) to be an ethical imperative. Conservation also supports ecological processes and functions that sustain and improve human well-being, known collectively as ecosystem services (Daily 1997). Ecosystem services can be divided into four categories: (1) *provisioning services,* providing food, timber, medicines, and other useful products; (2) *regulating services,* such as flood control and climate stabilization; (3) *supporting services,* such as pollination, soil formation, and water purification; and (4) *cultural services,* including aesthetic, spiritual, or recreational assets that provide both intangible and tangible benefits such as ecotourism attractions (Kremen and Ostfeld 2005). "Provisioning" has historically been seen as the highest-priority service provided by agricultural landscapes. But it is now recognized that even the "bread baskets" and "rice bowls" of the world also provide other ecosystem services, such as water supply and quality, or pest and disease control, that are critically important (Wood and Scherr 2000).

Agricultural Landscapes Provide Critical Habitat

The conservation community is moving toward an "ecosystem approach" to conserving biodiversity, in light of the dependence of protected areas on a supportive matrix of land and water use, and creation of biological corridors (CBD 2000). The international community has set a goal of having at least 10% of every habitat type under effective protection by 2015 (The Nature Conservancy 2004). This strategy, if successful, will protect many species and ecological communities. But some estimates suggest that more than half of all species exist principally outside protected areas, mostly in agricultural landscapes (Blann 2006). For example, conservation of wetlands within agricultural landscapes is critical for wild bird populations (Heimlich et al. 1998). Protecting such species requires initiatives by and with farmers. The concept of agriculture as ecological "sacrifice" areas is no longer valid in many regions because agricultural lands both perform services and provide essential habitat to many species. Thus the Convention of Biological Diversity agreed in 2002 to aim for 30% of agricultural lands worldwide to be managed to protect wild flora by 2010 (CBD 2002).

Agricultural Landscapes Provide Critical Watershed Functions

Many of the world's most important watersheds are densely populated and farmed, and most others are landscape mosaics where crop, livestock, and forest production influence hydrological systems (Wood et al. 2000). In such regions, agriculture can be managed for critical watershed functions, such as maintaining water quality, regulating water flow, recharging underground aquifers, mitigating flood risks, moderating sediment flows, and sustaining freshwater species and ecosystems. This has led to the concept of "green water"—an understanding that terrestrial land, soil, and vegetation management have critical roles in the hydrological cycle (Penning de Vries et al. 2003). Effective management of green water means using water-conserving crop mixtures, managing soil and water (including irrigation), maintaining soils to facilitate rainfall infiltration, creating vegetation barriers to slow movement of water down slopes, ensuring year-round soil vegetative cover, and maintaining natural vegetation in riparian areas, wetlands, and other strategic areas of the watershed. Well-managed agricultural landscapes can also provide protection against extreme natural events. With increased water scarcity and more frequent extreme weather events predicted in coming decades, the capacity of agricultural systems to sustain watershed functions is likely to be a priority consideration in agricultural investment and management.

Agricultural Landscapes Maintain "Green Space," Recreational Opportunities, Healthy Habitats, and Aesthetic Beauty in Human Settlements

With accelerating urbanization worldwide, the loss of natural habitats and natural features has become a central concern for planners and residents as well as for farmers operating in periurban areas. Agriculture can protect green spaces for aesthetic and recreational values and can help to finance the maintenance of green space for wildlife habitat and ecosystem services. Overall positive outcomes for human habitat and aesthetics require adequate management of crop and livestock wastes, air pollution (smoke, dust, odors), and polluting runoff.

Ecoagriculture: Integrating Production and Conservation at a Landscape Scale

The challenges described in this chapter are unlikely to be met by the solutions advocated most widely today: industrial agriculture, the Green Revolution, sustainable agriculture and natural resource management (with its important but limited focus on sustaining the resources underpinning production), or even agroecology or ecotechnology approaches Swaminathan (1994) (with their focus on the farmer's field), although all of these have major elements to contribute. Approaches to biodiversity conservation also need to move beyond the wild biodiversity focus of strictly protected areas and the modest goals of integrated conservation and development projects. Rather, many regions need ecoagriculture—a fully integrated approach to agriculture, conservation, and rural livelihoods, within a landscape or ecosystem context.

Ecoagriculture explicitly recognizes the economic and ecological relationships and mutual interdependence among agriculture, biodiversity, and ecosystem services (Fig. 1.1). Effective ecoagriculture systems rely on maximizing ecological, economic and social synergies among them, and minimizing the conflicts.

The term "landscape" itself is functionally defined, depending upon the spatial units needed or managed by the group of stakeholders working together to achieve biodiversity, production, and livelihood goals. Ecoagriculture landscapes are land-use mosaics consisting of the following:

- "Natural" areas (high-quality habitat niches to ensure ecosystem services that cannot be provided in areas under production), which are also managed to benefit agricultural livelihoods, either through positive synergies with pro-

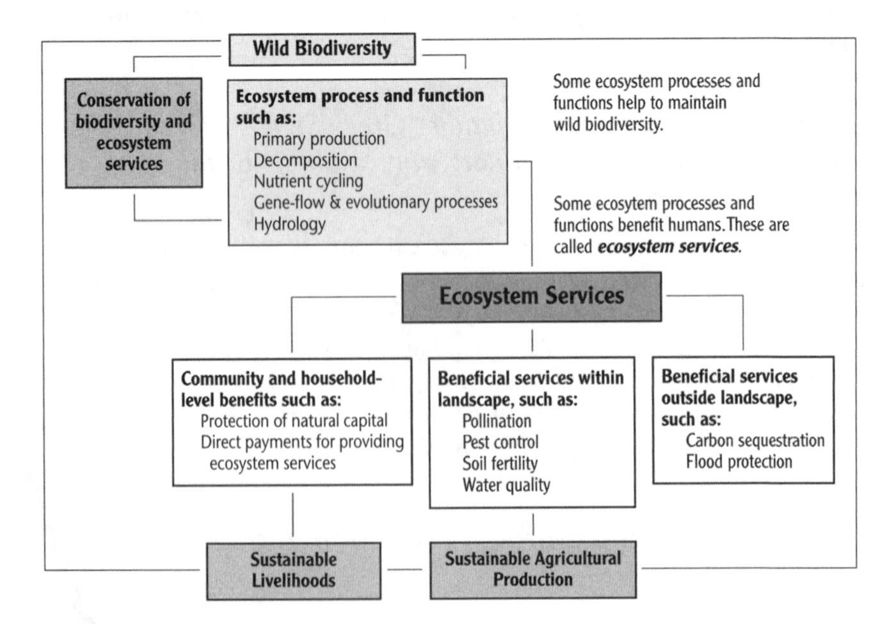

Figure 1.1. Links between the goals of ecoagriculture and ecosystem services (Buck et al. 2006)

duction or through providing other livelihood benefits such as firewood or clean water

- Agricultural production areas (productive, profitable, and meeting food security, market, and livelihood needs), which are also configured and managed to provide a matrix with benign or positive ecological qualities for wild biodiversity and ecosystem services
- Institutional mechanisms to coordinate initiatives to achieve production, conservation, and livelihood objectives at landscape, farm, and community scales, by exploiting synergies and managing trade-offs among them

The concept of ecoagriculture recognizes that agriculture-dependent rural communities are important (and often the principal) stewards of biodiversity and ecosystem services. Although protected "natural" areas are essential in ecoagriculture landscapes to ensure critical habitat for vulnerable species, maintain water sources, and provide cultural resources, these resources may often be owned or managed by local communities and farmers.

Biodiversity and Ecosystem Services in Ecoagriculture Landscapes

Conservation of biodiversity in ecoagriculture landscapes embraces all three elements of agricultural biodiversity defined by the Convention on Biological Diversity (CBD): genetic diversity of domesticated crops, animals, fish, and trees; diversity of wild species on which agricultural production depends (such as wild pollinators, soil microorganisms, and predators of agricultural pests); and diversity of wild species and ecological communities that use agricultural landscapes as their habitat (CBD 2002).

Although wild biodiversity and ecosystem services are closely linked, they are not synonymous. A landscape with a high degree of wild biodiversity is likely to provide many ecosystem services. However, ecosystem services can also be provided by nonnative species, or by combinations of native and nonnative species in heavily managed settings such as permanent farms. Even where wild biodiversity has been significantly reduced to make way for food and fiber production, high levels of ecosystem services can still be provided through land management practices. On the other hand, managing an ecoagriculture landscape for ecosystem services does not automatically ensure that wild biodiversity will be protected adequately. Thus wild biodiversity and ecosystem services both require explicit consideration in ecoagriculture systems.

Ecoagriculture Approaches

Broadly, ecoagriculture landscapes rely on six basic strategies of resource management, three focused on the agricultural parts of the landscape and three on natural areas. In production areas, farmers can achieve "intensification without simplification" by increasing agricultural outputs and reducing costs in ways that enhance the habitat quality and ecosystem services. Successful techniques include:

- Minimizing agricultural wastes and pollution
- Managing resources in ways that conserve water, soils, and wild flora and fauna
- Using crop, grass, and tree combinations to mimic the ecological structure and function of natural habitats

Farmers or other conservation managers protect and expand natural areas in ways that also provide benefits for adjacent farmers and communities by:

- Minimizing or reversing conversion of natural areas
- Protecting and expanding larger patches of high-quality natural habitat
- Developing effective ecological networks and corridors (McNeely and Scherr 2003)

The relative area and spatial configuration of agricultural and natural components (as well as physical infrastructure and human settlements) are key landscape design issues (Forman 1995). The conservation of wild species sensitive to habitat disturbance, including some of those most endangered or rare globally, requires large, well-connected patches of natural habitat. But many wild species, including some that are threatened and endangered, can coexist in compatibly managed agricultural landscapes, even in high-yielding systems. Ecosystem services can also be considered under a wide range of land uses, under good management.

The outcomes of planning and negotiations among multiple stakeholders in any particular landscape will take diverse forms depending on the context of local cultures and philosophies of land management. Identifying and managing potential synergies and trade offs at different scales will be essential, and calls for new approaches to knowledge-sharing organization and research. Examples of ecoagriculture landscapes with documented joint benefits for agricultural production, biodiversity conservation, and rural livelihoods include the following.

Kalinga Province, Philippines

For centuries, the Kalinga indigenous peoples of the Philippines have supported themselves and conserved mountain biodiversity through integrated landscape management. Communities manage watersheds to ensure a continual supply of water to communal irrigation systems, and in recent years over 150 ha of integrated rice terraces (including fish and vegetable production) have been rehabilitated. They manage indigenous forests for sustainable harvest of wild animals for protein, leading to an 81% rate of intact forest in Kalinga Province (Gillis and Southey 2005).

Transboundary Comanagement in Costa Rica and Panama

The Grandoca-Manzanillo National Wildlife Refuge on Costa Rica's Caribbean coast connects with Panama's San Pondsak National Wildlife Refuge. This 10,000 ha refuge is comanaged by local communities, nongovernmental organizations, and government agencies. Small farm agroecosystems are integral to regional biodiversity conservation. Over 300 farmers hold secure land titles in the refuge's buffer zone. A regional small farmers' cooperative (Smallholder Association of Talamanca) supports over 1500 small farmers, making it Central America's largest-volume organic producer and exporter, generating 15 to 60% increases in small-farmer revenue. Conservation-based carbon offset schemes are being developed to provide additional revenue for stewardship-focused farming.

Community Dryland Restoration in Rajasthan, India

For most of the past century, drought and environmental degradation severely impaired the livelihood security of local communities within Rajasthan's Avari Basin. Twenty years ago, the Tarun Bharat Sangh, a voluntary organization based in Jaipur, India, initiated a community-led watershed restoration program. The program reinstated *johads*, a traditional indigenous technology for water harvesting. Johads are simple concave mud barriers, built across small, uphill river tributaries to collect water. As the water drains through the catchment area, johads encourage groundwater recharge and improve hillside forest growth while providing water for irrigation, wildlife, livestock, and domestic use. More than 5000 johads now serve over 1000 villages in the region and are coordinated by village councils. Landscape changes include restoration of the Avari River, which had not flowed since the 1940s, and the return of native bird populations (Narain et al. 2005).

Where Ecoagriculture Approaches Are Needed

Ecoagriculture approaches can be relevant to all agricultural landscapes, in light of their focus on improving landscape performance vis-à-vis three goals (agricultural production, biodiversity conservation, and livelihoods). Synergies may be most apparent, and trade-offs least difficult, in areas with less productive agricultural lands (where the opportunity costs of protecting or restoring habitats are lower), and in heterogeneous areas where farms are already interspersed with hills, forests, and abandoned farms (Jackson and Jackson 2002). Nonetheless, the need to reconcile increased agricultural productivity and livelihoods with effective conservation of biodiversity and ecosystem services is widely found in both high- and low-income countries. Ecoagriculture approaches offer opportunities for integrated action, at a lower overall cost, to achieve Millennium Development Goals for poverty, hunger, water and sanitation, and environmental sustainability (Rhodes and Scherr 2005). Ecoagriculture also provides a strategy for implementing national commitments to multilateral environmental conventions, including the CBD, the Framework Convention on Climate Change, Ramsar, and the Convention to Combat Desertification.

But it is important to consider where integrated versus segregated land use is likely to be advantageous, and the scale at which integration is desirable (Balmford et al. 2001; Green et al. 2005). For example, if most biodiversity is likely to be lost in the transition from pristine to extensive systems or if key species are very sensitive to fragmentation, then segregated systems might be indicated at a coarser grain. But if the transition from extensive to intensive agriculture will result in greater biodiversity loss, then low-intensity agriculture finely interspersed with natural areas may be most desirable.

Real costs are associated with the cross-sectoral planning and coordination and technical innovations needed to achieve impacts at a landscape scale. These must be considered in prioritizing private, public, and civic ecoagriculture investments. There are four top priorities:

1. Agricultural landscapes located in or around critical habitat areas for wild species of local, national, or international importance
2. Degraded agricultural landscapes where restored ecosystem services will be essential to achieve both agricultural and biological diversity
3. Agricultural landscapes that must also function to provide critical ecosystem services
4. Periurban agricultural systems, where careful management is required to protect ecological, wildlife, and human health

Geographic scale and location of such priority areas for ecoagriculture development strategies (as distinct from agriculture-led or conservation-led development) have not been assessed. Undertaking such analyses is a critical step to guide policy action. There will be diverse "entry points" for ecoagriculture sometimes led by farmers and investment keepers, sometimes by conservationists, and sometimes by rural development leaders or even consumers.

Book Objectives and Structure

Farmers, conservationists, researchers, leaders in rural development, entrepreneurs, and policymakers in many parts of the world have begun to develop and promote ecoagriculture. But adoption of ecoagriculture is essential on a much larger scale to achieve the Millennium Development Goals on hunger, poverty, and environmental sustainability in developing countries, and to sustain ecosystems in strong rural economies in industrialized countries. This book assesses the current state of ecoagriculture systems and practices and begins developing a strategy to promote and support ecoagriculture development throughout the world.

The audience for this book includes the full spectrum of stakeholder groups who are, and must be, engaged to organize and manage the various elements of an ecoagriculture landscape:

- Land use planners, producer and conservation organizations, and resource managers responsible for conservation and production outcomes at a landscape scale
- Scientists and other innovators who study how agroecosystems work and who generate improvements in agricultural production and conservation management
- Agricultural and community enterprises, the food industry, other market players, and policymakers responsible for shaping the financial and livelihood incentives for land and resource management
- Grassroots practitioners responsible for production and for conservation management at farm and community scales

Structure of This Book

The book is divided into three sections.

Agricultural Production. This section presents current knowledge about agricultural production systems that have benign or positive impacts on biodiversity and ecosystem services. Chapters address annual crops, perennial crops, and livestock, as well as associated species, such as wild pollinators and soil microorganisms, and how diversity of domesticated species interacts with overall biodiversity. The section identifies barriers, gaps, and opportunities for increasing sustainable agricultural production in ways compatible with biodiversity conservation.

Biodiversity and Ecosystem Management. This section presents what has been learned about managing landscapes to achieve biodiversity and ecosystem objectives within mixed production–conservation mosaics. Chapters examine ecosystem "design" principles for terrestrial and freshwater biodiversity conservation, watershed management, and describe new developments in adaptive management, research, and monitoring at the landscape scale.

Institutional Foundations for Ecoagriculture. This section focuses on community- to policy-level action across ecosystems and farming systems to develop, implement, manage, and scale up successful ecoagriculture approaches. Chapters examine the central role of rural community leadership in developing ecoagriculture landscapes and overcoming barriers. Institutional approaches that effectively support communities and engage key stakeholders in planning and implementation are described, as are research approaches to support ecoagriculture development. Chapters also discuss new market-based approaches to increasing the financial viability of ecoagriculture, and policy actions needed to benefit livelihoods and ecosystems at a meaningful scale.

The ecoagriculture innovators whose work is reviewed and synthesized in this book come from strikingly diverse backgrounds, cultures, professions, and ecosystems. Their perspectives and philosophies are the result of many years of experience and consideration. But ecoagriculture is an emerging challenge that is still in the early years of its development; its implementation is an evolving and dynamic process. The pathways to ecoagriculture that are described may not only differ but may sometimes conflict. Nonetheless, the process of systematically sharing experiences and findings in a spirit of mutual respect will, it is hoped, broaden readers' understanding of the values, assumptions, and empirical evidence underlying the diversity of views and approaches. Solutions will be site specific; what works well in one setting may fail in another, emphasizing again the need for diversity.

References

Acharya, K.P. 2006. Linking trees on farms with biodiversity conservation in subsistence farming systems in Nepal. *Biodiversity and Conservation* 15:631–646.

Balmford, A., J.L. Moore, T. Brooks, N. Burgess, L.A. Hansen, P. Williams, and C. Rahbek. 2001. Conservation conflicts across Africa. *Science* 291(5513):2616–2619.

Blann, K. 2006. *Habitat in Agricultural Landscapes: How Much Is Enough? A State-of-the-Science Literature Review.* Defenders of Wildlife, West Linn, Oregon.

Breckwoldt, R. 1983. *Wildlife in the Home Paddock: Nature Conservation for Australian Farmers.* Angus and Robertson, Sydney.

Brown, L.R. 2004. *Outgrowing the Earth: The Food Security Challenge in an Age of Falling Water Tables and Rising Temperatures.* W.W. Norton, New York.

Buchmann, S.L. and G.P. Nabhan. 1996. *The Forgotten Pollinators.* Island Press, Washington, DC.

Buck, L.E., J.C. Milder, T.A. Gavin, and I. Mukherjee. 2006. Discussion paper 2—*Understanding Ecoagriculture: A Framework for Measuring Landscape Performance.* Ecoagriculture Partners, Washington, DC.

Carpenter, S.R., P. Pingali, E. Bennett, and M. Zurek (eds.). 2005. *Ecosystems and Human Well-Being: Scenarios.* Island Press, Washington, DC.

CBD (Convention on Biological Diversity). 2000. Decision V/5. Agricultural Biological Diversity: Review of Phase I of the Programme of Work and Adoption of Multi-year Work Programme. Conference of the Parties of the Convention on Biological Diversity. Nairobi, Kenya.

CBD. 2002. Decision VI/5. Agricultural Biological Diversity. Conference of the Parties of the Convention on Biological Diversity. The Hague, the Netherlands.

CBD. 2006. *Decision VI/9: Global Strategy for Plant Conservation.* http://www.biodiv.org/decisions/default.asp?dec=VI/9.

Cincotta, R.P. and R. Engelman. 2000. *Nature's Place: Human Population and the Future of Biological Diversity.* Population Action International, Washington, DC.

Cohen, J.E. 2003. Human population: the next half century. *Science* 302:1172–1175.

Daily, G.C. (ed.). 1997. *Nature's Services: Societal Dependence on Natural Ecosystems.* Island Press, Washington, DC.

Delgado, C., M. Rosegrant, H. Steinfeld, S. Ehui, and C. Courbois. 1999. 2020Vision. Discussion Paper No. 28. International Food Policy Research Institute, Washington, DC.

Diversitas. 2002. Diversitas Science Plan. http://www.diversitasinternational.org/pub_diversitas.html.

Forman, R.T. 1995. *Land Mosaic: The Ecology of Landscapes and Regions.* Cambridge University Press, Cambridge.

Gillis, N. and S. Southey. 2005. *New Strategies for Development: A Community Dialogue for Meeting the Millennium Development Goals.* Fordham University Press, New York.

Green, R.E., S. Cornell, J.P.W. Scharlemann, and A. Balmford. 2005. Farming and the fate of wild nature. *Science* 307:550–555.

Hassan, R., R. Scholes, and N. Ash (eds.). 2005. *Ecosystems and Human Well-Being: Current State and Trends, Volume 1.* Island Press, Washington, DC.

Heimlich, R.E., K.D. Wiebe, R. Klassen, and D. Gadsby. 1998. *Wetlands and Agriculture: Private Interests and Public Benefits.* US Department of Agriculture, Washington, DC.

IUCN (World Conservation Union). 1994. *Guidelines for Protected Area Management Categories.* IUCN, Gland, Switzerland.

Jackson, D.L. and L.L. Jackson (eds.). 2002. *The Farm as Natural Habitat: Reconnecting Food Systems with Ecosystems.* Island Press, Washington, DC.

Kremen, C. and R.S. Ostfeld. 2005. A call to ecologists: measuring, analyzing, and managing ecosystem services. *Frontiers in Ecology and the Environment* 3(10):540–548.

MA (Millennium Ecosystem Assessment). 2005. *Ecosystems and Human Well-Being: Synthesis.* World Resources Institute, Washington, DC.

Matthews, S. and K. Brand. 2004. *Africa Invaded: The Growing Danger of Invasive Alien Species.* Global Invasive Species Programme, Cape Town.

McCalla, A.F. 2000. Agriculture in the 21st century. CIMMYT Economics Program Fourth Distinguished Economist Lecture. CIMMYT (International Maize and Wheat Improvement Center), Mexico City.

McNeely, J.A. and S.J. Scherr. 2003. *Ecoagriculture: Strategies for Feeding the World and Conserving Wild Biodiversity.* Island Press, Washington, DC.

Mooney, H.A., R.N. Mack, J.A. McNeely, L.E. Neville, P.J. Schei, and J.K. Waage (eds.). 2005. *Invasive Alien Species: A New Synthesis.* Island Press, Washington, DC.

Myers, N.A., R.A. Mittermeier, G.C. Mittermeier, G.A.B. da Fonseca, and J. Kent. 2002. Biodiversity hotspots for conservation priorities. *Nature* 403:853–858.

Nairain, P., M.A. Khan, and G. Singh. 2005. Potential for water conservation and harvesting against drought in Rajasthan, India. Working Paper 104, Drought Series: Paper 7. International Water Management Institute, Colombo, Sri Lanka.

Nelson, M., R. Dudal, H. Gregersen, N. Jodha, D. Nyamia, J.-P. Groenewold, F. Torres, and A. Kassam. 1997. Report of the Study on CGIAR Research Priorities for Marginal Lands. Technical Advisory Committee, Consultative Group on International Research, and FAO, Rome.

NRC (National Research Council). 2002. *Environmental Effects of Transgenic Plants.* Pew

Initiative on Food and Biotechnology: Issues in the regulation of genetically engineered plants and animals.

OECD-FAO. 2005. *OECD-FAO Agricultural Outlook: 2005–2014.* OECD, Paris, and FAO, Rome.

Oksman-Caldentey, K.-M. and W.H. Barz. 2002. *Plant Biotechnology and Transgenic Plants.* Marcel Dekker, New York.

Omamo, S.W. and K. von Grebmer (eds.). 2005. *Biotechnology, Agriculture, and Food Security in Southern Africa.* IFPRI, Washington, DC.

Paoletti, M.G. 2005. *Ecological Implications of Mini-livestock: Potential of Insects, Rodents, Frogs and Snails.* Science Publishers Inc., Enfield, NH.

Penning de Vries, F.W.T., H. Acquay, D. Molden, S.J. Scherr, C. Valentin, and O. Cofie. 2003. *Integrated Land and Water Management for Food and Environmental Security.* Comprehensive Assessment of Water Management in Agriculture, Research Report 1, Colombo, Sri Lanka.

Pretty, J.N., A.D. Noble, D. Bossio, J. Dickson, R. Hein, F. Penning De Vries, and J. Morison. 2006. Resource-conserving agriculture increases yields in developing countries. *Environmental Science and Technology* 40(4):114–119.

Rabalais, N.N., R.E. Turner, W.J. Wiseman. 2002. Gulf of Mexico hypoxia, aka "The Dead Zone." *Annual Review of Ecology and Systematics* 33:235–263.

Rhodes, C. and S.J. Scherr (eds.). 2005. *Ecoagriculture: Integrating Strategies to Achieve the Millennium Development Goals.* Ecoagriculture Partners, Washington, DC.

Rhodes, C. and S. Scherr (eds.). 2005. *Developing Ecoagriculture to Improve Livelihoods, Biodiversity Conservation and Sustainable Production at a Landscape Scale: Assessment and Recommendations from the First International Ecoagriculture Conference and Practitioners' Fair, Sept. 25–Oct. 1, 2004.* Ecoagriculture Partners, Washington, DC.

Rosegrant, M.W., X. Cai, and S.A. Clein. 2002. *World Water and Food to 2025: Dealing with Scarcity.* International Food Policy Research Institute, Washington, DC.

Rosegrant, M.W. and S.A. Clein, T. Sulser, and R. Valmonte-Santos. 2005. *Long-Term Prospects for Africa's Agricultural Development and Food Security.* International Food Policy Research Institute, Washington, DC.

Runge, C.F., B. Senauer, P.G. Pardey, and M.W. Rosegrant. 2003. *Ending Hunger in Our Lifetime: Food Security and Globalization.* Johns Hopkins University Press, Baltimore.

Sanderson, E.W., M. Jaiteh, M.A. Levy, K.H. Redford, A.V. Wannebo, and G. Woolmer. 2002. The human footprint and the last of the wild. *Bioscience* 52(10):891–904.

Scherr, S.J. and J.A. McNeely. 2007 forthcoming. (Royal Society paper).

Swaminathan, M.S. (ed.). 1994. *Ecotechnology and Rural Employment: A Dialogue.* Macmillan India, Madras.

Tan, G. and R. Shibasaki. 2003. Global estimation of crop productivity and the impacts of global warming by GIS and EPIC integration. *Ecological Modeling* 168:357–370.

The Nature Conservancy. 2004. The Nature Conservancy's 2015 Goal. http://sites-conserveonline.org/gpg/projects/tnc2015goal.html.

Tilman, D. 2001. Forecasting agriculturally driven global environmental change. *Science* 292(13):281–284.

UN Millennium Project. 2005. *Investing in Development: A Practical Plan to Achieve the Millennium Development Goals.* United Nations, New York.

Wood, S., K. Sebastian, and S. Scherr. 2000. *Pilot Analysis of Global Ecosystems: Agroecosystems.* IFPRI and WRI, Washington, DC.

Part I

Agricultural Production in Ecoagriculture Landscapes

Overview

Conventional wisdom assumes that increasing agricultural production and productivity necessarily reduces biodiversity and ecosystem services. But state-of-the-art reviews of both science and field practice, synthesized in the next six chapters of this book, challenge that assumption. New and traditional approaches to annual and perennial crop and livestock production have been documented to sustain or increase production, reduce production costs or risks, or otherwise benefit producers, while at the same time having benign or positive impacts on wild species and ecosystems. The chapters in this section assess these diverse approaches, and identify major gaps in knowledge and action that must be addressed for agricultural systems to contribute to biodiversity conservation, as well as agricultural supply and rural livelihoods.

Louise E. Buck, Thomas A. Gavin, Norman T. Uphoff, and David R. Lee provide a comprehensive overview of scientific literature on ecological crop production systems, mainly for conventional annual crops, that benefit or are compatible with wild biodiversity. The authors evaluate a subset of studies that quantify the relationships among the three "legs of the stool" of ecoagriculture—sustainable production, rural livelihoods, and conservation of biodiversity and ecosystems. They highlight the major gaps in knowledge and methodological challenges of developing and assessing production systems for ecoagriculture landscapes.

A major aim of ecoagriculture is to enhance the diversity of both domesticated and wild species within a landscape. **Judith Thompson, Toby**

Hodgkin, Kwesi Atta-Krah, Devra Jarvis, Coosje Hoogendoorn, and Stefano Padulosi focus on agricultural biodiversity. They discuss the role of diversity in sustainable production, in contributing toward pest and disease control, in ensuring proper use and management of neglected and underutilized species, and in buffering livelihoods against environmental changes and challenges. They point out that agricultural diversity depends on wild sources of genes in order to maintain the productivity and adaptability of domesticated species. They advocate for a stronger integration between initiatives to conserve agricultural biodiversity and wild biodiversity.

Lee R. DeHaan, Tomas.S. Cox, David L. Van Tassel, and Jerry.D. Glover look at ecoagriculture innovations for the main food grains that sustain modern society. They argue that converting annual grains to perennials will make a significant contribution to ecoagriculture by enhancing production sustainability, enabling integrated pest management approaches, nurturing greater biodiversity in soils, and providing better grassland habitat for wildlife than is currently supported by the production systems that characterize most of the world's "breadbaskets" and "rice bowls." Their chapter describes research challenges and the state of perennial grains development around the world.

Roger Leakey examines another promising strategy for enhancing the diversity of ecoagriculture landscapes—domestication of new indigenous tree species and herbaceous plants for foods, pharmaceuticals, timbers, fibers, gums, and dyes. In addition to evaluating the benefits for farming systems and rural livelihoods of integrating these species, the author presents case studies where domestication of indigenous trees has a direct benefit on wildlife conservation. Leakey explains the continuum from wild to domestic that provides numerous options for future development.

Götz Schroth and Maria do Socorro S. da Mota evaluate evidence on diverse roles that agroforestry systems may play in supporting biodiversity conservation in human-dominated landscapes in the tropics. First, establishment of agroforestry systems can reduce the conversion of primary habitat and provide ecological synergies that further protect such habitat. Second, agroforestry plantings can provide secondary habitat. Third, landscapes with agroforestry can offer a more benign matrix for "islands" of primary habitat in the agricultural landscape, especially by buffering forest edges and creating biological corridors. Scaling up successful agroforestry approaches requires both improving livelihood and biodiversity impacts at the plot scale, and strategically placing and managing agroforestry plots within the landscape, to provide ecosystem services, such as watershed protection or wildlife habitat connectivity.

Conventional livestock production systems, both extensive and intensive, have an even larger ecological "footprint" than do crops, which is exacerbated by the "Livestock Revolution" that has increased meat and dairy consumption. But **Constance L. Neely and Richard Hatfield** identify a wide range of innovations in the livestock sector that not only improve the productivity and sustainability of production but also conserve plant and animal biodiversity. The authors contend that to achieve sustainable benefits from these technical approaches will require more widespread use of community-based approaches, targeted efforts to conserve livestock and grassland genetic diversity, the development of market chains for livestock products in which buyers recognize biodiversity benefits, and policies that support producers who conserve biodiversity.

Together, this set of chapters provides a broad, yet detailed review of production options for ecoagriculture landscapes and suggests new ways of thinking about sustaining agricultural production while conserving a full range of biodiversity and ecosystem services.

Chapter 2

Scientific Assessment of Ecoagriculture Systems

Louise E. Buck, Thomas A. Gavin, Norman T. Uphoff, and David R. Lee

Introduction

When *Ecoagriculture: Strategies to Feed the World and Save Biodiversity* was published in 2003, the concept was widely received as a promising and innovative approach to addressing environmental, food security, and livelihood goals. Emerging from conclusions derived from a review of spatial information developed through the Millennium Ecosystem Assessment, authors Jeffrey McNeely and Sara Scherr distilled principles, case studies, and policy recommendations that supported the proposition that both conservation and productivity can be mutually enhanced in many areas. As the concept of ecoagriculture entered forums of debate about international development-assistance priorities, however, various challenges emerged from specialists, scholars, and skeptics concerning some of the assumptions and evidence that had been brought to bear on the subject. At the same time, proponents were beginning to promote ecoagricultural practices in diverse settings, expecting to achieve the prospective benefits that ecoagriculture could offer. Thus the need for having more solid scientific bases for ecoagriculture became evident.

The basic challenge is how to support both agriculture and biodiversity conservation in areas where both are important and where segregating the two land uses may not meet their respective objectives or be economically and politically feasible. Would it be possible to create entirely new habitats on agricultural land that encompass both cultivated and wild plants, both herbaceous and woody plants, in edible and medicinal forms, with both highly cultivated and less intensely managed areas and a diversity of domesticated livestock along with wild vertebrates and invertebrates? Even more challenging, would it be possible to work with landscapes that are nearly devoid of life, as found in many degraded agricultural environments, and purposefully design landscapes that include both wild and agrobiodiversity, plus humans who live off the products of this design?

Achieving this Edenesque result would require community and population ecologists, agronomists, landscape ecologists, landscape architects, horticulturists, city and regional planners, and indigenous knowledge specialists, as well as local residents with practical knowledge, to join together in imagining and creating some novel outcome that might protect more biodiversity than the piecemeal, "hold onto what you can afford to" approach commonly espoused now. Many people spend their lives trying to design towns and cities that function efficiently and are physically and psychologically healthy for their inhabitants. Why should a diversity of scientists and agriculturalists not attempt to design agricultural landscapes from the ground up with the same general goals?

This idea is actually decades old, having been advocated as a way to at least think about designing future nature reserves in a degraded world. Wilson and Willis (1975) termed this approach "applied biogeography." To initiate such a process would require a great deal of knowledge about species interactions, the compatibility of domestic and wild species of plants and animals, the appropriate spatial configuration of land uses for maintaining viable populations of wild species in an agricultural matrix, and so on. The questions that need answers are not entirely new, but the novel context within which these questions would be asked could lead in unexpected and fruitful directions.

Responding to this vision, a study was designed around the following objectives:

1. To evaluate claims that have been made about the benefits of ecoagriculture on the basis of scientific evidence
2. To characterize methodologies for evaluating, with scientific rigor, the tradeoffs and synergies among objectives that are addressed within an ecoagriculture framework
3. To identify significant gaps in knowledge and corresponding methodological and institutional limitations in the generation of such knowledge
4. To identify promising opportunities and methods for improving knowledge and understanding about ecoagriculture

Method

A previous, in-depth assessment covered literatures from the fields of production agriculture, agricultural and natural resources economics, conservation biology, social science, and natural resources planning and management (Buck et al. 2004). Some 75 experts in these fields were consulted. The authors also examined evidence from project initiatives that are related to ecoagriculture and consistent with its precepts. Reviews of key literatures were updated for this chapter, and findings and conclusions of the assessment were revisited.

The study was not framed as an evaluation of ecoagriculture per se. The authors considered this to be too new an area to warrant any conclusive evaluation, positive or negative, at this point in time. The approach in the review was to bring together ideas and examples that could help others engage intellectually and practically with this emerging concept, to assess what is already known, and to suggest some ways forward to accumulate more useful experience and knowledge.

Concepts and Definitions

The feasibility of ecoagriculture strategies requires that there be compatibility between economic and environmental goals. By definition, ecoagriculture requires systems of sustainable agricultural and natural resource management that "simultaneously achieve improved livelihoods, conservation of biodiversity (genetic resources, ecosystem services, and wild flora and fauna), and sustainable production at a landscape scale" (Ecoagriculture Partners 2004).

The image of a three-legged stool provides a visual aid for working with the concept of ecoagriculture, as depicted in Figure 2.1. An effective ecoagriculture system needs to satisfy the three interdependent objectives in space and over time. Institutional and policy underpinnings affect the overall success of ecoagriculture systems, like the bars connecting and stabilizing the legs.

Synergies and Tradeoffs

Because ecoagriculture has multiple objectives that are not necessarily closely correlated, the challenge is to find accommodations and synergies among them. Where are there conflicts or contradictions among goals? What tradeoffs can give the best joint outcome? Given the significant role that synergies play in the success of ecoagriculture, what is the record regarding the simultaneous accomplishment of these joint objectives?

Synergistic outcomes from ecoagricultural practices may stem from a variety of different sources, including the following:

- Increased efficiency of input use
- Synergies among uses of component inputs
- Substitution among sources of capital (natural, financial, human, etc.)
- Accrual of natural capital
- More efficient spatial organization

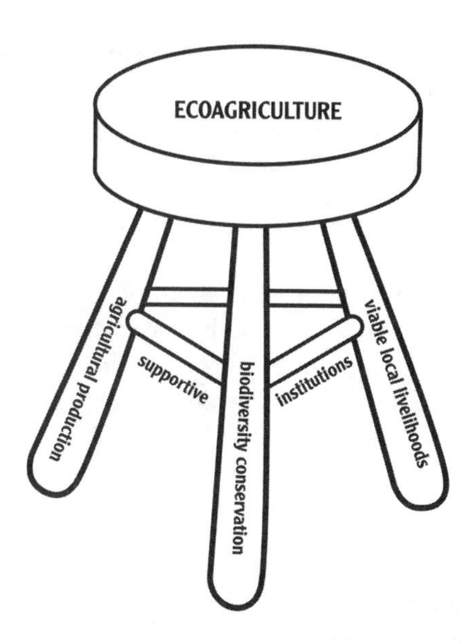

Figure 2.1. Ecoagriculture envisioned as a three-legged stool

- Pricing of services and collection of revenue from these
- Benefits to farming from improved use of biodiversity (e.g., through integrated pest management [IPM] or agroforestry)
- Economies of scale, including those through farmer collaboration (Scherr 2003)

Biodiversity

The authors define biodiversity in terms of the multiplicity of all species of native organisms, plant and animal, vertebrate and invertebrate, macro- and micro-, as well as the genetic and morphological variation within the populations of each of those species. In addition, the conservation of biodiversity includes maintaining the biological processes (e.g., predator–prey and plant–pollinator relationships) that produced those organisms and of which those organisms are a part. The assessment refers to the foregoing as "wild biodiversity," and to the variety of crop and livestock organisms, landraces, and species that contribute to food and agricultural production as "agrobiodiversity."

Biodiversity has commonly been understood in quantitative terms, perhaps because of the emphasis on preservation of certain endangered species. But there are "scale" questions and "value" questions that quickly make evaluation more complex. Most biologists are reluctant to make judgments that would diminish the importance of any particular species, yet almost all agree that, in a particular location, one species may deserve more attention than others because of its tenuous status or functional role in global or ecosystem terms. Thus a practice or farming system that would preserve a "more critical" species, even if local species diversity were diminished, could be positively evaluated.

Spatial Scale

The conservation of biodiversity is unlikely to be assured by sustainable agriculture practiced only at the farm scale because demographic units of plants and animals usually encompass larger areas than individual farms. Spatial scale is also important in economic terms. Individual farming operations may experience benefits and externalities that differ from those experienced by large aggregations of small producers.

Intensification

Agricultural "intensification" typically refers to an increase of use of inputs per unit of production output, but there are different frameworks within which this is assessed. The term "intensification" therefore must specify *in regard to what resource.* External-input-intensive agriculture is very different from labor-intensive production, having different implications for soil and plant biodiversity. Therefore, the *fact* of intensity (usually, by definition, one or more kinds of resources will be used more intensively in production than others) may be more important than the *kind* of intensity in production processes.

Evidence of Biodiversity Conservation in Agricultural Systems

It is clear that agricultural practices have been instrumental in reducing the amount of high-quality habitat for wild biodiversity all around the world (Alkorta et al. 2003). Yet examples are proliferating of agricultural production systems, both traditional and science-based, for both smallholder and large-scale commercial farming, that may have benign or positive impacts on wild biodiversity (Clay 2004; McNeely and Scherr 2003). Such practices are being used, together with improved management of natural areas, to develop ecoagriculture landscapes. However, the biodiversity benefits of most of these practices have not been rigorously documented. This section reviews some of the available quantitative evidence in order to explore trends in the findings.

Quantitative Studies

The authors explored the agricultural and biological literature for studies that document a correlation, positive or negative, between the implementation of a particular agricultural strategy, or suite of practices, with some measure of wild biodiversity or its habitat. They surveyed peer-reviewed journals, government reports, in-house documents of experiments or field trials, and other sources of evidence that confirms, or does not confirm, the claim that some particular agricultural method has enhanced biodiversity conservation. Only results were included where the investigators actually quantified the amount or kinds of biodiversity. These studies were organized within a matrix that depicts the range of both agricultural approaches and kinds of biodiversity (e.g., taxa or habitat conserved). Table 2.1 lists the number of studies found that apply to each cell; some studies apply to more than one cell. For a complete list of references found in the table, go to http://ecoagriculturepartners.org/resources/publications.php.

The sample includes a wide-ranging set of 82 studies that met these criteria, but this set is by no means exhaustive. It should be viewed as representative of the kinds of results that are being produced by researchers who study agriculture and its effects on wild biodiversity. The table is organized according to the main crop production practice examined by each study, which is the approach generally used in field studies. It is common to try to isolate or control for all other variables and to determine, for example, the effect on bird populations of having perennial crops versus annual crops. The number of practices or variables covered in the analysis, (18, the number of column headings), suggests the complexity of this issue. At any given location in the world, several of these variables may be extant and interacting in complex ways on the same site. To make matters more complicated, the particular set of interacting variables is likely to have different effects on different taxonomic or functional groups of wild biodiversity.

Nevertheless, there are some identifiable patterns. The authors found that 12 of the 18 ecoagricultural strategies surveyed were found by at least three researchers to conserve at least three taxa. The strategy most often correlated with the conservation of wild biodiversity was the *maintenance of adjacent hedgerows or woodlots;* 21 studies documented positive correlations with eight taxa plus the conservation of natural habitat (see also Benton et al. 2003). *Organic agriculture* was correlated with an increase in seven taxa plus habitat in eight studies. *Shaded tropical agricultural production*—especially of coffee and cacao—had higher species richness of three taxa according to 11 studies.

The results of studies listed show that some agricultural approaches are more beneficial than others in conserving wild biodiversity, and it is possible to document beneficial effects from certain kinds of agricultural practice. As already noted, most studies that documented conservation benefits were assessing the value of maintaining hedgerows, windbreaks, or natural habitat adjacent to agricultural fields. Similarly, shaded tropical agriculture (Donald 2004) and organic agriculture have been well documented to maintain more species at all, or nearly all, taxonomic levels than the standard modern practices that these ecoagriculture strategies replace (although see Hole et al. 2005 for a somewhat

Table 2.1. Studies that examine the relationships between various agricultural practices and wild biodiversity or habitat

Taxa or surrogate of diversity measured		*Agricultural practices claimed to conserve biodiversity*							
		Shaded tropical agriculture	*Trees in pastures*	*Other agroforestry*	*Other mixed cropping*	*Perennial crops*	*Low-tillage*	*No-tillage*	*Increased fallow*
Species	Mammals			1					
abundance	Birds	6	2		1	1	2		
or other	Ants	5	1						
measure	Other vertebrates								
of on-site	Soil invertebrates			1	1				1
diversity	Other invertebrates	2	1	3	2		1		
of	Soil microbes				1		1	1	
organisms	Soil biomass						1		
	Soil organic matter	1				2			
	Trees	1		1					
	Other plants	1	1			1	1		
Habitat (acre)									

Taxa or surrogate of diversity measured		*Agricultural practice claimed to conserve biodiversity*							
		Other grassland management	*Reduced chemicals*	*Organic*	*Other European innovation*	*Postharvest treatment*	*Managed flooding*	*Ecocertification*	*Lower-intensity agriculture*
Species abundance or other measure of on-site diversity of organisms	Mammals								
	Birds		1	3	1	3	4		1
	Ants			1		1			
	Other vertebrates		1						
	Soil invertebrates				1		1		1
	Other invertebrates	3	2	1	1				
	Soil microbes		3	1					1
	Soil biomass			1					
	Soil organic matter		1						
	Trees							1	
	Other plants	2		1	1				2
Habitat (area)			1	1	1			1	1

Table 2.1. (*Continued*)

Taxa or surrogate of diversity measured		*Off-plot management*: *Hedgerow or forest adjacent to plot*	*Off-plot management*: *Landscape-level effects*
Species abundance or other measure of on-site diversity of organisms	Mammals	4	
	Birds	5	3
	Ants	2	
	Other vertebrates	1	
	Soil invertebrates	8	8
	Other invertebrates		
	Soil microbes		
	Soil biomass		
	Soil organic matter	1	
	Trees	4	1
	Other plants	4	4
Habitat (area)		2	

cautious endorsement of organic farming in lowland Europe). Because these are also the practices that have been best studied, there could be others with similar or even greater conservation benefits.

The matrix highlights some patterns in research and reveals some important gaps. The authors believe the patterns can support and focus research on the probable conservation benefits of sustainable agriculture practices. However, many of these prescriptions are likely to be criticized as site specific, taxon specific, or agronomic specific, and some of those that measured effects across taxa have found inconsistent or even inverse correlations with biodiversity. Few studies included economic or productivity data or could be linked directly to parallel studies with such data. Moreover, most studies relied on local species richness as a criterion for evaluation without any regional or global measure of impacts on biodiversity.

More broadly, a few high-priority knowledge gaps merit mention. First, the current debate over the linkage between diversity and ecosystem functioning or ecosystem stability will not likely be settled soon (Swift et al. 2004), and yet important advances are being made in knowledge. One of the most important aspects of this debate, regarding the nexus between agriculture and conservation, is the agricultural value of below-ground biodiversity (Brussaard et al. 2004). Several studies have demonstrated that the application of agricultural chemicals significantly reduces soil biodiversity, while the reduction or elimination of those chemicals results in higher below-ground biodiversity. There is some evidence to suggest that gains in crop productivity and biodiversity can be obtained from reduced chemical use (Pretty 2005a).

Missing Elements

Most of these studies still rely on simple on-site species counts and, therefore, they are missing two key components for evaluating the overall impact on biodiversity: (1) the impact on regional or global diversity, and (2) the biological significance of finding an individual of some species at some site. To evaluate agricultural practices that maintain or enhance as much of the biodiversity that exists locally, regionally, or globally, quantity versus quality of wild biodiversity needs clarification: which species or species assemblages, and in what abundance, indicate that the landscape is "headed in the right direction"?

The hierarchy of local to global diversity is important. For example, it is possible to increase aggregate biodiversity locally through various land-use practices and the introduction of nonnative species while simultaneously making it impossible for certain "sensitive" species to survive there (Sax and Gaines 2003). If these sensitive species are the ones that tend to be lost in nearly every local situation, they become rare regionally (e.g., gray wolves, *Canis lupus*) or globally (e.g., Spix's Macaws, *Cyanopsitta spixii*). Therefore, if one takes a larger spatial perspective, more biodiversity may be conserved globally if certain local practices encourage certain sensitive species at the expense of other, more common species (Lennon et al. 2004). The claim that certain agricultural practices actually increase local biodiversity may be true, but in some cases the biota that benefit from the practices may not be biota that need benefiting. So claims that particular agricultural practices "increase biodiversity" need to be evaluated relative to which goal is selected: (1) to enhance local biodiversity that may be compatible with local agricultural productivity only, or (2) to conserve particular species of biodiversity locally that contribute to global or regional biodiversity and are in need of protection.

Determining what levels of biodiversity are present at local sites is often difficult, and determining the biological meaning of those levels is even more difficult. Yet effecting any positive changes is not an easy proposition either. In the Netherlands, for example, various agri-environment schemes have been established. But the overall landscapes have already been so affected by intensive agricultural use that some studies suggest the schemes have little positive impact on wild biodiversity relative to nearby fields not under the conservation schemes (Kleijn et al. 2004).

There are pitfalls in concentrating too intently on individual fields, or plots, or crops. The context within which those units exist is extremely important, and the principles involved have been well articulated in recent reviews sanctioned by the Ecological Society of America (Christensen et al. 1996; Dale et al. 2000). Understanding what needs to be done to conserve wild biodiversity in agricultural landscapes is difficult enough, but enacting the appropriate changes on the ground without compromising economic and productivity gains will be even more challenging. Gains in conserving wild biodiversity without significant losses in the other two spheres constitute the greatest information gap in the evaluation and promotion of ecoagriculture.

The Productivity of Crop Production Practices That Exhibit Environmental and Livelihood Benefits

The economic and environmental costs of producing more food to meet higher levels of demand can be substantial and need to be reckoned with. The situation of the poor will not necessarily be improved in this case by producing more food at higher cost, because the necessary technology requires more capital investment and uses marginal, less accessible land and water resources. Moreover, in terms of economic growth and livelihood expansion, if food production becomes more costly, this will direct resources away from investment in creating jobs and wealth and meeting other needs. So finding ways to raise *system productivity* in the agricultural sector is essential to be able to meet all three ecoagriculture objectives in positive-sum (i.e., in synergistic ways).

To assess opportunities for capitalizing on such positive-sum dynamics, the authors reviewed the state of knowledge and practice in a number of areas that can be characterized as integrated agricultural management practices. Literature on these types of practice offers insight into possibilities for achieving increases in productivity that focus on combinations and communities of plants, animals, and soil organisms, rather than on one particular species at a time. The practices are rooted in an agroecological approach that seeks to achieve and benefit from synergies between objectives and components of production systems. The goals include having benign impacts on ecosystems and contributing to social and cultural values, rather than directing all efforts toward yield-oriented outcomes (Altieri 1995).

When the assessment was conducted in 2004 the authors found relatively few studies that provided quantitative evidence of gains in the productivity or value of crops produced in agroecological practices. Generally, studies published before 2004 tended to focus on documentation of environmental and/or social benefits, making inferences about prospective productivity gains. During the past three years, however, many more studies have appeared that quantify effects on agronomic and economic productivity in conservation-oriented agricultural systems. These findings are presented in Table 2.2.

The table identifies 25 studies that quantify gains in agroecological practices, compared with conventional practices, in terms of yield per area and/or profitability per area. The table also indicates environmental and/or livelihood benefits from the respective practices that are documented in the studies.

Three of the 25 studies include an indicator of biodiversity conservation among their performance criteria. Two of the three report on shaded coffee systems and highlight the need for ecocertification-based price premiums to offset productivity losses under shade and to justify the cost of the conservation benefit. The other suggests the potential for greater biodiversity from reduced ecotoxicity from pesticides, though it does not quantitatively demonstrate this outcome. To date, studies remain to be done that rigorously examine trade-offs and potential synergies between agroecological practice and biodiversity conservation.

Productivity gains are possible in practices that also benefit the environment in ways that are likely to help sustain production over time and/or that limit negative off-site impacts. In numerous cases, livelihood benefits such as income diversification and risk mitigation can be achieved simultaneously (see the next section for further discussion).

There are substantial *opportunities* for agriculture to become more productive, profitable, sustainable, and environmentally friendly by relying more on energy and nutrients made available through biological processes (Uphoff et al.

Table 2.2. Survey of studies that compare benefits of integrated crop management practices

	Resource-conserving agriculture; ecological agriculture	*Shaded tropical crop agriculture*	*Organic agriculture*	*Conservation agriculture*
Yield increase	Pretty et al. 2006			Bakhsh et al. 2005; Gregory et al. 2005
Yield competitive with conventional alternative	Rasul and Thapa 2004		Delate et al. 2003; Pimentel et al. 2005	
Yield profitable		Bacon 2005; Perfecto et al. 2005	Delate et al. 2003; Pimentel et al. 2005	Bakhsh et al. 2005; Janosky et al. 2002
Soil benefits	Rasul and Thapa 2004		Pimentel et al. 2005	Gregory et al. 2005; Janosky et al. 2002
Pesticide reduction	Pretty et al. 2006; Rasul and Thapa 2004			
Water-use efficiency	Pretty et al. 2006			
Energy-use efficiency			Pimentel et al. 2005	Lithourgidis et al. 2006
Biodiversity benefits		Bacon 2005; Perfecto et al. 2005		
Livelihood benefits		Bacon 2005		Bakhsh et al. 2005; Lithourgidis et al. 2006

Agroecological practice → performance indicator	*Integrated water management*	*Raised beds*	*Rotational cropping*	*Alley cropping*
Yield increase	Jat et al. 2005; Narayanamoorthy 2004; Sayre and Hobbs 2004; Thadchayini and Thiruchelvam 2005	Jat et al 2005; Sayre and Hobbs 2004	Gregory et al. 2005	Ajayi et al. 2006
Yield competitive with conventional alternative	Berg 2002; Frei and Becker 2005	Govaerts et al. 2005	Govaerts et al. 2005	
Yield profitable	Berg 2002; Frei and Becker 2005; Jat et al. 2005; Narayanamoorthy 2004; Sayre and Hobbs 2004; Thadchayini and Thiruchelvam 2005	Jat et al. 2005; Sayre and Hobbs 2004		Ajayi et al. 2006
Soil benefits	Sayre and Hobbs 2004	Sayre and Hobbs 2004	Gregory et al. 2005	Ajayi et al. 2006
Pesticide reduction	Berg 2002			
Water-use efficiency	Jat et al. 2005; Narayanamoorthy 2004; Sayre and Hobbs 2004; Thadchayini and Thiruchelvam 2005	Jat et al. 2005; Sayre and Hobbs 2004		
Energy-use efficiency				
Biodiversity benefits				
Livelihood benefits	Frei and Becker 2005			

Table 2.2. (*Continued*)

Agroecological practice ⟶ performance indicator	*Integrated pest management*	*System of rice intensification*
Yield increase		Uphoff 2003
Yield competitive with conventional alternative	Berg 2002; Hassan and Bakshi 2005	
Yield profitable	Berg 2002; Hassan and Bakshi 2005	Uphoff 2003
Soil benefits		Uphoff 2003
Pesticide reduction	Berg 2002; Hassan and Bakshi 2005	
Water use efficiency		Uphoff 2003
Energy use efficiency		
Biodiversity benefits		
Livelihood benefits		

2006). These processes can be favorably affected by changes in management practices for soil, water, plants, animals, and nutrients. This trend may or may not have a significant impact on the use of chemical amendments in the production of specific crops in particular settings, but in many of the cases cited, and others, the presumption appears to be shifting from chemicals *substituting for* or *compensating for* biology, to chemicals *supplementing* biology, informed by better scientific understanding of microbiology, ecology, and their derived discipline of microbial ecology. Advances in scientific knowledge in the practices reviewed in the assessment and presented in Table 2.2 demonstrate that there is more scope for compatibility between agricultural production increases and benign environmental impacts than seen previously. This evidence provides a theoretical and practical foundation for existing and potential success of ecoagriculture systems.

Enhancing Livelihoods through Biodiversity-Friendly Agricultural Practices

A central premise of the ecoagriculture approach is that conservation and production goals can, to a significant extent, be jointly achieved—at the landscape scale—while at the same time enhancing household and community welfare. Many of these livelihood benefits stem from products harvested from natural areas in the landscape, or from the ecosystem services produced jointly by natural and agricultural areas (e.g., clean and secure access to water). But production systems remain a chief contributor to livelihoods in most ecoagriculture landscapes. This section reviews the evidence of livelihood benefits from biodiversity-friendly production.

Agroecological practices, as just reviewed, are commonly characterized as leading to diversified incomes, improved access to assets (natural, physical, human, social, and/or financial forms of capital), and reduced vulnerability to food insecurity and other shocks and threats to local household production systems. This is consistent with the focus on "means, capabilities, assets and activities" associated with sustainable livelihood systems (Carney et al. 1999; Chambers and Conway 1992). In this view, the diversification of a household's asset base, the heterogeneity of and complementarity among those assets, and the existence of multiple "entry points" for rural development programs all assume a prominent role (de Janvry and Sadoulet 2003).

Recent research documents the potential for positive interactions and benefits where rural livelihoods are involved. Table 2.2 cites some typical studies, although the cases cited are illustrative only and far from complete. Bakhsh et al. (2005) find that zero tillage of wheat (versus conventional tillage) in the Pakistani Punjab increases yields and profits, improves livelihoods, and reduces poverty. Similarly, Hobbs and Gupta (2003) find that the use of conservation technologies in South Asian agriculture—including zero and reduced tillage and bed planting—improve input-use efficiency, reduce costs, improve farmer livelihoods, and provide environmental benefits. Hassan and Bakshi (2005), in a study of IPM in rice farming in Bangladesh, find that IPM "reduces pesticide costs with no countervailing loss in production," with evidently greater profits, and concomitant health and environmental benefits for rural communities.

Other studies have a broader, more inclusive scope. Frei and Becker (2005) and Pretty (2005b) identify broad-based benefits of integrated rice–fish farming, including improved land-use efficiency and total productivity; enhanced management of weeds, pests, and water; and increased nutrient cycling and availability. Rasul and Thapa (2004) examine conventional versus "ecological" farming systems in Bangladesh and find that, although the latter exhibit improvements in such measures as crop diversification, soil fertility management, pest and disease management, and agrochemical use, there are no major differences in key indicators such as land-use patterns, crop yield and stability, farmer risk exposure, and food security. Lee (2005) provides a critical review of several dozen recent studies of the factors influencing adoption of low-input sustainable agricultural practices. In a particularly ambitious study of a different type, Pretty et al. (2003) reviewed more than 200 sustainable agriculture development projects in 52 countries, involving nearly 9 million producers farming 29 million hectares, and find significant weighted average yield increases—37% per farm and 48% per hectare—compared to preadoption yield levels.

With specific regard to livelihood improvements, their results are less conclusive, but selected case studies do provide some evidence of positive capital accumulation (for diverse sources of capital). The specific outcomes of similar technologies, practices, and systems introduced in different locations can vary widely due to the influences of a wide variety of conditioning factors: biophysical, capital endowments (for all types of capital), market access, property rights, policy-related factors, and institutional and collective-action factors. In addition, poverty concerns are likely to weigh significantly in livelihood outcomes because rural poverty is centered in many of the same regions where wild biodiversity is richest and also under threat (Nelson et al. 1997, cited in McNeely and Scherr 2003). Equity concerns may also be important because efforts to conserve biodiversity may prove unsustainable if benefits are distributed unequally and there is persistent conflict among stakeholders.

Ecoagriculture and Long-Term Livelihood Strategies

A key question is the feasible scope of adoption of ecoagriculture production practices. Ecoagriculture production systems for niche markets do have positive attributes with long-term potential advantages. It has long been evident that the globe as a whole has more than enough food production capacity; long-term real price declines for most basic food and feed grains are but one sign of that capacity. So the agricultural and rural growth strategy of encouraging producers, where possible, to move out of commodity-type production of undifferentiated products and into the production of higher-value differentiated products that are likely to face more promising long-term price and market trends is a positive strategy and will benefit ecoagriculture systems.

But the specific technologies and systems currently being used for ecoagriculture are still very limited. Although some are broad based and widely applicable—conservation tillage being the most notable example—many are applicable only to agroecological "niches." Organic coffee, rice–fish systems, selected high-value horticultural products, agroforestry systems designed for particular biophysical environments—these and other such systems may represent viable diversification alternatives for producers in selected environments. But they have not yet demonstrated the capacity to generate production, livelihood, and biodiversity benefits across the board.

Indeed, scaling up those systems that depend on niche-market certification premiums or minor products with limited demand will often create price-depressing market effects that mitigate the gains achieved at more modest levels of production. Further, when considered at the aggregate level—the region or nation, for example—the benefits generated from these systems are typically dwarfed by the effects of crop or animal production in conventional commodity-type environments. It is thus essential that strategies to promote ecoagriculture focus greater attention on development of production methods (as well as associated natural area management strategies) that are suitable and economically feasible for commodity production (see the examples in Clay 2004) as well as for a much greater diversity of niche products for which market demand is increasing.

Conceptual and Empirical Challenges

The simultaneous achievement of biodiversity, productivity, and livelihood outcomes is a challenge, conceptually as well as in practice. A major source of the conceptual challenge arises in considering the gains from rural households' focusing on a single goal (e.g., maximizing production, biodiversity, or economic value) versus their targeting a set of outcomes in which no single outcome is maximized but where the outcomes, considered together, represent a joint improvement across multiple criteria. Whether specific technologies or agroecological system strategies are able to generate synergistic solutions, rather than trade-offs, becomes essentially an empirical question. For this reason, previous research has focused a great deal on the practical challenges revealed in empirical case studies in order to determine patterns and "lessons learned" (Lee and Barrett 2001).

The empirical literature examining the relationships between biodiversity conservation, sustainable production, and economic livelihood enhancement remains partial in nature, rarely analyzing measurable outcomes *across multiple criteria,* and results are often inconclusive. Furthermore, many of the examples cited in the literature that pertain to agroforestry, organic production systems, and related practices are premised on their benefits to local livelihoods but fail, upon closer inspection, to demonstrate economic competitiveness using quantitative measures and rigorous analysis. A prominent exception to these partial approaches is research conducted in three countries (Brazil, Cameroon, Indonesia) under the auspices of the Alternatives to Slash and Burn Program sponsored by the Consultative Group on International Agricultural Research, which remains one of the few cases of applied research measuring and documenting multiple system outputs across multiple sites using a consistent methodology. It thus affords the basis for directly comparing results across sites (see Tomich et al, chapter 18 this volume).

Modeling Trade-Offs and Synergies

The challenges posed by ecoagriculture raise the stakes for researchers when it comes to constructing and estimating analytical models. To evaluate which ecoagriculture production and land-use strategies may lead to synergies—or conversely, to trade-offs—among economic, production, and environmental goals, models need to identify specific measurable indicators and outcomes across multiple criteria. This is not an easy task, particularly with regard to biodiversity and other environmental indicators. In addition, meaningful evaluation of the feasibility of ecoagriculture practices and technologies means moving beyond the multitude of one-time studies conducted in specific agroecological and economic environments to examine the performance of these systems in different sites and facing different biophysical and socioeconomic conditions. This in turn puts a premium on models and analytical approaches that can flexibly evaluate these alternatives. Finally, the additional demand on ecoagriculture practices that they be sustainable creates further data and modeling requirements, namely, the desirability of monitoring and evaluating results over time, or at a minimum, being able to perform long-term simulations to assess potential sustainability.

A variety of quantitative and mixed modeling approaches can prove useful in assessing ecoagriculture strategies. Bioeconomic models are among the most promising vehicles because they are capable of explicitly linking biophysical and economic interactions and can formally incorporate parameters representing soil and water conditions, land suitability, measures of biological diversity, and diverse farm household variables. Systems analysis approaches have many of the same advantages. Moreover, both these types of modeling approaches are capable of directly assessing trade-off relationships among alternative outcomes. Spatially explicit models incorporating geographic information system (GIS)-based approaches can also be useful in situations where spatial dimensions are key, such as in evaluating changes in land-use patterns. Modeling approaches are reviewed in detail in Buck et al. (2004), and examples are provided in van Noordwijk et al. (2002) and in Tomich et al. (chapter 18 this volume).

Conclusions

As reviewed here, many of the elements of ecoagriculture strategies have sound foundations in scientific knowledge and understanding. The authors find that there are many specific examples and case studies in which achieving progress in the multidimensional outcomes of sustainable production, livelihood improvements, and biodiversity conservation—compared to conventional technologies and practices—has been documented. There is considerable evidence, some of it rigorously quantified, that a variety of agricultural practices throughout the world can provide habitat for locally and globally important species of wildlife at the same time that they produce food and other benefits, including livelihood support. Yet existing research does not fully clarify the conditions under which multiple goals can be achieved in different spatial settings.

A cornerstone in evaluating ecoagriculture strategies is assessing the multidimensional outcomes of empirical research. Yet most relevant research to date has yielded only partial results and outcomes, only infrequently addressing all three dimensions of ecoagriculture. This makes it difficult to assess comprehensively the trade-offs and complementarities that can arise in evaluating specific strategies. Another constraint that must be surmounted in future research includes incorporating better long-term indicators, activities, and constraints to address sustainability criteria.

It is also essential to understand the numerous scale issues posed by ecoagriculture. The authors' 2004 assessment noted the importance of accounting for scale effects in different ways: (1) *spatial scale:* the need to link scales of analysis and action and the corresponding units of measurement and management; (2) *aggregation and scaling up:* larger spatial units present challenges in ramping up smaller-scale efforts, including pricing problems and difficulties in collective action; and (3) *measurement scales:* recognizing that the relations among scales of analysis are not linear and must deal with emergent properties at the same time that well-conceived reductionist analysis is carried out. These issues will have to be reckoned with in forthcoming evaluations of land use employing ecoagriculture strategies.

The preponderance of empirical evidence suggests that tradeoffs among agricultural productivity, livelihood improvements, and biodiversity conservation are presently the rule in most agricultural landscapes. The challenge of ecoagriculture research and development is to find and develop synergistic effects and positive-sum interactions between and among the three legs of the ecoagricultural "stool." The evidence in this chapter suggests that there is greater scope for doing this at the scale of individual fields and farms than is commonly assumed. But in many situations, trade-offs will need to be addressed through solutions at a landscape scale (see Jackson et al., chapter 17 this volume).

Research Needs and Opportunities

Perhaps the greatest near-term research need is for improved multidimensional measures and indicators of the performance of ecoagriculture practices and systems that can be consistently used to compare experience across diverse landscapes and agroecosystem settings. Such tools would help researchers to surmount the limitations that currently exist in relating disparate results across many one-off studies. This would also help in sensitizing stakeholders to think about the multiple outcomes of production systems and to plan and monitor land-use strategies that aim to deliver "win-win-win" strategies. This obstacle must be overcome if the research evidence on ecoagriculture strategies is to be fully understood and a convincing case is to be made for the general feasibility of the approach.

Given our limited understanding of the role of biodiversity in agricultural landscapes and how biodiversity can be conserved in these settings, developing indicators and means of measuring wild biodiversity, with respect to its variety and abundance, is one of the highest priorities. There are increasingly sophisticated methodologies for evaluating biodiversity within agricultural systems (Thies and Grossman 2006). In addition, more needs to be known about the relationships between below-ground and above-ground biodiversity.

Related to this priority, we need general guidelines for farm management that are reliably correlated with enhancement of wild biodiversity. To what extent are fewer chemicals, less tillage, better soil health, and more tree cover measurably correlated with wild biodiversity? With more time and resources, it should be possible to construct data-based relationships documenting how examples of wild biodiversity specifically contribute to or enhance crop and animal production, and vice versa, as well as a protocol for documenting additional examples from around the world. Building up a systematic knowledge base on these relationships would strengthen the case for making ecofriendly transitions in farming systems.

Given that any agricultural practice is bound to help some species and displace others, it would be instructive to identify species that are known to be highly important in terms of ecosystem services, or that are rare regionally or globally, or that are good indicator species for the presence and survivability of other groups of species. At the same time, we need to identify which agricultural practices and agricultural landscape-design features can actually help promote sustainable populations of these selected species. It will be further desirable then to be able to identify how indicators of livelihood performance correspond to these relationships.

As a general matter, we need to know more about the specific conditions under which multiple goals can or must be achieved at the field, farm, or local landscape scale, and when more spatially segregated approaches are appropriate. Given that the critical factors conditioning specific empirical production–livelihood–biodiversity outcomes differ across space, we need to achieve a better understanding of the empirical regularities determining which of these conditioning factors are critical and under what specific circumstances.

The formation and mission of Ecoagriculture Partners is an important development in this direction. The organization is building a large network of agricultural practitioners and researchers around the world who are allied with ecoagriculture thinking. If these leaders can work together to create a strategy for systematically measuring and evaluating important ecoagriculture phenomena, we can anticipate that governments, donor agencies, nongovernmental organizations, local government bodies, and, most of all, rural communities and resource users will make better-informed decisions about how to move landscapes toward realizing the important joint objectives of ecoagriculture.

References

Ajayi, O.C., F. Place, F. Kwesiga, and P. Mafongoya. 2006. Impact of natural resource management technologies: Fertilizer tree fallows in Zambia. World Agroforestry Centre (Occasional paper no. 5).

Alkorta, I., I. Albizu, and C. Garbisu. 2003. Biodiversity and agroecosystems. *Biodiversity and Conservation* 12:2521–2522.

Altieri, M.A. 1995. *Agroecology: The Scientific Basis of Alternative Agriculture.* Westview Press, Boulder, CO.

Bacon, C. 2005. Confronting the coffee crisis: Can fair trade, organic, and specialty coffees reduce small-scale farmer vulnerability in Northern Nicaragua. *World Development* 33(3): 497–511.

Bakhsh, K., I. Hassan, and A. Maqbool. 2005. Impact assessment of zero-tillage technology in rice–wheat system: a case study from Pakistani Punjab. *Electronic Journal of Environmental, Agricultural and Food Chemistry* 4(6): 1132–1137. http://ejeafche.uvigo.es/4(6)2005/007462005.pdf.

Benton, T.G., J.A. Vickery, and J.D. Wilson. 2003. Farmland biodiversity: is habitat heterogeneity the key? *Trends in Ecology and Evolution* 18:182–188.

Berg, H. 2002. Rice monoculture and integrated rice-fish farming in the Mekong Delta, Vietnam: economic and ecological considerations. *Ecological Economics* 41:95–107.

Brussaard, L., T.W. Kuyper, W.A.M. Didden, R.G.M. de Goede, and J. Bloem. 2004. Biological soil quality from biomass to biodiversity importance and resilience to management stress and disturbance. In: *Managing Soil Quality: Challenges in Modern Agriculture,* ed. P. Schjønning, S. Elmholt, and B.T. Christensen. CABI Publishing, New York: 139–161.

Buck, L.E., T.A. Gavin, D.R. Lee, and N.T. Uphoff with D.C. Behr, L.E. Drinkwater, W.D. Hively, and F.R. Werner. 2004. *Ecoagriculture: A Review and Assessment of Its Scientific Foundations.* Ecoagriculture Discussion Paper No. 1, Ecoagriculture Partners, Washington, DC.

Carney D., M. Drinkwater, T. Rusinow, K. Neefjes, S. Wanmali, and N. Singh. 1999. *Livelihoods Approaches Compared: A Brief Comparison of the Livelihoods Approaches of the UK Department for International Development (DFID), CARE, Oxfam and the United Nations Development Programme (UNDP).* Department for International Development, London.

Chambers, R. and G. Conway. 1992. Sustainable rural livelihoods: practical concepts for the 21st century. IDS Discussion Paper 296, University of Sussex, UK, Institute for Development Studies.

Christensen, N.L., A.M. Bartuska, J.H. Brown, S. Carpenter, C. D'Antonio, R. Francis, J.F. Franklin, J.A. MacMahon, R.F. Noss, D.J. Parsons, C.H. Peterson, M.G. Turner, and R.G. Woodmansee. 1996. The report of the Ecological Society of America Committee on the Scientific Basis for Ecosystem Management. Ecological Applications 6:665–691.

Clay, J. 2004. *World Agriculture and the Environment: A Commodity-by-Commodity Guide to Impacts and Practices.* Island Press, Washington, DC.

Dale, V.H., S. Brown, R.A. Haeuber, N.T. Hobbs, N. Huntly, R.J. Naiman, W.E. Riebsame, M.G. Turner, and T.J. Valone. 2000. Ecological principles and guidelines for managing the use of land. *Ecological Applications* 10:639–670.

De Janvry, A. and E. Sadoulet. 2005. Achieving success in rural development: toward implementation of an integral approach. *Agricultural Economics* 32(S1):75–89.

Delate, K., M. Duffy, C. Chase, A. Holste, H. Friedrich, and N. Wantate. 2003. An economic comparison of organic and conventional grain crops in a long-term agroecological research (LTAR) site in Iowa. *American Journal of Alternative Agriculture* 18(2):59–69.

Donald, P.F. 2004. Biodiversity impacts of some agricultural commodity production systems. *Conservation Biology* 18:17–37.

Ecoagriculture Partners. 2004. Defining Ecoagriculture: The Nairobi Declaration. www.ecoagriculturepartners.org.

Frei, M. and K. Becker. 2005. Integrated rice–fish culture: coupled production saves resources. *Natural Resources Forum* 29:135–143.

Gregory, M.M., K.L. Shea, and E.B. Bakko. 2005. Comparing agroecosystems: Effects of cropping and tillage patterns on soil, water, energy use, and productivity. *Renewable Agriculture and Food Systems* 20(2):81–90.

Govaerts, B., K.D. Sayre, and J. Deckers. 2005. Stable high yields with zero tillage and permanent bed planting? *Field Crops Research* 94:33–42.

Hassan, A.R. and K. Bakshi. 2005. Pest management, productivity and environment: a comparative study of IPM and conventional farmers of northern districts of Bangladesh. *Pakistan Journal of Social Sciences* 3(8):1007–1014.

Hobbs, P. and R.K. Gupta.2003. Resource-conserving technologies for wheat in the rice–wheat system. In: *Improving the Productivity and Sustainability of Rice–Wheat Systems: Issues and Impacts,* ed. J.K. Ladha, J.E. Hill, R.K. Gupta, J. Duxbury, and R.J. Buresh. ASA Special Publication 65. ASA, Madison, Wisconsin: 149–171.

Hole, D.G., A.J. Perkins, J.D. Wilson, I.H. Alexander, P.V. Grice, and A.D. Evans. 2005. Does organic farming benefit biodiversity? *Biological Conservation* 122:113-130.

Janosky, J.S., D.L. Young, and W.F. Schillinger. 2002. Tillage: Economics of conservation tillage in a wheat-fallow rotation. *Agronomy Journal* 94:527–531.

Jat, M.L., S. Singh, H.K. Rai, R.S. Chhokar, S.K. Sharma, and R.K. Gupta. 2005. Furrow irrigated raised bed (FIRB) planting technique for diversification of rice-wheat system in Indo-Gangetic Plains. *Japan Association for International Collaboration of Agriculture and Forestry* 28(1):25–42.

Kleijn, D., F. Berendse, R. Smit, N. Gilissen, J. Smit, B. Brak, and R. Groeneveld. 2004. Ecological effectiveness of agri-environment schemes in different agricultural landscapes in the Netherlands. *Conservation Biology* 18:775–786.

Lee, D.R. 2005. Agricultural sustainability and technology adoption: issues and policies for developing countries. *American Journal of Agricultural Economics* 87(5):1325–1334.

Lee, D.R. and Barrett, C.B., (eds.). 2001. *Tradeoffs or Synergies? Agricultural Intensification, Economic Development and the Environment.* CABI Publishing, Wallingford, UK.

Lennon, J.J., P. Koleff, J.J.D. Greenwood, and K.J. Gaston. 2004. Contribution of rarity and commonness to patterns of species richness. *Ecology Letters* 7(2):81–87.

Lithourgidis, A.S., K.V. Dhima, C.A. Damalas, I.B. Vasilakoglou, and I.G. Eleftherohorinos. 2006. Tillage effects on wheat emergence and yield at varying seeding rates, and on labor and fuel consumption. *Crop Science* 46:1187–1192.

McNeely, J.A. and S.J. Scherr. 2003. *Ecoagriculture: Strategies to Feed the World and Save Wild Biodiversity.* Island Press, Washington, DC.

Narayanamoorthy, A. 2004. Impact assessment of drip irrigation in India: A case for sugarcane. *Development Policy Review* 22(4):443–462.

Nelson, M., R. Dudal, H. Gregeresen, N. Jodha, D. Nyamia, J.-P. Groenewold, F. Torres, and A. Kassam. 1997. *Report of the Study on CGIAR Research Priorities for Marginal*

Lands. Technical Advisory Committee, Consultative Group on International Research, and FAO, Rome.

Perfecto, I., J. Vandermeer, A. Mas, and L. Soto-Pinto. 2005. Biodiversity, yield, and shade coffee certification. *Ecological Economics* 54:435–446.

Pimentel, D., P. Hepperly, J. Hanson, D. Douds, and R. Seidel. 2005. Environmental, energetic, and economic comparisons of organic and conventional farming systems. *BioScience* 55(7):573–582.

Pretty, J. (ed.). 2005a. *The Pesticide Detox.* Earthscan, London.

Pretty, J. 2005b. Sustainability in agriculture: recent progress and emergent challenges. *Environmental Science and Technology* 21:1–15.

Pretty, J.N., Morison, J.I.L. and R.E. Hine. 2003. Reducing food poverty by increasing agricultural sustainability in developing countries. *Agriculture, Ecosystems and Environment* 95(1):217–234.

Pretty, J.N., A.D. Noble, D. Bossio, J. Dixon, R.E. Hine, and F.W.T. Penning de Vries. 2006. Resource-conserving agriculture increases yields in developing countries. *Environmental Science and Technology* 40(4):114–1119.

Rasul, G. and G.B. Thapa. 2004. Sustainability of ecological and conventional agricultural systems in Bangladesh: An assessment based on environmental, economic, and social perspectives. *Agricultural Systems* 79:327–351.

Sax, D.F. and S.D. Gaines. 2003. Species diversity: from global decreases to local increases. *Trends in Ecology and Evolution* 18(11):561–566.

Sayre, K.D., and P. Hobbs. 2004. The raised-bed system of cultivation for irrigated production conditions. In R. Lal (ed.), *Sustainable Agriculture and the International Rice-Wheat System* (pp. 337–355). New York: Marcel Dekker.

Scherr, S.J. 2003. Ecoagriculture: Landscapes that Integrate Agriculture, Biodiversity, and Ecosystem Services. Seminar presented to the Cornell International Institute for Food, Agriculture, and Development, Cornell University, New York, November 2003.

Swift, M.J., A.-M.N. Izac, and M. van Noordwijk. 2004. Biodiversity and ecosystem services in agricultural landscapes: are we asking the right questions? *Agriculture, Ecosystems and Environment* 104:113–134.

Thadchayini, T., and S. Thiruchelvam. 2005. An Economic Evaluation of a Drip Irrigation Project for Banana Cultivation in Jaffna District. Water Professionals' Day Symposium.

Thies, J.E. and J.G. Grossman. 2006. The soil habitat and soil ecology. In: *Biological Approaches to Sustainable Soil Systems,* ed. N. Uphoff, A. Ball, E. Fernandes, H. Herren, O. Husson, M. Laing, C. Palm, J.N. Pretty, P.A. Sanchez, N. Sanginga, and J. Thies. CRC Press, Boca Raton, Florida: 59–78.

Uphoff, N. 2003. Higher yields with fewer external inputs? The system of rice intensification and potential contributions to agricultural sustainability. *International Journal of Agricultural Sustainability* 1:38–50.

Uphoff, N., A. Ball, E. Fernandes, H. Herren, O. Husson, M. Laing, C. Palm, J.N. Pretty, P.A. Sanchez, N. Sanginga, and J. Thies (eds.). 2006. *Biological Approaches to Sustainable Soil Systems.* CRC Press, Boca Raton, Florida.

Van Noordwijk, M. 2002. Scaling trade-offs between crop productivity, carbon stocks and biodiversity in shifting cultivation landscape mosaics: the FALLOW model. *Ecological Modelling* 149(1–2):113–126.

Wilson, E.O. and E.O. Willis. 1975. Applied biogeography. In: *Ecology and Evolution of Communities,* ed. M.L. Cody and J.M. Diamond. Belknap Press, Cambridge, Massachusetts: 522–534.

Chapter 3

Biodiversity in Agroecosystems

Judith Thompson, Toby Hodgkin, Kwesi Atta-Krah, Devra Jarvis, Coosje Hoogendoorn, and Stefano Padulosi

Introduction

This chapter presents agricultural biodiversity in the context of ecoagriculture from a crop production perspective. It describes examples of agricultural biodiversity's role in and potential for ensuring stability and productivity of agriculture while benefiting the environment and wild biodiversity.

Agricultural Biodiversity

Agricultural biodiversity comprises all the elements, from genes to agricultural ecosystems, that are used in food production. These include crops, trees, fish, and livestock, as well as interacting species of pollinators, pests, parasites, and predators (Qualset et al. 1995.) Maintaining diversity and the balance among its various components enhances agricultural productivity, food security, socioeconomic and nutritional value, and environmental sustainability.

These benefits are central to the achievement of the Millennium Development Goals that were adopted by the United Nations in 2000 and were recognized in the Convention on Biological Diversity Program of Work on Agricultural Biodiversity.

Agriculture has often been viewed as in conflict with environmental concerns and is still frequently portrayed as an "enemy" to many of those involved in conservation and the environment. In reality, improved management and the use of diversity for production are necessary components of an agriculture that both meets human needs and contributes to environmental improvement and conservation. The importance of the various components of agricultural biodiversity and the contribution they make to sustainable production, livelihoods, and ecosystem health are now widely appreciated. Thus crop and tree diversity (within and between crop variation) can help farmers avoid risk, increase food security, and generate income, as well as optimize land use and increase adaptability to changing conditions (Brush 1995). Livestock and fish diversity has the same functions and provides similar benefits in many farming and aquatic systems. Soil organisms contribute a wide range of essential services to the sustainable functioning of agroecosystems through nutrient cycling, regulation of the dynamics of soil carbon sequestration and greenhouse gas emission, effects on soil physical structure and water regimes, and influences on plant life (e.g., nitrogen fixation and the interactions in the soil of pests, predators, and other organisms) (Swift et al. 2003). Pollinators are essential for seed and fruit production, and their number and diversity can profoundly affect crop production levels. However, knowledge is still limited on the interactions and synergies among these different systems (crop, livestock, and aquatic) and their associated biodiversity (e.g., soil organisms, pollinators).

The diversity of plant and animal species maintained in traditional farming systems over many centuries and the knowledge associated with managing these resources constitute key assets of rural people. The management and use of these assets, and the practices that maintain pollinators and associated belowground diversity, provide the natural capital of rural livelihoods. In marginal and difficult farming conditions this is especially true. In these circumstances, diversity management can become a central part of the livelihood management strategies of farmers and pastoralists in stress-prone production areas.

Historically, a substantial decline in biodiversity in agroecosystems has tended to follow agricultural intensification and development, and these trends continue. The consequences of such a decline for small-scale farmers in developing countries can be devastating, decreasing the resilience of agroecosystems and increasing their vulnerability. Resilience is "the capacity to absorb shocks while maintaining function," providing the components for renewal and reorganization when change occurs (Folke et al. 2002). Inability to cope with risks, stresses, and shocks, be they economic or environmental, undermines small-scale farmers' livelihoods substantially.

Biodiversity carries additional values because of its role in providing ecosystem services. These services are conventionally not assigned any value and are often, erroneously, regarded as "free" resources. Humans exploit biodiversity's ecosystem functions, such as nutrient cycles and water regimes, biological control of diseases and pests, and regulation of greenhouse gases (MA 2005). These ecosystem services can be sustained in biologically diverse agricultural systems. The biodiversity maintained by farmers is therefore of benefit not only to the farmers but also to society as a whole. Different sectors of society perceive these values in distinct ways. The farmer depends for livelihood on the services with a direct agricultural production function such as nutrient cycles. Neighboring communities benefit from disease control and water quality regulation. The global community benefits from greenhouse gas regulation and preservation of biodiversity itself.

Diversity in the different components of agriculture contributes to an increased but sustainable level of agricultural production. Although the three main levels of biodiversity (ecosystem, species, and genetic) are all important for sustainable production, the focus here is mainly on genetic and species diversity, using examples from work over the past 10 years on the role of crop diversity in production. These examples show how within- and between-crop diversity is used to maintain and improve production and what practical approaches are being developed with farmers in developing countries for the management and maintenance of genetic diversity.

Genetic Diversity for Sustainable Production and Productivity

Traditional low-input agricultural systems are still common in many parts of the developing world. They are often characterized by high levels of diversity, both within and between crops. Numerous varieties of different crops are maintained by individual communities, and the varieties themselves often contain considerable genetic variation. Millions of small-scale farmers depend on within- and between-variety genetic diversity for their livelihoods and food security strategies. Such diversity is also seen in traditional agroforestry systems such as shifting cultivation, home gardens (Kumar and Nair 2004), and agroforests/forest gardens (Kaya et al. 2002).

Although traditional varieties (synonymous to landraces, e.g., Harlan 1975) are generally thought to have lower yield potential than modern ones, they offer many advantages to the farmers who continue to use them. They also have inherent potential for increased productivity, although this potential is largely untapped because of neglect by research and development concerns. They are often better adapted to local climatic conditions, cultural practices, and pests and diseases; provide resilience and stability under variable and difficult cropping conditions; and meet different special use requirements (Bellon 1995; Brush 1995). During the 1970s and 1980s the general feeling was that these varieties would rapidly disappear, but they have proved remarkably resilient. Even where newer high-yielding varieties of a crop have been introduced, local traditional varieties continue to be used where they meet the needs (agronomic, economic, social, or cultural) of the farmers who grow them.

Maintenance of Crop Diversity in Production Systems

Over the past decade, understanding of the ways in which genetic diversity within and between crop varieties is maintained and used has improved considerably (e.g., Teshome et al. 2001). Studies on major food crops like sorghum in Ethiopia (Teshome et al. 1997) and Burkina Faso, rice in Nepal (Bajracharya et al. 2005), and potatoes in the Andes (Zimmerer 2003) have shown that large numbers of cultivars can often be maintained by communities of only a few thousand households. Although these varieties often lack uniformity, they are agromorphologically distinguishable and characterized by specific genetic traits. They are maintained because they meet specific needs and farmers are familiar with their traits. Thus Teshome et al. (1999) found that farmers had a good knowledge of resistance to pests associated with sorghum varieties in Ethiopia and that increasing abiotic stress risk was correlated with increased diversity in terms of numbers of varieties maintained. These findings have been borne out for many other crops in diverse situations throughout the world (e.g., Brush 1999; Jarvis et al. 2000).

The traditional varieties that farmers maintain meet a range of needs in production and use that act as primary drivers for diversity maintenance in these production systems. A few varieties are usually grown in relatively large areas, often by many farmers, to meet general subsistence requirements. Others are grown in much smaller areas by many of the farmers and usually meet specific needs or opportunities. These can include relatively low-yielding but high-value varieties or varieties for special occasions or of particular cultural significance. However, by far, the largest number of varieties are grown in relatively small areas by only one or two farmers. The reasons for the maintenance of these varieties are more difficult to determine, and they may be particularly likely to disappear as development proceeds.

The maintenance of diversity in traditional farming systems depends on farmers maintaining their own seed or obtaining seed of the varieties that they need in other ways. Studies of seed supply systems have shown that, although most seed is saved by individual farmers, significant amounts may be exchanged in informal networks of friends and relatives, sold, or obtained from other sources such as local markets (e.g., Louette et al. 1997; Balma et al. 2004). Extremely complex matrices of seed supply have been identified that depend on informal mechanisms and effective community institutions (Subedi et al. 2003). These seem to play a key part in ensuring farmers have the seed they need and that varietal diversity is maintained within production systems. Healthy seed supply systems resemble healthy ecosystems in that they have relatively greater stability and resilience (McGuire 2001).

Diversity against Pests and Diseases

Up to 30% of the world's potential annual harvest is lost to pest damage and diseases, with developing countries experiencing the greatest devastation. The resulting economic and food resource costs are one consequence of the continuing evolution of new races of pests and pathogens that are able to overcome resistance genes introduced by modern breeding. New cultivars are produced with new resistance genes, but these are frequently overcome in only a few cropping seasons by new virulent races of the pathogen. Indeed, host resistance breeding and pesticide use remain the most common strategies to protect crops against pest and disease pressures. In most cases, however, these responses provide only temporary solutions.

Loss of genetic choices, reflected as loss of local crops or cultivars, diminishes farmers' capacities to cope with changes in pest and disease infestations and leads to yield instability and loss and taking more, often marginal, land for production. Pesticides are prohibitively expensive for small-scale farmers and threaten human health and ecosystem stability. Farmers and plant breeders have long selected for pest and pathogen resistance in agriculture (Finckh and Wolfe 1997; Thinlay et al. 2000) and in developing systems for reducing the impact of diseases on their crops. Farmers have local preferences for growing mixtures of cultivars that provide resistance to local pests and diseases and enhance yield stability (Trutmann et al. 1993; Youyong et al. 2000). Within-crop diversity (through variety mixtures, multilines, or the planned deployment of different varieties in the same production environment) reduces damage by pests and diseases, so local farmers often use traditional varieties and mixtures of traditional and modern varieties to achieve reduced pest and disease damage. Management of intraspecific diversity is, therefore, a potentially important option for pest and disease control in developing economies, where many resource-poor smallholders practice low-input agriculture.

Mechanisms for Diversity Maintenance

Given the value of diversity for agroecosystem management and sustainable production, what mechanisms can help farmers to maintain their traditional varieties where they wish to do so? A wide range of actions (reviewed in Jarvis et al. 2004) are providing many useful, tested approaches (CIP-UPWARD 2003).

Local diversity fairs provide an opportunity for farmers to display the different varieties they manage, to exchange information about them, and to buy, sell, or exchange seed (Sthapit et al. 2003). Seed fairs, using a similar approach, are providing ways of ensuring that farmers' seed supplies can be reestablished following wars or other disasters such as drought or floods (Sperling and Longley 2002). Other ways of ensuring those seed supplies include community gene banks in Ethiopia (Tesema 2004), and gene–seed–grain banks in India (Bala Ravi, pers. comm.). Such banks often require some physical infrastructure and considerable community commitment to maintenance and management. A practical alternative can be the maintenance of a Community Biodiversity Register (Anil Kumar et al. 2003) that helps farmers to keep track of who is growing what variety.

More ambitious activities, often carried out as part of specific development projects, involve farmer review of available materials by on-farm testing of diversity blocks, combined with participatory variety selection or participatory plant breeding (Cleveland and Soleri 2002; Sthapit et al. 1996). Borrowing from the concept of farmer field schools, established for integrated pest management, farmer diversity fora are being tested in West Africa to enable farmers to review the diversity in their area (or from nearby areas) and work together to plan how to maintain supplies of desired types.

Crop and Species Diversity and the Role of Neglected and Underutilized Crops

Local species and varieties maintained in traditional, small-scale farming communities are being replaced by high-yielding varieties of just a few crops (Altieri and Merrick 1988). Causes include globalization trends such as the strengthening of major commodity markets, globalization of food habits accompanied by a standardization of diets across regions and cultures, loss of agriculture-related traditional knowledge as its transmission from one generation to another is reduced, abandonment of traditional cultivating practices, and urbanization and abandonment of rural areas and related lifestyles.

More specific constraints that hamper the continued deployment of some 7000 less-common species, both cultivated and wild (Wilson 1992), include the lack of germplasm required to support crop improvement programs (Wallis et al. 1989), poor market organization and poor competitiveness of products, inadequate cultivation and postharvest practices, and poor human and physical capacities to address research and development bottlenecks (Padulosi et al. 2002). These "neglected and underutilized species" (NUS) are challenged by marginalization (i.e., neglected by researchers and conservationists) and economic and social underuse (Padulosi and Hoeschle-Zeledon 2004). Much research is still needed on the role of NUS and other locally important agricultural biodiversity in support of the multifunctionalities and complexities of local agroecosystems (see Leakey, chapter 5), but a few generalizations can already be made:

Diversity Is Important in Low-Input Agriculture in Marginal Landscapes

One of the strongest arguments in favor of increasing the conservation and use of NUS and landraces of major crops is the fact that they have been selected to thrive in marginal lands and in low-input agroecosystems where they have evolved comparative advantages over major crops during centuries of natural and human selection. Examples include quinoa (*Chenopodium quinoa*) growing in the highlands of the Andes up to 3800 meters above sea level (Partap et al. 1998; Risi and Galwey 1989); minor millets (such as *Eleusine coracana*), cultivated in the poorest soils of high and hilly lands of India and Nepal (Bhag Mal 1994); and teff (*Eragrostis tef*), grown at high altitudes in Ethiopia (Cheverton and Chapman 1989). Most NUS are cultivated without resort to pesticides because farmers are either too poor to afford these or because their products are gathered directly in the wild. The fact that they are grown organically increases the market value of these plants, particularly for the production of baby food, herbal tea, and medicinal remedies.

NUS Enhance Sustainable Exploitation of Soil Nutrients

NUS are part of a broad portfolio of species characterized by lower nutritional requirements than those of commodity crops or high-yielding varieties. Including NUS in agricultural systems increases complementarity in nutrient absorption and hence reduces the risks of production loss associated with their depletion in the soil (including the buildup of pests and diseases) (NRC 1993).

NUS Buffer against Environmental Changes

Although greater species diversity usually requires more labor, the greater diversity of crops deployed in the field minimizes risks associated with unpredictable changes of weather and other environmental factors (pest and disease attacks, shortage of water, etc.) (Nuez et al. 1997). The fact that many NUS are highly adapted to marginal lands and present in areas prone to drought, water shortages, and other abiotic stresses makes this buffering role extremely valuable for sustainable livelihoods.

NUS Maximize Returns from Production Inputs

Traditional agricultural systems are characterized by small plots of land grown with a diverse portfolio of species meant to satisfy many needs of the grower's farm and community. Such plots may contain starch and protein crops, aromatic plants for condiments, vegetables, fruit trees, fiber crops, and species planted for their aesthetic, cultural, or social values. The diversity deployment is highest in home gardens (Chweya and Eyzaguirre 1999), which are important throughout the world, especially in the tropics and in marginal lands. In vulnerable areas (e.g., flood-prone areas of Bangladesh) they can be critical to survival because they withstand periodic disturbance (e.g., Oakley 2004).

Home gardens are productive ecosystems extremely rich in genetic diversity as well as species diversity. Small in size (usually 0.1–1.0 ha.) and adjacent to the homestead, they sometimes harbor more than 100 species to meet the food, fodder, green manure, building, firewood, and medical needs of the family. They conserve agricultural biodiversity, creating microenvironments for crops and crop varieties that were once more widespread in the larger agroecosystem.

These multistory, multispecies microenvironments support a richly productive combination of locally adapted cultivars, newly domesticated wild species, and introduced cultivars arising from exchange and interactions between cultures and communities. This combination increases the overall productivity of the agroecosystem (Watson and Eyzaguirre 2002) and helps maximize returns to farmers from use of land, labor, and other inputs. Useful wild species are often moved into home gardens when their natural habitats are threatened, as in the case of loroco (*Fernaldia pandurata*) following the high rate of deforestation in Guatemala (Leiva et al. 2002). Home garden species in Guatemala, Ghana, Vietnam, Venezuela, and Cuba contain a significant amount of crop genetic diversity and can constitute a sustainable in situ conservation system (Watson and Eyzaguirre 2002).

Links between Crops and Other Components of the Agroecosystem

Crop production and the maintenance of diversity within cropping systems are closely related to the ways in which wild diversity is maintained within and around production systems. Good examples of some possible relationships are the use of wild indicators to help manage production, the maintenance and use of crop wild relatives, and the role of agroforestry.

Natural Ecosystem Components and Crop Production

Of increasing interest to researchers and farmers alike is the link between crop diversity and the surrounding natural ecosystem. Natural ecosystem components are used by most farming communities as an agricultural almanac, pointing to the beginning and end of the rainy season, and thus influencing decisions on the timing of planting and harvesting of local crop cultivars (Sawadogo et al. 2001). In Burkina Faso, farmers observe indicators in the natural ecosystem such as the appearance of new leaves, yellowing of leaves, flowering, and ripening of fruit in several wild tree species: *Lannea microcarpa*, *Sclerocarya birrea*, and *Butyrospermum paradoxum*. These plants, together with the appearance of certain migratory birds, types of bird calls, birds building their nests, lizards shedding skin, toads going from bush to ponds, star constellations, weather signs, and dates on the traditional lunar calendar, are used to predict the characteristics and timing of seasons.

Crop Wild Relatives

Many species characteristic of secondary or disturbed vegetation, and which occur in agroecosystems, are themselves closely related to crop species. Often, as in the case of wild barley (*Hordeum spontaneum*), they are direct progenitors of the cultivated species. In other cases they are members of the primary, secondary, or tertiary gene pool (Harlan and de Wet 1971) and are used in crop improvement. They have provided genes conferring improved disease resistance, yield potential, nutritional qualities, and even medicinal benefits such as the anticarcinogenic traits found in wild *Brassica* spp. from Sicily (Meilleur and Hodgkin 2004). *Lactuca serriola*, a common weed found in cultivated land in many European countries, provides the primary source for disease resistance to lettuce mildew (Crute 1992). *Oryza rupifogon*, a weedy relative of rice, can improve rice yields by up to 17% (Xiao et al. 1998), and wheat relatives characteristic of grasslands in West and Central Asia are a primary source of resistance to biotic and abiotic stress resistance in wheat (Appels and Lagudah 1990). This is not to say that weeds are desirable, but the crop-related species need to be maintained as a major source of new genes that could reduce inputs of pesticides and fungicides and raise yields in sustainable ways.

Wider Considerations

Plant diversity, particularly crop plant diversity, is essential to ecoagricultural systems. Equally good cases could have been made for the importance of maintaining diversity of soil biota (Giller et al. 1997), pollinators (Buchmann and Nabhan 1996; Kevan 1999), or livestock (Steinfeld et al. 1997), none of which have been addressed in this chapter. One of the real challenges is to bring work on different components together in an integrated whole, just as the components themselves form an integrated whole on the farm.

Ecoagriculture seeks to increase global food security while conserving biodiversity. This requires considering both wild and cultivated biodiversity and the dynamic interaction between them in food production, ecosystem services, and human livelihoods. Through sustainable management of agricultural systems, which are irretrievably linked to natural ecosystems, wild biodiversity can be conserved and protected in the face of changing climate, global conflict, and diminishing possibilities for expanding agricultural land. This requires making conservation of agricultural biodiversity a prominent component of agricultural management practices, and ensuring that agricultural biodiversity considerations are built into the broader biodiversity debate and discourse. The benefits derived from integrating into the ecoagriculture framework a direct concern with the maintenance and use of biodiversity within the production system include food security, land-use optimization, improved livelihoods for and conservation of indigenous cultures, and ultimately a strengthening of ecoagriculture itself.

Food Security

Agricultural biodiversity can also provide insurance against crop failure. Both production stability and crop adaptability, and thus ultimately productivity, strengthen the resilience of crops and their ability to withstand stress.

Optimization of Land Use

The ecosystem services that agricultural biodiversity provides protect the environment from degradation. Traditional crops and landraces, for example, often require less input in terms of fertilizers and pesticides that can build up dangerous residues in the soil and water table. Crop varieties that are highly adapted to local or marginal conditions allow farmers to achieve larger and more varied production without encroaching on protected areas and forests. Improvements in management of agricultural land can reduce the harvesting of wild plant species for food, fuel, and medicine. Intercropping, such as planting coffee under shade trees that yield fruit or nontimber forest products or interspersing banana and cassava among coconut trees, makes optimal use of land for multiple crops.

Improvement of Livelihoods and Conservation of Indigenous Cultures

A mixture of crops and uses, as in home gardens, can improve the nutritional quality of diets as well as providing a source of income for households. Preserving traditional crops means preserving the traditional knowledge associated with those crops and the agricultural (or agroforestry) practices that created and nurtured them. Along with this indigenous knowledge, associated social practices and their cultural context are conserved, thus adding value to the conservation process itself. NUS, for example, contribute significantly to keeping alive centuries-old traditions, such as the agricultural fairs organized in the Andes for quinoa, cañihua, and amaranth (Woods and Eyzaguirre 2004); social rituals, such as preparations of elaborate dishes like the Italian "pistic"—a soup made of 56 different herbs (Paoletti et al. 1995); and religious celebrations.

Strengthening Ecoagriculture

Agricultural management practices that use biodiversity to foster the functions of food security, land-use optimization, and livelihood improvement will have long-term positive effects on wild biodiversity but will require global attention and action. Agricultural research institutes need to explore ways to improve access to genetic material and to information on that material. Breeders need to improve material for marginal lands through participatory plant breeding, and farmers need to work together to adopt ways to reduce agricultural inputs. In addition, national and local organizations should look for ways to increase market availability for underutilized crops and products and improve market and postproduction crop management.

Challenges and Opportunities

Agricultural biodiversity in production systems can contribute to high productivity, in terms of both product quantity and financial returns. Productivity gains based on judicious use of biodiversity can greatly contribute to the sustainability of production levels. Using biodiversity to achieve this is by no means old-fashioned or synonymous with underdevelopment, although high-diversity production systems are often equated with developing countries and traditional systems. In fact, increasing productivity by optimizing diversity rather than using blanket pesticide and fertilizer application on single varieties is very knowledge- and technology-intensive and can be done on a wider scale only if optimal use is made of disciplines such as modeling, precision farming, geographical information systems, genetics, and, particularly in the case of NUS, postharvest technology, processing, and marketing. Because each farm is different, such disciplines must be combined so that better tools and methodologies can be developed and made widely available to help farming communities and farmers optimize biodiversity management in their production systems. The ecoagriculture initiative provides a framework to help the international agricultural research system and other partners work on the missing elements of such a toolkit.

A major challenge remains: bringing together and integrating information from different areas, on different aspects of production, and concerned with different components (e.g., crops, livestock, agroforesty, soils, pollinators, etc.) of the agricultural system. The recently formed Platform for Agrobiodiversity Research seeks to address this challenge. Developed from ideas discussed at various international meetings, the idea of the Platform was endorsed by the 7th Conference of the Parties (COP) to the Convention on Biological Diversity (CBD) in Malaysia (Decision VII/3). The Platform proposes to provide ways of integrating, sharing, and mobilizing knowledge and research findings on sustainable management and use of agrobiodiversity. Its objectives include identifying the contribution agrobiodiversity can make to major global issues and promoting new partnerships to undertake innovative research.

The ecoagriculture approach is based on both a concern for wild biodiversity and the recognition that agriculture needs to provide food and other products for a growing world population in a sustainable way. Although farmers and wildlife protection services are often perceived to be in opposing camps, wild biodiversity as found on and around farms can be very important for most farmers. The increasing call for farming communities to provide ecosystem services and to be paid for doing so recognizes that farmers are influential ecosystem managers. Therefore, farmers need to be provided with the tools to be able to use both farm and wild biodiversity to increase global food security today while protecting and conserving this biodiversity for use and enjoyment by future generations.

References

Altieri, M.A. and L.C. Merrick. 1988. In situ conservation of crop genetic resources through maintenance of traditional farming systems. *Economic Botany* 41:86–96.

Anil Kumar, N., V. Balakrishnan, G. Gigiran, T. Raveendran. 2003. Peoples' biodiversity registers: a case from India. In: *Conservation and Sustainable Use of Agricultural Biodiversity: A Sourcebook,* CIP-UPWARD. International Potato Center—Users' Perspectives with Agricultural Research and Development, Los Banos, California: 236–244.

Appels, R., E.S. Lagudah. 1990. Manipulation of chromosomal segments from wild wheat for the improvement of bread wheat. *Annual Review of Plant Physiology and Plant Molecular Biology* 7:253–266.

Bajracharya J., K.A. Steele, D.I. Jarvis, B.R. Sthapit, and J.R. Witcombe. 2005. Rice landrace diversity in Nepal: variability of agromorphological traits and SSR markers in landraces from a high altitude site. *Field Crops Research* 95:327–335.

Balma, D., T.J. Ouedraogo, M. Sawadogo. 2004. On-farm seed systems and crop genetic diversity. In: *Seed Systems and Crop Genetic Diversity On-Farm: Proceedings of a Workshop 16–20 September 2003,* ed. D. Jarvis, R. Sevilla, and T. Hodgkin. International Plant Genetic Resources Institute, Rome: 51–56.

Bellon, M.R. 1995. The dynamics of crop infraspecific diversity: a conceptual framework at farmer level. *Economic Botany* 50:26–39.

Bhag Mal. 1994. *Underutilized Grain Legumes and Pseudo Cereals: Their Potentials in Asia.* RAPA Publication, FAO 1994/14. Bangkok, Thailand.

Brush, S.B., 1995. In situ conservation of landraces in centers of crop diversity. *Crop Science* 35:346–354.

Brush, S.B., 1999. *Genes in the Field: On-Farm Conservation of Crop Diversity.* International Plant Genetic Resources Institute, Rome; International Development Research Center (IDRC), Ottawa; Lewis Publishers, Boca Raton.

Buchmann, S.E. and G.P. Nabhan. 1996. *The Forgotten Pollinators.* Island Press, Washington, DC.

Cheverton, M.R. and G.P. Chapman. 1989. Ethiopian t'ef: a cereal confined to its centre of variability. In: *New Crops for Food and Industry,* ed. G.E. Wickens, N. Haq, and P. Day. Chapman and Hall, London: 235–248.

Chweya, J.A. and P.B. Eyzaguirre (eds.). 1999. *The Biodiversity of Traditional Leafy Vegetables.* International Plant Genetic Resources Institute, Rome.

CIP-UPWARD. 2003. *Conservation and Sustainable Use of Agricultural Biodiversity: A*

Sourcebook. Vols. 1–3. International Potato Center—Users' Perspectives with Agricultural Research and Development, Los Banos, California.

Cleveland, D.A. and D. Soleri. 2002. *Collaborative Plant Breeding: Integrating Farmer and Scientist Knowledge and Practice.* CAB International, Wallingford, UK.

Crute, I.R. 1992. From breeding to cloning (and back again?): a case study with lettuce downy mildew. *Annual Review of Phytopathology* 30:485–506.

Finckh, M. and M. Wolfe. 1997. The use of biodiversity to restrict plant diseases and some consequences for farmers and society. In: *Ecology in Agriculture,* ed. L.E. Jackson. Academic Press, London: 203–237.

Folke, C. et al. 2002. Resilience and sustainable development: building adaptive capacity in a world of transformation. Scientific Background Paper on Resilience for the process of the World Summit on Sustainable Development, on behalf of the Environmental Advisory Council to the Swedish Government. http://www.resalliance.org/698.php.

Giller, K.E., M.H. Beare, P. Lavelle, A.-M.N. Izac, and M.J. Swift. 1997. Agricultural intensification, soil biodiversity and agroecosystem function. *Applied Soil Ecology* 6:3–16.

Harlan, J.R. 1975. *Crops and Man.* American Society of Agronomy, Madison, Wisconsin.

Harlan, J.R. and J.M.J. deWet. 1971. Toward a rational classification of cultivated plants. *Taxon* 20:509–517.

Jarvis, D., B. Sthapit, and L. Sears. 2000. *Conserving Agricultural Biodiversity In Situ: A Scientific Basis for Sustainable Agriculture.* International Plant Genetic Resources Institute, Rome.

Jarvis, D.I., V. Zoes, D. Nares, and T. Hodgkin. 2004. On-farm management of crop genetic diversity and the Convention on Biological Diversity's Programme of Work on Agricultural Biodiversity. *Plant Genetic Resources Newsletter* 138:5–17.

Kaya M., L. Kammesheidt, and H.J. Weidelt. 2002. The forest garden system of Saparua Island, central Maluku, Indonesia, and its role in maintaining tree species diversity. *Agroforestry Systems* 54:225–234.

Kevan, P.G. 1999. Pollinators as bioindicators of the state of the environment: species, activity and diversity. *Agriculture, Ecosystems and Environment* 74:373–393.

Kumar B.M. and P.K.R. Nair. 2004. The enigma of tropical homegardens. *Agroforestry Systems* 61:135–152.

Leiva, J.M., C. Azurdia, W. Ovando, E. Lopez, and H. Ayala. 2002. Contribution of home gardens to in situ conservation in traditional farming systems: Guatemalan component. In: *Home Gardens and In Situ Conservation of Plant Genetic Resources in Farming Systems,* ed. J.W. Watson and P.B. Eyzaguirre. Proceedings of the Second International Home Gardens Workshop, 17–19 July 2001, Witzenhausen, Federal Republic of Germany: 56–72.

Louette, D., A. Charrier and J. Berthaud. 1997. In situ conservation of maize in Mexico: genetic diversity and maize seed management in a traditional community. *Economic Botany* 51:20–38.

MA (Millennium Ecosystem Assessment). 2005. *Synthesis Report.* Kuala Lumpur, Malaysia. www.maweb.org.

McGuire, S.J. 2001. A note on health indicators. In: *Targeted Seed Aid and Seed System Interventions: Strengthening Small Farmer Seed Systems in East and Central Africa,* ed.

L. Sperling. Proceedings of a workshop in Kampala, 21–24 June 2000. CGIAR Systemwide Program on Participatory Research and Gender Analysis (PRGA), International Center for Tropical Agriculture (CIAT), and International Development Research Center (IDRC): 1–9.

Meilleur, B.A. and T. Hodgkin. 2004. In situ conservation of crop wild relatives. *Biodiversity and Conservation* 13:663–684.

NRC (National Research Council). 1993. *Sustainable Agriculture and the Environment in the Humid Tropics.* National Academy Press, Washington, DC.

Nuez F., J.J. Ruiz and J. Prohens. 1997. Mejora genetica para mantener de la diversidad en los cultivos agricolas. *Documento informativo de estudio No. 6.* Food and Agricultural Organization of the United Nations (FAO), Rome.

Oakley, E. 2004. Home gardens: a cultural responsibility. *LEISA* 20(1):22–23.

Padulosi, S. and I. Hoeschle-Zeledon. 2004. Underutilized plant species: what are they? *LEISA* 20(1):5–6.

Padulosi, S., T. Hodgkin, J.T. Williams, and N. Haq. 2002. Underutilized crops: trends, challenges and opportunities in the 21st century. In: *Managing Plant Genetic Resources,* ed. J.M.M. Engels et al. CAB International Plant Genetic Resources Institute, Rome: 323–338.

Paoletti, M.G, A.L. Dreon, G.G. Lorenzoni. 1995. *Pistic,* traditional food from Western Friuli, N.E. Italy. *Journal of Economic Botany* 49(1):26–30.

Partap, T., B.D. Joshi, and N.W. Galwey. 1998. Chenopods: *Chenopodium* spp. In: *Promoting the Conservation and Use of Underutilized and Neglected Crops.* Institute of Plant Genetics and Crop Plant Research, Gatersleben/International Plant Genetic Resources Institute, Rome: 22.

Qualset, C.O., P.E. McGuire, and M.L. Warburton. 1995. Agrobiodiversity: key to agricultural productivity. *California Agriculture* 49:45–49.

Risi, J.C. and N.W. Galwey. 1989. *Chenopodium* grains of the Andes: a crop for temperate latitudes. In: *New Crops for Food and Industry,* ed. G.E. Wickens, N. Haq, and P. Day. Chapman and Hall, London: 222–234.

Sawadogo, M., D. Balma, J.B. Ouedrago, O.M. Belem, B. Dossou, and D.I. Jarvis. 2001. Sustainability of in situ conservation on-farm: Burkina Faso case study. Paper presented for the workshop, In Situ Conservation of Agrobiodiversity: Scientific and Institutional Experiences and Implications for National Policies, August 14–17, International Potato Centre (CIP), La Molina, Peru.

Sperling, L. and C. Longley. 2002. Beyond seeds and tools: effective support to farmers in emergencies. *Disasters* 26(4):283.

Steinfeld, H., C. De Haan, and H. Blackburn. 1997. *Livestock and the Environment: Issues and Options.* European Commission/FAO/World Bank, Brussels.

Sthapit, B., K.D. Joshi, and J.R. Witcombe. 1996. Farmer participatory crop improvement, III: Participatory plant breeding: a case study for rice in Nepal. *Experimental Agriculture* 32:479–496.

Sthapit, B., D. Rijal, Nguyen Ngoc De, and D. Jarvis. 2003. A role for diversity fairs: experiences from Nepal and Vietnam. In: *Conservation and Sustainable Use of Agricultural Biodiversity: A Sourcebook,* CIP-UPWARD, International Potato Center—Users' Perspectives with Agricultural Research and Development, Los Banos, California: 271–276.

Subedi, A., P. Chaudhary, B. Baniya, R. Rana, R.K. Tiwari, D. Rijal, D.I. Jarvis, and B.R. Sthapit. 2003. Who maintains genetic diversity and how? Policy implications for agro-biodiversity management. In: *Agrobiodiversity Conservation On-Farm: Nepal's Contribution to a Scientific Basis for Policy Recommendations,* ed. D. Gauchan, B.R. Sthapit, and D.I. Jarvis. International Plant Genetic Resources Institute, Rome: 24–27.

Swift, M.J., A.M.N. Izac, and M. Van Noordwijk. 2003. Biodiversity and ecosystem services in agricultural landscapes: are we asking the right questions? *Agriculture, Ecosystems and Environment* 104:113–134.

Tesema, T. 2004. Genetic diversity and on-farm seed systems in Ethiopia. In: *Seed Systems and Crop Genetic Diversity On-Farm,* ed. D. Jarvis, R. Sevilla, and T. Hodgkin. Proceedings of a Workshop, 16–20 September 2003, Pucallpa, Peru.

Teshome, A., B.R. Baum, L. Fahrig, J.K. Torrance, T.J. Arnason, and J.D. Lambert. 1997. Sorghum (*Sorghum bicolor* (L.) Moench) landrace variation and classification in north Shewa and south Welo regions of Ethiopia. *Euphytica* 97:255–263.

Teshome, A., A.H.D. Brown, and T. Hodgkin. 2001. Diversity in landraces of cereal and legume crops. *Plant Breeding Reviews* 21:221–261.

Teshome, A., J.L. Fahrig, J.K. Torrance, J.D. Lambert, T.J. Arnaason, and B.R. Baum. 1999. Maintenance of sorghum (*Sorghum bicolor*, Poaceae) landrace diversity by farmers' selection in Ethiopia. *Economic Botany* 53:79–88.

Thinlay, J., R. Zeigler, and M.R. Finkh. 2000. Pathogenic variability of the blast pathogen *Pyricularia grisea* from the high and mid-altitude zones of Bhutan. *Phytopathology* 90:621–628.

Trutmann P., J. Voss, and J. Fairhead. 1993. Management of common bean diseases by farmers in the Central African highlands. *International Journal of Pest Management* 39(3):334–342.

Wallis, E.S., I.M. Wood, and D.E. Byth. 1989. New crops: a suggested framework for their selection, evaluation and commercial development. In: *New Crops for Food and Industry,* ed. G.E. Wickens, N. Haq, and P. Day. Chapman and Hall, London: 36–52.

Watson, J.W. and P.B. Eyzaguirre (eds.). 2002. *Home Gardens and In Situ Conservation of Plant Genetic Resources in Farming Systems.* Proceedings of the Second International Home Gardens Workshop, 17–19 July 2001, Witzenhausen, Federal Republic of Germany.

Wilson, E.O. 1992. *The Diversity of Life.* Penguin, London.

Woods, A. and P. Eyzaguirre. 2004. Cañahua deserves to come back. *LEISA* 20(1):11–13.

Xiao J., J. Li, S. Grandillo, S.N. Ahn, L. Yuan, S.D. Tanksley, and S.R. McCouch. 1998. Identification of trait-improving quantitative trait loci alleles from a wild rice relative, *Oryza rufipogon*. *Genetics* 150:899–909.

Youyong Zhu, H. Chen, J. Fan, Y. Wang, Y. Li, J. Chen, J. Fan, S. Yang, L. Hu, H. Leung, T. Mew, P. Teng, Z. Wang, C. Mundt. 2000. Genetic diversity and disease control in rice. *Nature* 406:718–722.

Zimmerer, K. 2003. Geographies of seed networks for food plants (potato, ulluco) and approaches to agrobiodiversity conservation in the Andean countries. *Society and Natural Resources* 16:583–601.

Chapter 4

Perennial Grains

Lee R. DeHaan, Thomas S. Cox, David L. Van Tassel, and Jerry D. Glover

Introduction

Annual grain production dominates more than two-thirds of global cropland. Despite recent improvements, grain production continues to cause erosion, nitrogen loss, and pesticide contamination. These problems are major threats to human health, biodiversity, and sustained food production. Organic and no-till methods are unable to remedy these problems inherent in annual grain agriculture. Conversion to perennial crops (i.e., crops that live for more than two years) addresses all three of these major issues, but current perennial crops are unable to provide the grain required to feed the human population.

Developing perennial grains is an obvious solution. Plant breeders are increasingly aware of the promise that perennial grains hold, and work has begun with wheat, triticale, intermediate wheatgrass, rye, sorghum, sunflower, Illinois bundleflower, flax, chickpea, rice, pearl millet, and hybrid hazelnuts. Potential exists to develop the following perennial grain crops: beach wildrye, oats, maize, Indian ricegrass, barley, soybean, buckwheat, pigeonpea, and wild senna. This list is not exhaustive, but it provides an idea of the wide range of possibilities.

Perennial grain development depends upon support from researchers in diverse fields, but the work of plant breeders is primary. The sustained funding of long-term perennial grain breeding programs is a unique challenge. Endowments, the work of private nongovernment organizations, and participation of major public research institutions will be critical. If perennial grain development is supported, humanity may at last achieve a means of sustaining food production without compromising human health or wild biodiversity.

Perennial grain crops, were they to become invaders of natural ecosystems, could be a threat to biodiversity. However, there is no reason to believe that perennial grain crops are more likely to become invasive species than are current annual grain crops. Domestication typically results in noncompetitive, short plants that produce large, nondormant seeds that are held on the plant indefinitely. These traits are in direct contrast with the traits typical of good invaders: competitiveness, abundant vegetative growth, many small seeds, effective seed dispersal, and lengthy seed dormancy.

Rationale

Agriculture has been at odds with the preservation of biodiversity for millennia. Although techniques and strategies are now available to produce food while preserving biodiversity (Jackson and Jackson 2002; McNeely and Scherr 2003), the most successful current approaches have been limited to grazing systems, agroforestry, and small-scale fruit and vegetable production. Despite successes in these areas, robust strategies for reconciling grain production and biodiversity conservation are sorely lacking. We are faced with the stark reality that more than two-thirds of global cropland is dedicated to annual grain crops largely produced in monocultures (FAO 2003). Although small improvements to annual grain production systems have been developed, most systems remain directly opposed to the preservation of wild biodiversity. Grain agriculture is driven by its extensive scale and complete reliance upon annual plants, so the only viable long-term approach to meeting the grain needs of humanity and conserving biodiversity lies in the development of new and more adaptive perennial grain crops. Fortunately, breeding perennial grain crops is a reasonable, albeit long-term, objective (Cox et al. 2006).

The necessity of perennial grain crops cannot be appreciated without an understanding of the differences between the perennial and annual habits. To illustrate, we will consider a summer annual cropping system in the temperate region. The annual crop matures approximately in September. From crop maturity until planting time in the following May, the agricultural landscape is mostly devoid of living root, stem, or leaf tissue. After planting in May, the annual crop has to start the new season's growth from a seed. Importantly, seeds cannot be planted until the minimum soil temperature required for germination is reached and the soil is dry enough for field operations. Growing from a seed requires substantial lag time as a new root system is developed and leaves are deployed to harvest solar energy. It is not until July that the annual crop has enough leaf tissue to maximize harvest of incoming solar energy and a deep enough root system to maximize uptake of the available water and nutrients.

Now consider the growth cycle of a prairie, which is dominated by perennial plants. The perennial plants of the prairie regrow every year from vegetative storage organs like crowns, tubers, and rhizomes. Even in midwinter, prairie soils are permeated by a thick mass of living roots. When the soil warms and the first spring rains come in March and April, cool season perennial plants use their substantial carbohydrate reserves to rapidly deploy leaves that harvest solar energy. Furthermore, their root systems are using available water and nutrients from the soil. In summer, warm-season plants dominate the prairie and extract water and nutrients from great depth with their deep roots. In September when annual cropping systems are mostly devoid of life, the cool-season prairie plants are again growing vigorously and continue to grow until the soil is frozen.

Soil Erosion

The differences between perennial and annual plants as just described translate into vastly different ecosystem functions. We will begin by considering the effect upon soil erosion. In natural ecosystems dominated by perennials, the dense mat of fibrous roots present in the soil year-round holds the soil tightly in place even in the most sloping conditions. With annual crops, living root systems are present in their fully developed state for only a few weeks during the growing season. Combined with tillage, which buries surface residue and damages soil structure, annual cropping can result in extremely high soil erosion rates. After 100 years of annual cropping, up to 66% of topsoil has been lost when compared to continuous perennial cover (Gantzer et al. 1990). Topsoil loss of this magnitude (22 Mg/ha/yr) can translate into a loss of productivity potential of 60%, despite the addition of fertilizers (Gantzer et al. 1990). As soil erosion reduces the productive capacity of land, conversion of existing wildlands to agriculture is necessary to meet human food needs. Eroded soil and associated phosphorus entering aquatic systems also have a direct negative impact on aquatic life (Tilman et al. 2001; Uri 2001).

In the United States, soil erosion has been substantially reduced, from an estimated 3.1 billion tons in 1982 to 2.1 billion tons in 1992 (Uri 2001). A portion of this reduction is due to improved land management practices, such as no-till. By 1995, no-till and mulch-till were used on 31% and 24%, respectively, of northern US fields (Illinois Agricultural Statistics Service 1998). But about 60% of the reduction in soil erosion since 1982 is accounted for by conversion from annual crops to perennials via government programs (Brady and Weil 1999). Conversion of highly erodable land to perennial cover through the Conservation Reserve Program (CRP) reduces soil erosion by an average of 38.6 Mg/ha/yr, whereas better management of annual crops had only reduced erosion by 6.3 Mg/ha/yr between 1982 and 1992 (Uri 2001). In other words, conversion to perennials was six times as effective at controlling erosion as was improved management of annual crops. Despite progress, soil erosion continues to be a major problem, causing the United States an estimated $37.6 billion in social costs annually (Uri 2001).

Nitrogen Loss

Nitrogen fertilization of agriculture is now about half of the total fixed by all bacteria in all natural terrestrial ecosystems; in local regions, inputs from human activity can surpass natural flows by more than 10-fold (Smil 2001). Nitrogen enrichment of native ecosystems can have a detrimental effect on their species composition (Perry et al. 2004). Nitrogen lost from annual cropping systems can endanger aquatic biodiversity thousands of kilometers distant, illustrating the seamlessness of managed and natural ecosystems (Burkhart and James 1999; Turner and Rabalais 2003).

The key to preventing nitrogen loss and the associated environmental degradation is efficient nitrogen use by agricultural crops. Annual crops have an inherent timing problem when it comes to using nitrogen efficiently. In temperate regions, thawing soils and spring rainfall flush nitrogen from the soil and into ground and surface waters well before annual crops are planted (Dinnes et al. 2002). After annual crops mature in the fall, a second round of nitrate leaching typically occurs (Dinnes et al. 2002). Due to the lack of year-round vegetative cover, annual cropping systems can lose five times the water and 35 times the nitrogen to leaching as perennial systems (Randall et al. 1997). Globally, about 30 to 50% of applied nitrogen fertilizer is taken up by annual crops (Tilman et al. 2002), whereas some perennials can be fertilized at a rate of 200 kg N/ha/yr and lose only 1 kg N/ha/yr to leaching (Andrén et al. 1990; Paustian et al. 1990).

Thus far, technological solutions for nitrogen loss from annual crops have been inadequate (Dinnes et al. 2002). Annual cover crops, which are used in an attempt to mimic the perennial habit with annuals, are one of the more promising solutions. In Minnesota, fall-seeded cereal rye on average reduced nitrogen leaching into tile drains by 11% (Strock et al. 2004). This reduction is modest compared to the gains achieved with perennial crops, and the authors cautioned that stand establishment of the winter cover will only be achieved about one year in four. Reduced tillage practices have almost no impact on nitrogen losses (Dinnes et al. 2002). Fertilization with legume-derived green manures (as is common in organic systems) rather than chemical fertilizer is also not likely to dramatically increase nitrogen use efficiency by annual crops (Crews and Peoples, 2005).

Pesticide Use

In 2000, global pesticide production was 3,750,000 Mg, and if current trends continue it will be 1.7 times that by 2020 (Tilman et al. 2001). The effects of sublethal doses of pesticides on humans and other life forms are currently the topic of much investigation. Recent findings suggest that deformities and population declines of amphibians have been due to pesticide exposure (Hayes et al. 2002; Sparling et al. 2001). In humans, pesticides have been linked to profound learning disorders (Guillette et al. 1998), childhood leukemia (Reynolds et al. 2002), birth defects and shifts in sex ratios (Garry et al. 2002), and reduced sperm counts and quality (Swan et al. 2003). Endocrine disruption effects can occur at very low doses and are likely to be caused by largely unregulated "inactive ingredients" (Lin and Garry 2000).

Given that pesticides can have detrimental impacts on human and animal health at very low doses, the question is whether nontarget organisms, including humans, are being exposed. Strong evidence now suggests that no pesticide-free zones are left on the planet, including the arctic, which is now contaminated with numerous pesticides (Arctic Monitoring and Assessment Programme 1998). Due to the inevitable contamination of nontarget organisms, pesticide-free approaches to controlling insects, pathogens, and weeds are desperately needed. Strategies are available for reducing pesticide use in current systems (McNeely and Scherr 2003), but perennial crops offer expanded opportunities. Perennial crops planted in diverse mixtures that mimic the structure of natural ecosystems are expected to function much like the natural ecosystem (Ewell 1999). Because the prairie has less susceptibility to pest outbreaks than annual monoculture cropping systems, it is hoped that perennial grain polycultures managed with fire, mowing, and grazing will be productive without widespread use of pesticides (Cox et al. 2005; Soule and Piper 1992). Because perennials promote a healthier soil system than annuals, the perennial crop system may be suppressive to many diseases when compared to annual crop systems (Cox et al. 2005). Perennial plants often have very high levels of disease and insect resistance, and we have noted that the perennial grains being developed at the Land Institute in Kansas have high levels of disease resistance when compared to their annual relatives.

Herbicides are the most commonly used pesticide globally, making up about 45% of the dollar value of all pesticides sold (Yudelman et al. 1998). This major category of pesticide could be sharply reduced if cropping systems could be made less susceptible to weed invasion. Prairie restoration has been shown to reduce weed biomass by 94% (Blumenthal et al. 2003). Possible mechanisms of weed reduction in perennial polycultures (prairie) include greater aboveground biomass throughout the season reducing light availability to weed seedlings, establishment limitation, the presence of mutualists or antagonists, and diverse plant communities increasing competitive effects. In our perennial grain breeding plots we have observed a trend similar to that with prairie restoration. Initially, weed pressure increases. But once perennial grain prototypes are established, weed populations are readily managed without herbicides.

The Annual Conundrum

Fundamentally, the problem with annual grains is the scale at which they are grown to meet human food needs. In sparsely distributed garden-sized patches, annual grains would have limited negative impact. However, the human population's demand for cereal grains combined with social and economic pressures will make such an arrangement extremely unlikely in most situations. When annual grains dominate the landscape, as they do in many agricultural regions, the impacts of soil erosion, nutrient leaching, and pesticide contamination are severe. Current agricultural technologies typically address only one of these major problems and remain unable to remedy the other two. For instance, organic production practices eliminate pesticide contamination but have little impact on nutrient leaching or soil erosion. No-till agriculture limits soil erosion but has little impact on nutrient leaching or pesticide contamination. Conversion to perennials is the only proven method of addressing all three of these major impacts of grain agriculture.

Understanding Annual versus Perennial Trade-Offs

One of the strongest barriers to perennial grain crop development has been conceptual in nature. Because perennials use carbohydrate reserves to overwinter in temperate environments, it is often assumed that seed production must necessarily be reduced. Wagoner (1990), a strong proponent of perennial grains, has stated, "The photosynthetic energy assimilated by the perennial plant . . . must be divided among its perennating structures and seeds. . . . Consequently, the resources available for seed production in a perennial appear to be less than in an annual." Although resources in a perennial plant must obviously be divided between vegetation, perennating structures, and seeds, the resources available for seed production need not be less for the perennial. DeHaan et al. (2005) have given a thorough response to this trade-off argument. In short, a robust understanding of trade-offs acknowledges that they are not absolute. They readily shift in response to environmental or genetic changes. Perennial grains have access to resources unavailable to annuals in both space and time. Therefore, total carbon fixation in a perennial system can be greater than in an annual system, even with reduced inputs. Indeed, the productivity of managed systems is on average reduced compared to the natural, largely perennial, systems they have replaced (Field 2001). With more carbon available in the perennial crop, breeding can then be used to make the genetic changes that will allocate a larger fraction of the plant's resources to seed.

A second major conceptual advance recently achieved is a clear understanding of the time scales required. In the past, many perennial grain breeding attempts were poorly funded and relatively short-lived (Cox et al. 2002; Wagoner 1990). In many cases, perennial grain breeding has been a side project. When resources get tight, annual by perennial hybrids are used to improve resistance in the annual crops and the original objective of developing a perennial grain is abandoned (Cox et al. 2002). Having learned this lesson, researchers are hopeful that current and future perennial grain development programs will be initiated under the assumption that the project must be sustained for decades rather than years. By actively engaging in and supporting perennial grain breeding, private organizations like the Land Institute may be able to sustain breeding work despite the employee turnover and political changes that jeopardize long-term projects in public research institutions.

Broad-Scale Approaches

Domesticating promising wild perennials is one of two major strategies for achieving perennial grain crops. The first step is to evaluate the diversity present within wild populations for important traits. The process will then require selection for many traits, including seed yield, seed size, resistance to shattering, resistance to lodging, synchronous maturing of seed, ease of threshing, and food or milling quality. Throughout the process, desirable characteristics of the original species, such as perennialism, must be maintained. Some traits, such as shattering resistance, are often under control of a few genes and will therefore be easily selected once they are identified in wild populations. Seed size and yield are generally controlled by many genes. Increasing traits such as these may therefore require decades of work to reach economic levels if values in the original wild populations are low. Therefore, wild perennials that are candidates for domestication should have reasonably high values for these important quantitative traits. Legume species, moreover, require inoculation with the appropriate bacteria to enable nitrogen fixation. For human consumption, processing or breeding of some species will be required to eliminate an undesirable flavor from the seed.

The second approach is to use genes from existing annual crops as a sort of shortcut to breeding perennial grains. Many annual crops have perennial relatives with which they can be hybridized. In these cases we can attempt to combine the perennial habit with the desirable domestication traits already present within the annual crop. Unfortunately, the hybrids between crops and their wild relatives are often mostly sterile. Many years of work and the use of special techniques can be required to obtain fully fertile hybrids. Therefore, the amount of genetic variability available for breeding can be limiting to future breeding progress.

We do *not* see transgenic technology as a viable approach to developing perennial grains. Many genes controlling functions such as regrowth, disease resistance, insect resistance, cold tolerance, carbohydrate storage, and dormancy are required for a plant to be perennial. Therefore, adding a few genes to an existing crop plant will not be a viable method of creating a perennial crop. Similarly, the number of genes required to obtain high yield precludes the use of transgenic technology to develop a perennial grain. However, there may be a minor role for transgenic improvement of crops developed via standard plant breeding techniques.

Perennial grains are being developed by researchers at several institutions. The following section highlights ongoing perennial grain breeding projects. Note that wheat and rice are, today, by far the most important grains for food consumption. But many of the minor grains discussed below have potential for greater use for human consumption, processing, and animal feed.

Wheat

Wheat (*Triticum* spp.) has been crossed with numerous wild relatives, many of which are perennial (Sharma 1995). Perennial wheat breeding had its origin in the Soviet Union, where work was conducted from the 1930s through the 1950s (Tsitsin 1960), although a truly perennial wheat was never released. Perennial wheat breeding was begun at the Land Institute in 2001. In the first year about 500 hybrids were made between wheat and intermediate wheatgrass (*Thinopyrum intermedium*). Crosses were also made with the wild perennials *Th. elongaum* and *Th. ponticum*. Although the hybrids were completely male sterile, more than 500 progeny were obtained by applying pollen either from wheat or from the perennial parent. In the following season, these 500 progeny set a total of 850 selfed seed, evidence that fertility is being restored. In the coming seasons, researchers will continue to advance generations to restore fertility and to broaden the gene pools by making new hybrids. We have also been diversifying our approach by hybridizing wheat with *Leymus racemosus* and *Agropyron repens*. The ultimate objective is to obtain diverse, fully fertile gene pools from which breeders will be able to develop perennial wheat varieties adapted to specific geographical regions. This process will be made possible through recent advances in cytogenetic (such as the ability to readily identify specific chromosomes) and molecular techniques that were unavailable to the earlier perennial wheat breeders (Cox et al. 2002).

Perennial wheat for the US Northwest is currently being developed by researchers at Washington State University. The program was begun in 1991 and now contains more than 2000 lines that are derived from hybrids between wheat and *Thinopyrum* spp. (Scheinost et al. 2001). This group of researchers is also investigating perennial wheat using cytogenetic approaches (Cai et al. 2001). Currently, they are engaged in a molecular project aimed at mapping the major gene responsible for regrowth in perennial hybrids (M. Arterburn, pers. comm., 2004).

Rice

The International Rice Research Institute (IRRI) had a perennial rice breeding program from 1995 to 2001. Despite very promising results, the program was discontinued and the breeding populations were distributed to cooperators in China, where perennial rice breeding efforts may still continue (Cox et al. 2002).

Rice (*Oryza sativa*) can be readily crossed with one of its perennial ancestors, *O. rufipogon* (Majumder et al. 1997). Upland field performance of *O. sativa/ O. rufipogon* hybrids and progenies has been evaluated in the Philippines (Sacks et al. 2003a). After one year, average survival of the hybrids was 30.6%. Yields of many of the hybrids were large, indicating a good potential for perennial rice to exceed yields of 1 to 2 Mg/ha in Southeast Asia upland conditions. Some of the most strongly perennial families yielded best, and there was not a negative correlation between yield and survival. Particularly promising was the ability of the perennial hybrids to produce a dry season crop, which could be extremely important to the welfare of farmers in Southeast Asia. Ultimately, the researchers concluded that "breeding perennial cultivated rice should be feasible but it will likely take five to ten more years" (Sacks et al. 2003a). Crossing with *O. rufipogon* holds promise for developing perennial rice for temperate regions because an ecotype of the species known as Dongxiang can persist in regions where temperatures below −10°C are common (He et al. 1996).

Rice can also be hybridized with *O. longistaminata*, a vigorous perennial that spreads and persists through underground rhizomes. Hybrids producing rhizomes have been obtained, but severe interspecific crossing barriers result in persistent sterility (Sacks et al. 2003b).

The development of perennial rice may be aided by the detailed molecular map of rice and knowledge of genomic regions influencing important domestication traits (Cai and Morishima 2000; Xiong et al. 1999). Furthermore, markers or cloned genes for rhizome production in other species may be readily transferable to rice (Hu et al. 2003).

Triticale

Triticale (X *Triticosecale*) is a humanmade hybrid between wheat (*Triticum*) and rye (*Secale*). Although commercialization of the species remains limited, there is still room for widespread adoption, particularly in marginal environments (Skovmand et al. 1984). One approach to developing perennial triticale is to hybridize wheat with the wild perennial grass *Secale montanum*. Although fertile hybrids have been obtained, strongly perennial plants have not yet been identified (Cox et al. 2002). Researchers at the Land Institute have been crossing winter triticale varieties adapted to Kansas with *Thinopyrum* species. The resulting hybrids have been vigorous perennials, and progress parallels the wheat breeding program already described.

Intermediate Wheatgrass

Intermediate wheatgrass (*Thinopyrum intermedium*) is a perennial grass used mostly for forage in the United States. For a wild grass, it has many desirable agronomic and culinary qualities (Wagoner 1990). Domestication of intermediate wheatgrass was initiated by the Rodale Institute in Pennsylvania in 1987 and has been continued by the US Department of Agriculture–Natural Resources Conservation Service (USDA-NRCS) Big Flats Plant Materials Center in New York (Cox et al. 2002; Wagoner 1990). One cycle of selection increased seed yield by a remarkable 25% (Wagoner et al. 1996). However, with each selection cycle taking four to five years, an economically viable crop could take several decades to develop. Combining direct domestication with domestication genes obtained from crossing with wheat as a "donor parent" could be a remedy to the long time scale.

Rye

Rye (*Secale cereale*) is a promising candidate for perennialization because it can readily be crossed with its direct perennial ancestor, *S. montanum*. Perennial rye breeding began in the Soviet Union, but a perennial rye was never widely grown (Tsitsin 1960). Similarly, a perennial rye program in Germany was only partially successful (Reimann-Philipp 1995). Perennial rye breeding has been plagued by cytogenetic difficulties causing sterility or loss of perenniality, but some solutions have been proposed (Cox et al. 2002; Reimann-Philipp 1995).

Perennial rye has shown greater promise as forage than as a grain crop. In Australia, cereal rye has been used to improve *S. montanum*, and the cultivar Black Mountain has been released (Oram 1996). In Canada, the perennial rye cultivar AC E-1 has been released for use as forage (S. Acharya, pers. comm., 2003). In Lethbridge, Alberta, stands persist for three to four years. The biggest difficulty has been sterility, which results in severe ergot infestations.

At the Land Institute, perennial rye breeding using *S. montanum* as the perennial parent has been mostly discontinued. We have yet to identify an *S. montanum* line that will persist through the hot, dry Kansas summers on clay soils. We have taken a new approach by making hybrids between cereal rye and *Th. intermedium*. Although the F_1 plants are perennial, they have low vigor and are completely sterile. Thus far, we have had no success restoring fertility by chromosome doubling or backcrossing.

Sorghum

Grain sorghum (*Sorghum bicolor*) is an annual grain that is an excellent candidate for perennialization through wide hybridization. Although grain sorghum is a diploid, induced tetraploid lines can be easily crossed with the perennial weed johnsongrass (*S. halapense*) (Casaday and Anderson 1952). Researchers have been evaluating the progeny of hybrids first crossed in 1983 and then again in 2001 after a period of inactivity, first to identify plants or families that are perennial (in Kansas, "perennial" translates as rhizomatous and winter hardy).

Then, from among winter hardy families they will identify the most croplike plants. From previous crosses, they have selected tetraploid plants that are rhizomatous, early, fertile, and relatively large seeded; some have survived one to two winters, and their progeny are being selected for winter hardiness, productivity, and increased seed size. None of those plants are winter hardy, however, so scientists are crossing them back to winter hardy but less "domesticated" plants.

Because of excessive height and asynchronous tillering among both winter hardy and non–winter hardy plants, these populations are not suitable for yield testing. Researchers intermated these populations in 2004 in an effort to produce populations that will segregate and can be evaluated for grain yield, seed size, winter hardiness, and molecular markers associated with those traits

Molecular markers may be an important tool for future perennial sorghum breeding work. Markers associated with rhizome production and growth in sorghum hybrids have been identified (Paterson et al. 1995) and further efforts are under way at Cornell University to identify rhizome genes (S. Murray, pers. comm., 2004).

Sunflower

Perennial sunflower species are widely distributed in North America. The Land Institute has been investigating one Kansas-adapted species, *Helianthus maximiliani*, for several years, collecting wild populations, identifying variation in key traits of interest, and evaluating performance in two intercropping systems (Jackson and Jackson 1999; Van Tassel, unpublished data). Other perennial *Helianthus* species could also be directly domesticated (Cox et al. 2002) and many have important traits such as disease resistance or desirable fatty acid composition (Seiler 1992), or they may even produce renewable resources such as rubber (Seiler et al. 1991). Other genera in the sunflower family may also hold promise for domestication. The Land Institute now has common-garden observation plots for species from the genera *Silphium* and *Liatris*. Observed in the wild, these genera appear to be extremely drought tolerant, relatively large seeded, and slow to shatter. They are reasonably short and erect and bear easily harvested terminal inflorescences.

An interspecific route to a perennial sunflower oilseed crop is also possible. Numerous crosses between perennial sunflower species and crop sunflower (*H. annuus*) have been achieved. Only the few polyploid perennial species, such as the tuber crop Jerusalem artichoke (*H. tuberosus*) reliably produce fertile hybrids with the annual, diploid crop sunflower without the use of embryo rescue (a technique that often allows even distantly related species to be hybridized).

The Land Institute has attempted a similar strategy using the locally adapted perennial species *H. maximiliani*, *H. salicifolius*, and *H. grosseseratus*. Embryo rescue is required for these crosses, the few surviving hybrid plants are highly sterile, and attempts to restore fertility by chromosome doubling have failed thus far. Future efforts will focus on the more easily crossed perennial species. The Land Institute is beginning to develop a population of these doubled hybrids (amphiploids) with enough genetic diversity and local adaptation to allow selection for both yield and longevity.

Illinois Bundleflower

Illinois bundleflower (*Desmanthus illinoensis*) is an herbaceous legume native to the prairie regions of the United States. It is a promising candidate for domestication due to its high seed yields, which have exceeded 1700 kg/ha in Kansas (Kulakow et al. 1990). In Florida, annual seed yield with two harvests has exceeded 3000 kg/ha (Adjei and Pitman 1993). Furthermore, the seed has a crude protein content of about 38% (Kulakow et al. 1990).

With Illinois bundleflower, wild population diversity has already been evaluated for important traits. Kulakow (1999) identified populations with large seed, nonshattering characteristics, and high seed yield when grown in Kansas. DeHaan et al. (2003) evaluated the performance of populations collected from the northern United States when grown in Iowa and Minnesota.

The Land Institute's primary objective is to transfer the genes conferring nonshattering from low-yielding lines to Kansas-adapted, high-yielding lines. To achieve this goal, we have been working to develop a method for making controlled crosses. Thus far, researchers have had difficulty achieving emasculation (the removal of the pollen-producing anthers) followed by effective pollination. In the future, we may rely on spontaneous outcrossing (which occurs at a rate of about 20%) and attempt to identify hybrids based on their phenotype.

Researchers at the University of Minnesota are conducting agronomic and utilization research with Illinois bundleflower. Preliminary findings indicate extremely high levels of antioxidant compounds, and ongoing hog feeding trials suggest that the grain could be used in animal feed (C. Sheaffer, pers. comm., 2003). These researchers have selected superior strains of nitrogen-fixing bacteria that can increase Illinois bundleflower growth by up to three times that of uninoculated plants (E. Beyhaut, pers. comm., 2004).

Flax

Researchers at the University of Minnesota plan to develop perennial flax by hybridizing cultivated flax (*Linum usitatissimum*) with wild perennial relatives (D. Wyse, pers. comm., 2004). Although many of the perennial species are robust plants with relatively large seed, shattering of seed is a problem. If interspecific hybrids can be made, the nonshattering trait could be readily introgressed from cultivated flax.

The University of Minnesota program has obtained the following perennial and biennial flax species: *Linum altaicum, L. austriacum, L. baicalense, L. bienne, L. flavum, L. hirsutum, L. lewisii, L. perenne, L. tauricum, L. campanulatum, L. sulcatum, L. tenufolium*, and *L. thracium*. These accessions and elite lines of the annual domestic flax *L. usitatissimum* are now being grown in evaluation plots. Hybridization will be attempted, using embryo rescue if necessary (D. Wyse, pers. comm., 2004).

Chickpea

Domestic chickpea (*Cicer arietinum*) is one of nine annual species in the genus *Cicer*, which also contains 34 perennial species (van der Maesen 1972). Two perennial species, *C. anatolicum* and *C. songaricum*, have survived for 10 years in the field at the Western Regional Plant Introduction Station in Pullman, Washington (Cox et al. 2002). Preliminary evaluations of 34 perennial chickpea accessions at Pullman indicated that the necessary diversity for domestication of perennial chickpea exists (C. Coyne, pers. comm., 2002).

Pearl Millet

The prospects for breeding perennial pearl millet are reasonable (Cox et al. 2002). The grain crop species (*Pennisetum glaucum*) is diploid, and the tetraploid, perennial forage crop napiergrass (*P. purpureum*) contains a similar and compatible genome but also a second incompatible genome. These species can be crossed, and napiergrass is considered a good source of genetic variation for pearl millet (Hanna 1990). A useful strategy may be to use pearl millet as a source of "yield genes" for perennial species (National Research Council 1996), but no one has reported attempting backcrosses to the perennial. Chromosome doubling can be used to generate fertile perennial plants with the full genomes of both of the parents (Gonzalez and Hanna 1984; Hanna 1981). If many of these plants were generated using genetically diverse parents, a population—essentially a new species—could be created and selected for high yielding types.

A winter hardy perennial pearl millet is a more distant prospect. Accessions of at least three species, *P. orientale, P. flaccidum*, and *P. alopecuroides*, are winter hardy in Kansas. The first two of these are considered to be more useful for improving pearl millet (Dujardin and Hanna 1989). Hybrids between pearl millet and *P. orientale* are difficult to produce (Kaushal and Sidhu 2000), but Dujardin and Hanna (1987) obtained partially fertile hybrids and backcrosses and predicted that with techniques to increase recombination between genomes these hybrids could serve as a "valuable bridge" for transferring to pearl millet "desirable genes for perenniality, pest resistance, drought tolerance, and winter-hardiness." As usual, attempts to backcross to the perennial have generally not been attempted.

Hybrid Hazelnuts

The European hazelnut tree (*Corylus avellana*) has been crossed with wild North American hazelnut shrub species, *Corylus americana* and *Corylus cornuta*. After decades of breeding, the result is a domestic shrub with the disease resistance and winter hardiness of North American hazelnuts and the productivity and nut quality approaching that of the European hazelnut (Rutter 1994). In Minnesota and surrounding states, the crop is moving toward full commercialization. The American Heartland Hazelnut Association was recently formed to facilitate adoption of the new crop (P. Rutter, pers. comm., 2004).

Although hybrid hazelnuts are woody shrubs, they are worthy of consideration as potential grains. Nuts have typically been considered a high-value specialty item due to the labor requirements involved in their culture and harvest. Hybrid hazels may ultimately require labor inputs more similar to grain crops than other nuts. The shrubs do not require careful pruning and both planting and harvesting can now be mechanized (P. Rutter, pers. comm., 2003). These factors indicate the potential for hazelnuts to move from the category of specialty crop to a staple in the human diet, possibly replacing soybeans.

Other Potential Perennial Grains

Beach wildrye, or lyme grass (*Leymus arenarius*) has been studied as a potential perennial grain crop in Iceland. The substantial genetic variation among accessions of lyme grass could be exploited in a direct domestication project (Anamthawat-Jonsson et al. 1999). The species can be hybridized with wheat, which makes the domestication genes in wheat available to lyme grass breeders (Anamthawat-Jonsson 1996).

The potential for developing perennial oats by crossing domestic oat (*Avena sativa*) with a wild, perennial relative, *A. macrostachya*, may be limited by winter hardiness and sterility due to chromosomal instability (Cox et al. 2002).

Maize (*Zea mays*) can be readily hybridized with its close perennial relative *Z. mays* ssp. *diploperennis*. However, recovering perennial plants as a result of this cross has been difficult (Srinivasan and Brewbaker 1999). Cox et al. (2002) have suggested that molecular markers could greatly assist with perennial maize breeding. Both species are tropical, so breeding perennial maize for temperate climates could be very difficult. Eastern gamagrass (*Tripsacum dactyloides*), a native species in midwestern US prairies, could be a source of winter hardiness in perennial corn (Cox et al. 2002).

Indian ricegrass (*Oryzopsis hymenoides*) is a wild perennial grass native to the western United States. In Montana, this species is currently being harvested and sold as a gluten-free grain under the name Montina™ (Cox et al. 2002). Beyond selection for shatter resistance, little breeding work has been done with the species, so it remains a strong candidate for domestication.

Barley (*Hordeum vulgare*) has several perennial relatives, *H. brachyantherum*, *H. bulbosum*, and *H. jubatum*, which are potential perennial parents in the breeding of perennial barley. Barley has also been hybridized with perennial *Leymus* species (Dewey 1984). Due to lack of wide hybridization work in *Hordeum*, the potential for perennial barley development remains an open question.

Soybean (*Glycine max*) would be among the most difficult annual crops to perennialize. The genus *Glycine* has 16 perennial species (Singh and Hymowitz 1999), but only one (*G. tomentella*) has been hybridized with soybean. Through treatment with colchicine, the chromosome number of the hybrids has been doubled, restoring some fertility (Singh et al. 1993). The hybrids tend toward perenniality even when backcrossed twice to annual soybean (Cox et al. 2002). Although perennial soybean for the tropics holds promise, winter hardiness does not exist in any species of *Glycine* (Cox et al. 2002).

Buckwheat (*Fagopyrum esculentum*) can be hybridized with perennial buckwheat, *F. cymosum*, using embryo rescue techniques, but the resulting hybrids have very low fertility (Woo et al. 1999). If the sterility barrier can be overcome, development of perennial buckwheat will likely be possible.

Many perennial legumes are potentially available for use in the tropics. A good example is pigeonpea (*Cajanus cajan*). In two studies in India, second-crop yields of this species were higher than the first (Newaj et al. 1996; Nimbole 1997).

Wild senna (*Senna marilandica* and *Senna hebecarpa*) is a wild perennial herbaceous legume native to the United States that produces abundant seed yields. Wild populations of *S. marilandica* grown in Kansas have produced, on average, more than 2000 kg seed/ha in the second year (Jackson and Jackson 1999). These species lack the ability of many other legumes to fix nitrogen via symbiosis with *Rhizobium* bacteria. Furthermore, unprocessed seed may be inedible due to the presence of chemicals with laxative properties.

The foregoing list of potential perennial grains includes many of the most promising species for use in the temperate United States, but it is by no means exhaustive. For instance, researchers in Australia are just beginning to evaluate native perennial plants that could be domesticated for use in their country (T. Lefroy, pers. comm., 2004). When the tropics are considered, the list of potential perennial grains may become long indeed. In summary, the species mentioned in this chapter are intended to expand, not limit, the reader's imagination.

Next Steps

The obvious next step is the initiation and expansion of perennial grain breeding programs globally. Although "development" in the context of perennial grains refers to much more than breeding, breeding is needed first. After several years to (in the case of some species) several decades, an interdisciplinary team will be needed to direct the breeding program and bring the new perennial grain into production. Such a team must include ecologists, weed scientists, soil scientists, entomologists, plant pathologists, seed scientists, food scientists, economists, sociologists, and others. However, the work of all these disciplines is secondary to that of the plant breeder. Without at least an early prototype of the perennial crop, the other scientists have little with which to work. Because the benefits of perennial grains would be so far reaching and the novelty of the approach so compelling, there will be numerous researchers in diverse fields willing to work on bringing a perennial grain into production once it has been bred.

For perennial grain development, the work of other scientists in most fields can be completed within several years and within the framework of current funding sources. Breeding of perennial grains, however, requires a long time scale and falls outside the realm of consideration by most funding sources. Therefore, the most critical step is to develop robust strategies for funding and executing the breeding programs. That public support for plant breeding programs has been drying up is a major concern (Knight 2003). If plant breeding in general becomes a lost art, there is little hope for success in perennial grain breeding.

Obtaining funding for perennial grain breeding begins with generating support for the idea. The Land Institute has been working toward this objective for decades, and the rationale presented earlier is derived from that work. But ultimately, success is required to bring greater success. If several perennial grain species can be developed, their benefits will be clear. Once perennial grain crops are actually being used by farmers, the support for developing additional perennial grains will likely be strong.

Success in breeding a new perennial grain will depend primarily upon continuity. Far too many breeding programs have been terminated due to changes in funding allocation, a retirement, or a grant proposal that wasn't funded. Therefore, the endowment of perennial grain development projects is a reasonable—possibly essential—objective. Furthermore, the work of private nongovernmental organizations like the Land Institute, which are buffered from the political and economic forces that can drive public institutions, is critical.

The model proposed at the Land Institute for a perennial grain breeding program on a national or perhaps even global scale is that of a center with numerous satellites. Plant breeding is inherently site specific. Therefore, each bioregion requires its own breeding program, or satellite, for every crop. In the United States, these breeding programs should be based at existing colleges and universities. A center is required so that work is not repeated unnecessarily in each of the satellites. The center can also serve as a clearing house for information and germplasm. Although satellite programs may come and go due to many factors, a center will provide the necessary continuity. Seed and information that is generated at a satellite can be preserved by the center until work at the satellite site can be resumed.

Economic forces currently support the perpetuation of annual crops, despite the problems with annual systems already described. Annual grain breeding programs have sustained internal rates of return on investment in the range of 10 to 40% (Marasas et al. 2003; Smale et al. 1998). These calculations consider yield increases and savings due to reduced pesticide consumption compared to the cost of variety development. In comparison, all new crops (including new annual crops) require initial investment in germplasm collection, agronomic research, and utilization research. New crops also typically have modest yields because they have not benefited from the decades of intense breeding and agronomic research dedicated to the established crops. Thus new crops are typically only successful when unique circumstances are combined with a sustained public and private investment (Paarlberg 1990). Perennial grains face a similar, if not greater, challenge. Evaluating yields of perennials takes a minimum of two years per cycle of selection, slowing progress compared to annuals. The initial investment is large, risk is moderate to high, and initially, grain yields will be moderate. For the economic picture to be positive, one must consider what are referred to in economics as externalities, such as soil erosion, pesticide contamination, and fertilizer runoff. An economic system that internalizes costs such as these and considers the long-term effects of depleting the nonrenewable resource of soil would also support the development of perennial grain crops.

Farmers will be the first to benefit economically from perennial grains. Perennial grains could reduce farmers' costs for fertilizer, pesticide, seed, tillage, and planting. According to one scenario, intermediate wheatgrass yielding 600 kg/ha, on average, over four years without resowing would have had a break-even price in North Dakota of $4.02 per bushel in 1989, similar to that of spring wheat at the time (Wagoner 1990). Unfortunately, farmers as a group have been difficult to organize, and their immediate problems often preclude support for long-term research. However, for those wishing to support systemic change resulting in sustained food production, environmental benefits, and increased economic return to farmers, perennial grains are an excellent "investment" opportunity.

References

Adjei, M.B. and W.D. Pitman. 1993. Response of *Desmanthus* to clipping on a phosphatic clay mine-spoil. *Tropical Grasslands* 27:94–99.

Anamthawat-Jonsson, K. 1996. Wide-hybrids between wheat and lymegrass: breeding and agricultural potential. *Buvusindi* 10:101–113.

Anamthawat-Jonsson, K., B.T. Bragason, S.K. Bodvarsdottir, R.M.D. Koebner. 1999. Molecular variation in *Leymus* species and populations. *Molecular Ecology* 8:309–315.

Andrén, O., U. Lindberg, U. Boström, M. Clarholm, A.C. Hansson, G. Johansson, J. Lagerlöf, K. Paustian, J. Persson, R. Pettersson, J. Schnürer, B. Sohlenius, M. Wivstad. 1990. Organic carbon and nitrogen flows. *Ecological Bulletins* 40:85–126.

Arctic Monitoring and Assessment Programme. 1998. AMAP assessment report: Arctic pollution issues. AMAP Secretariat, Oslo, Norway.

Blumenthal, D.M., N.R. Jordan, and E.L. Svenson. 2003. Weed control as a rationale for restoration: the example of tallgrass prairie. *Conservation Ecology* 7:6. http://www.consecol.org/vol7/iss1/art6.

Brady, N.C. and R.R. Weil. 1999. *The Nature and Properties of Soils*. 12th ed. Prentice-Hall, Upper Saddle River, New Jersey: 717–718.

Burkhart, M.R., and D.E. James. 1999. Agricultural-nitrogen contributions to hypoxia in the Gulf of Mexico. *Journal of Environmental Quality* 28:850–859.

Cai, X., S.S. Jones, and T.D. Murray. 2001. Molecular and cytogenetic characterization of *Thinopyrum* genomes conferring perennial growth habit in wheat-*Thinopyrum* amphiploids. *Plant Breeding* 120:21–26.

Cai, H.W., and H. Morishima. 2000. Genomic regions affecting seed shattering and seed dormancy in rice. *Theoretical and Applied Genetics* 100:840–846.

Casaday, A.J. and K.L. Anderson. 1952. Hybridization, cytological, and inheritance studies of a sorghum cross: autotetraploid sudangrass X (johnsongrass X 4n sudangrass) *Agronomy Journal* 43:189–194.

Cox, C.M., K.A. Garrett, and W.W. Bockus. 2005. Meeting the challenge of disease management in perennial grain cropping systems. *Renewable Agriculture and Food Systems* 20:15–24.

Cox, T.S., M. Bender, C. Picone, D.L. Van Tassel, J.B. Holland, E.C. Brummer, B.E. Zoeller, A.H. Paterson, and W. Jackson. 2002. Breeding perennial grain crops. *Critical Reviews in Plant Science* 21:59–91.

Cox, T.S., J.D. Glover, D.L. Van Tassel, C.M. Cox, and L.R. DeHaan. 2006. Prospects for developing perennial grain crops. *BioScience* 56:649–659.

Crews, T.E. and M.B. Peoples. 2005. Can the synchrony of nitrogen supply and crop demand be improved in legume and fertilizer-based agroecosystems? A review. *Nutrient Cycling in Agroecosystems* 72:101–120.

DeHaan, L.R., N.J. Ehlke, C.C. Sheaffer, R.L. DeHaan, and D.L. Wyse. 2003. Evaluation of diversity among and within accessions of Illinois bundleflower. *Crop Science* 43:1528–1537.

DeHaan, L.R., D.L. Van Tassel, and T.S. Cox. 2005. Perennial grain crops: a synthesis of ecology and plant breeding. *Renewable Agriculture and Food Systems* 20:5–14.

Dewey, D.R. 1984. The genomic system of classification as a guide to intergeneric hybridization with the perennial Triticeae. In: *Gene Manipulation in Plant Improvement*, ed. P.J. Gustafson. Plenum Press, New York: 209–279.

Dinnes, D.L., D.L. Karlen, D.B. Jaynes, T.C. Kaspar, J.L. Hatfield, T.S. Colvin, and C.A. Cambardella. 2002. Nitrogen management strategies to reduce nitrate leaching in tile-drained Midwestern soils. *Agronomy Journal* 94:153–171.

Dujardin, M. and W. Hanna. 1987. Inducing male fertility in crosses between pearl millet and *Pennisetum orientale* Rich. *Crop Science* 27:65–68.

Dujardin, M. and W.W. Hanna. 1989. Crossability of pearl millet with wild *Pennisetum* species. *Crop Science* 29:77–80.

Ewell, J.J. 1999. Natural systems as models for the design of sustainable systems of land use. *Agroforestry Systems* 45:1–21.

FAO. 2003. FAOSTAT agricultural data. http://apps.fao.org/cgi-bin/nph-db.pl?subset-agriculture.

Field, C.B. 2001. Sharing the garden. *Science* 294:2490–2491.

Gantzer, C.J., S.H. Anderson, A.L. Thompson, and J.R. Brown. 1990. Estimating soil erosion after 100 years of cropping on Sanborn Field. *Journal of Soil and Water Conservation* 45:641–644.

Garry V.F., M.E. Harkins, L.L. Erickson, L.K. Long-Simpson, S.E. Holland, and B.L. Burroughs. 2002. Birth defects, season of conception, and sex of children born to pesticide applicators living in the Red River Valley of Minnesota, USA. *Environmental Health Perspectives* 110:441-449.

Gonzalez, B. and W.W. Hanna. 1984. Morphological and fertility responses in isogenic triploid and hexaploid pearl millet X napiergrass hybrids. *Journal of Heredity* 75:317–318.

Guillette E.A., M.M. Meza, M.G. Aquilar, A.D. Soto, and I.E. Garcia. 1998. An anthropological approach to the evaluation of preschool children exposed to pesticides in Mexico. *Environmental Health Perspectives* 106:347-353.

Hanna, W.W. 1981. Method of reproduction in napiergrass and in the 3X and 6X alloploid hybrids with pearl millet. *Crop Science* 21:123–126.

Hanna, W.W. 1990. Transfer of germ plasm from the secondary to the primary gene pool in *Pennisetum*. *Theoretical and Applied Genetics* 80:200–204.

Hayes T.B., A. Collins, M. Lee, M. Mendoza, N. Noriega, A.A. Stuart, and A. Vonk. 2002. Hermaphroditic, demasculinized frogs after exposure to the herbicide, atrazine, at low ecologically relevant doses. *Proceedings of the National Academy of Sciences* 99:5476–5480.

He, G.C., L.H. Shu, Y.C. Zhou, and L.J. Liao. 1996. Overwintering ability of Dongxiang wild rice in Wuhan district, China. *International Rice Research Notes* 21:6–7.

Hu, F.Y., D.Y. Tao, E. Sacks, B.Y. Fu, P. Xu, J. Li, Y. Yang, K. McNally, G.S. Khush, A.H. Paterson, and Z.K. Li. 2003. Convergent evolution of perenniality in rice and sorghum. *Proceedings of the National Academy of Sciences* 100:4050–4054.

Illinois Agricultural Statistics Service. 1998. Annual summary: tillage systems. http://www.nass.usda.gov/il/1998/98116.htm.

Jackson, D.L. and L.L. Jackson. 2002. *The Farm as Natural Habitat: Reconnecting Food Systems with Ecosystems*. Island Press, Washington, DC.

Jackson, W. and L.L. Jackson. 1999. Developing high seed yielding perennial polycultures as a mimic of mid-grass prairie. In: *Agriculture as a Mimic of Natural Systems*, ed. E.C. Lefroy, R.J. Hobbs, M.H. O'Connor, and J.S. Pate. Kluwer Academic Publishers, Dordrecht, Netherlands: 1–37.

Kaushal, P. and J.S. Sidhu. 2000. Prefertilization incompatibility barriers to interspecific hybridizations in *Pennisetum* species. *Journal of Agricultural Science* 134:199–206.

Knight, J. 2003. A dying breed. *Nature* 421:568–570.

Kulakow, P.A. 1999. Variation in Illinois bundleflower (*Desmanthus illinoensis* [Michaux] MacMillan): a potential perennial grain legume. *Euphytica* 110:7–20.

Kulakow, P.A., L.L. Benson, and J.G. Vail. 1990. Prospects for domesticating Illinois bundleflower. In: *Advances in New Crops*, ed. J. Janick and J.E. Simon. Timber Press, Portland, Oregon: 168–171.

Lin, N. and V.F. Garry. 2000. In vitro studies of cellular and molecular developmental toxicity of adjuvants, herbicides and fungicides commonly used in Red River Valley, Minnesota. *Journal of Toxicology and Environmental Health* 60:423-439.

Majumder, N.D., T. Ram, and A.C. Sharma. 1997. Cytological and morphological variation in hybrid swarms and introgressed populations of interspecific hybrids (*Oryza rufipogon* Griff. X *Oryza sativa* L.) and its impact on evolution of intermediate types. *Euphytica* 94:295–302.

Marasas, C.N., M. Smale, and R.P. Singh. 2003. The economic impact of productivity maintenance research: breeding for leaf rust resistance in modern wheat. *International Association of Agricultural Economists* 29:253–263.

McNeely, J.A. and S.J. Scherr. 2003. *Ecoagriculture: Strategies to Feed the World and Save Wild Biodiversity*. Island Press, Washington, DC.

National Research Council. 1996. *Lost Crops of Africa Volume 1: Grains*. National Academy Press, Washington, DC.

Newaj, R., R.D. Roy, A.K. Bisaria, and B. Singh. 1996. Production potential and economics of perennial pigeonpea based on alley cropping systems. *Range Management Agroforestry* 17:69–74.

Nimbole, N.N. 1997. Maximum yield and survival of perennial pigeonpea (*Cajanus cajan*) by improving planting site, watering, and chemicals. *Indian Journal of Agricultural Science* 67:507–509.

Oram, R.N. 1996. *Secale montanum*: a wider role in Australasia? *New Zealand Journal of Agricultural Resources* 39:629–633.

Paarlberg, D. 1990. The economics of new crops. In: J. Janick and J.E. Simon (eds.). Advances in New Crops. Timber Press, Portland, OR: 2–6.

Paterson, A.H., K.F. Schertz, Y. Lin, S. Liu, and Y. Chang. 1995. The weediness of wild plants: molecular analysis of genes influencing dispersal and persistence of johnsongrass, *Sorghum halepense* (L.). *Proceedings of the National Academy of Sciences* USA 92: 6127–6131.

Paustian, K., L. Bergström, P. Jansson, and H. Johnsson. 1990. Ecosystem dynamics. *Ecological Bulletins* 40:153–180.

Perry, L.G., S.M. Galatowitsch, and C.J. Rosen. 2004. Competitive control of invasive vegetation: a native wetland sedge suppresses *Phalaris arundinacea* in carbon-enriched soil. *Journal of Applied Ecology* 41:151–162.

Randall, G.W., D.R. Hugins, M.P. Russelle, D.J. Fuchs, W.W. Nelson, and J.L. Anderson. 1997. Nitrate losses through subsurface tile drainage in CRP, alfalfa, and row crop systems. *Journal of Environmental Quality* 26:1240–1247.

Reimann-Philipp, R. 1995. Breeding perennial rye. *Plant Breeding Reviews* 13:265–292.

Reynolds, P., J. Von Behren, R.B. Gunier, D.E. Goldberg, A. Hertz, and M.E. Harnly. 2002. Childhood cancer and agricultural pesticide use: an ecological study in California. *Environmental Health Perspectives* 110:319–324.

Rutter, P.A. 1994. The potential of hybrid hazelnuts in agroforestry and woody agriculture systems. http://www.badgersett.com/basic%20haz.html.

Sacks, E.J., J.P. Roxas, and M.T. Sta. Cruz. 2003a. Developing perennial upland rice, I: Field performance of *Oryza sativa/O. rufipogon* F1, F4 and BC1F4 progeny. *Crop Science* 43:120-128.

Sacks, E.J., J.P. Roxas, and M.T. Sta. Cruz. 2003b. Developing perennial upland rice, II: Field performance of S1 families from an intermated *Oryza sativa/O. longistaminata* population. *Crop Science* 43:129–134.

Scheinost, P.L., D.L. Lammer, X. Cai, T.D. Murray, and S.S. Jones. 2001. Perennial wheat: the development of a sustainable cropping system for the U.S. Pacific Northwest. *American Journal of Alternative Agriculture* 16:147–151.

Seiler, G.J. 1992. Utilization of wild sunflower species for the improvement of cultivated sunflower. *Field Crops Research* 30:195–230.

Seiler, G.J., M.E. Carr, and M.O. Bagby. 1991. Renewable resources from wild sunflowers (*Helianthus* spp., Asteraceae). *Economic Botany* 45:4–15.

Sharma, H.C. 1995. How wide can a cross be? *Euphytica* 82:43–64.

Singh, R.J. and T. Hymowitz. 1999. Soybean genetic resources and crop improvement. *Genome* 42:605–616.

Singh, R.J., K.P. Kollipara, and T. Hymowitz. 1993. Backcross (BC2–BC4)-derived fertile plants from *Glycine max* and *G. tomentella* intersubgeneric hybrids. *Crop Science* 33:1002–1007.

Skovmand, B., P.N. Fox, and R.L. Villareal. 1984. Triticale in commercial agriculture: progress and promise. *Advances in Agronomy* 37:1–45.

Smale, M., R.P. Singh, K. Sayre, P. Pingali, S. Rajaram, and H.J. Dublin. 1998. Estimating the economic impact of breeding nonspecific resistance to leaf rust in modern bread wheats. *Plant Disease* 82:1055–1061.

Smil, V. 2001. *Enriching the Earth*. MIT Press, London.

Soulé, J.D. and J.K. Piper. 1992. *Farming in Nature's Image: An Ecological Approach to Agriculture*. Island Press, Washington, DC.

Sparling D.W., G.M. Fellers, and L.L. McConnell. 2001. Pesticides and amphibian population declines in California, USA. *Environmental Toxicology and Chemistry* 20:1591-1595.

Srinivasan, G. and J.L. Brewbaker. 1999. Genetic analysis of hybrids between maize and perennial teosinte, I: Morphological traits. *Maydica* 44:353–369.

Strock, J.S., P.M. Porter, and M.P. Russelle. 2004. Cover cropping to reduce nitrate loss through subsurface drainage in the northern U.S. corn belt. *Journal of Environmental Quality* 33:1010–1016.

Swan, S.H., R.L. Kruse, F.L. Dana, B. Barr, E.Z. Drobnis, J.B. Redmon, C. Wang, C. Brazil, and J.W. Overstreet. 2003. Semen quality in relation to biomarkers of pesticide exposure. *Environmental Health Perspectives* 111:1478–1484.

Tilman, D., K.G. Cassman, P.A. Matson, R. Naylor, and S. Polasky. 2002. Agricultural sustainability and intensive production practices. *Nature* 418:671–677.

Tilman, D., J. Fargione, B. Wolff, C. D'Antonio, A. Dobson, R. Howarth, D. Schindler, W.H. Schlesinger, D. Simberloff, and D. Swackhamer. 2001. Forecasting agriculturally driven global environmental change. *Science* 292:281–284.

Tsitsin, N.V. 1960. *Wide Hybridization of Plants*. Israel Program for Science Translation, Jerusalem.

Turner, R.E. and N. Rabalais. 2003. Linking landscape and water quality in the Mississippi River Basin for 200 years. *BioScience* 53:563–572.

Uri, N.D. 2001. A note on soil erosion and its environmental consequences in the United States. *Water, Air, and Soil Pollution* 129:181-197.

Van der Maesen, L.J.G. 1972. Cicer *L., a Monograph of the Genus, with Special Reference to the Chickpea (*Cicer arietinum *L.), Its Ecology and Cultivation*. Mededelingen Landbouwhogeschool, Wageningen, Netherlands.

Wagoner, P. 1990. Perennial grain development: past efforts and potential for the future. *Critical Reviews in Plant Science* 9:381–409.

Wagoner, P., M. van der Grinten, and L.E. Drinkwater. 1996. Breeding intermediate wheatgrass (*Thinopyrum intermedium*) for use as a perennial grain. *Agronomy Abstracts* 1996:93.

Woo, S.H., Y.J. Wang, and C.G. Campbell. 1999. Interspecific hybrids with *Fagopyrum cymosum* in the genus *Fagopyrum*. *Fagopyrum* 16:13–18.

Xiong, L.Z., K.D. Liu, X.K. Dai, C.G. Xu, and Q. Zhang. 1999. Identification of genetic factors controlling domestication-related traits of rice using an F2 population of a cross between *Oryza sativa* and *Oryza rufipogon*. *Theoretical and Applied Genetics* 98:243–251.

Yudelman, M., A. Ratta, and D. Nygaard. 1998. Pest Management and Food Production. 2020 Discussion Paper 25. International Food Policy Research Institute, Washington, DC.

Chapter 5

Domesticating and Marketing Novel Crops

Roger R.B. Leakey

Introduction

This chapter examines the impacts of land clearance and degradation, and the importance of perennial plants, especially trees and herbaceous shrubs, for agricultural sustainability and ecosystems, and the people who depend on them. New developments in agroforestry focus on the livelihood and income-generating opportunities flowing from the domestication of indigenous trees producing marketable products. This is seen as creating an incentive for farmers to plant a wide range of tree species and reestablish functional agroecosystems based on perennial vegetation.

Understanding the Cycle of Environmental Degradation

To develop practical interventions that promote sustainable agriculture, it is important to have some understanding of the cycle of environmental degradation. A generic scenario for the developing tropics (Fig. 5.1) illustrates that the problems of poverty, land degradation, loss of biodiversity, social deprivation, malnutrition and hunger, poor health, and declining livelihoods are all contributory, being interlinked and cyclical. Typically, the starting point is the result of the farmers' circumstances, induced by external factors (e.g., the weather, fire, social disputes, politics, or the overexploitation of natural resources by outsiders), population pressure, poor health, or crop failure. Such circumstances lead either to decisions to clear more forest, overstock pastures, or to grow crops in ways that

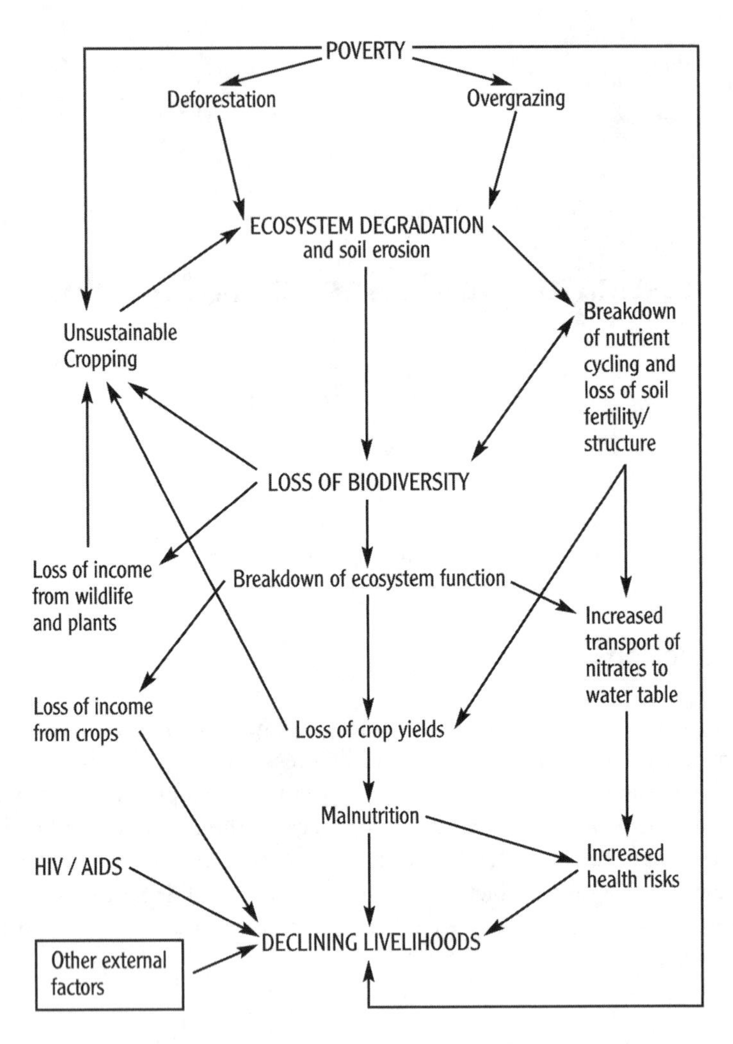

Figure 5.1. Biophysical and socioeconomic scenario for the developing tropics

deplete soil fertility or run the risk of pests and disease. After a period of time ecosystem degradation and soil erosion occur, which in turn lead to the loss of biological diversity above and/or below ground. Eventually, the resilience of the agroecosystem is overcome and there is a loss of ecosystem function, with detrimental consequences for crop yield and income. In arid areas in particular, this can lead to failure to recycle nutrients and their transport to depths below the rooting zone, with their eventual passage down to the water table where they pollute groundwater. The accumulation of these negative impacts leads to the next round of the degradation cycle. With each turn of the cycle the problems are compounded and the processes accelerated, reaching a point where all the problems occur simultaneously.

Given the complexity and interactive nature of this cycle of degradation, interventions to alleviate any one portion of the whole are unlikely to have any long-lasting success. Consequently, meaningful attempts to reverse the cycle have to simultaneously target a number of different points within the cycle (Leakey, Tchoundjeu, et al. 2005). Both agroforestry and ecoagriculture aim to provide this solution, each having a slightly different focus.

Sustainable Land Use: Contributions of Agroforestry to Ecoagriculture Landscapes

While maintaining or increasing productivity, ecoagriculture systems seek to make more space for wildlife and improve habitat quality of productive areas (www.ecoagriculturepartners.org). One way of achieving this is to integrate perennial trees, shrubs, and grasses into production systems to mimic natural vegetation and ecological functions.

Modern agroforestry has tended to be a set of stand-alone technologies, which together form various land-use systems. Trees can be sequentially or simultaneously integrated with crops and/or livestock in many ways (Nair 1989). However, agroforestry practices should create successional phases in the development of a productive and complex agroecosystem, akin to the succession of natural ecosystems (Leakey 1996). In this way, trees producing different agroforestry tree products (AFTPs) and providing different services can be used to create and fill niches in the farming system, as well as to develop a mosaic of patches in the landscape. In accordance with this concept, agroforestry has been defined as "a dynamic, ecologically based, natural resource management system that, through the integration of trees on farms and in the landscape, diversifies and sustains production for increased social, economic and environmental benefits" (Leakey 1996, 6). Based on this ecological approach, planting agroforestry species fitted for different niches in the agroforest canopy makes both farming systems and agricultural landscapes ecologically more stable and biologically more diverse. It is anticipated that this diversity would increase with each phase of the agroecological succession.

Since becoming a field of research (Sanchez 1995), agroforestry's prime objective has been to improve farming systems, making them more productive. In the early days, this research focused on the replenishment of soil fertility (Young 1997). These agronomic approaches to soil management using nitrogen-fixing trees basically maintained farmers' fields in early phases of agroecological succession. With the advent of the more ecological approach to agroforestry came the need to progress this succession to more mature (climax) phases. In many parts of the tropics, scattered indigenous trees producing indigenous fruits, nuts, and medicinal and other products were a common feature of farming systems, but these trees were virtually wild and had been overlooked by agricultural and horticultural science. These "Cinderella" species (called thus because they have been neglected, despite having great value) have become the focus of the next phase of agroforestry development (Leakey and Newton 1994a; Leakey, Tchoundjeu, et al. 2005). Since the early 1990s the domestication of indigenous trees that produce marketable forest products has been developed as a means to diversify and intensify agroecosystems and thereby create mature and diverse agroecosystems.

Agroforestry is a land-use option that promotes human welfare and environmental resilience (ICRAF 1997; Sanchez et al. 1997). The achievement of these twin aims is dependent on the development of agroforests that provide the following:

1. Tree products that both increase food and nutritional security and generate cash income for poverty alleviation
2. Enhanced biodiversity and other ecosystem services that support and promote ecosystem function

The domestication of a wide range of indigenous trees producing timber and nontimber AFTPs is seen as one of the major incentives for resource-poor subsistence farmers to practice this agroecological approach to agroforestry (Leakey 2001a,b; Leakey, Tchoundjeu, et al. 2005; Simons and Leakey 2004), which in turn maintains and enhances biodiversity in farming systems: ecoagriculture (Leakey 1999a).

Novel Crops for Ecoagriculture

Many plant species are useful to humankind, but only about 0.04% have been domesticated for food over the last 2000 to 3000 years, although 8% have edible parts. In agriculture, the "Green Revolution" has been the main vehicle for crop breeding and this has focused mainly on a few staple food crops (Tribe 1994). This narrow focus has been justified on the grounds that it has maximized the returns on limited resources in terms of the impact on the needs of the human population. In contrast to agriculture, much larger numbers of species (5%) have been domesticated for shelter fiber, medicinal, and aesthetic values.

In agroforestry, the relatively new programs of tree domestication have a very different approach from that of the intensive, plant-breeding model implemented on a few species in sophisticated international research stations. Domestication in agroforestry is now seeking to bring a wide range of Cinderella species into cultivation in partnership with farmers on their own farms. This farmer-oriented process is, however, market driven (Simons and Leakey 2004), and it seeks to utilize the intraspecific diversity of locally important tree species in ways that bring diversity to farming systems.

Domestication of Trees for Timber and Nontimber Products

Throughout the tropics numerous perennial woody species have provided indigenous peoples with many of their daily needs for millennia (e.g., Abbiw 1990; Irvine 1961 for Ghana). These species vary in stature and hence their location within a multistrata agroforest (Table 5.1).

Much has been written in recent years about the strategies, processes, and techniques for domesticating agroforestry tree species, and policies required for their implementation (reviewed by Leakey, Tchoundjeu, et al. 2005). Domestication spans activities from ethnobotany, market studies, gene conservation, genetic selection, propagation, and breeding to sustainable cultivation and interactions with the environment, market adoption, and the assessment of environmental, social, and economic impacts (Leakey and Newton 1994a,b).

The domestication strategy will involve the development of a combination of vegetative propagation, selection, and genetic resource strategies (Fig. 5.2), which will vary depending on the value of the products, the extent of intraspecific variation, appropriate propagation methods, and many other factors (Leakey, Tchoundjeu, et al. 2005). However, for high-value species, such as those producing AFTPs, the vegetative propagation of superior genotypes identified from within the existing wild populations will be appropriate (Leakey and Simons 2000). This is the approach generally followed in horticulture and now being adopted in agroforestry (reviewed by Leakey, Tchoundjeu, et al. 2005). Typically, the really superior individuals appropriate for cultivar development are very rare, and thus it is essential to either screen large numbers of trees or seek the assistance of local people with knowledge of the trees in the area. Approaches to screening populations for the selection of elite trees include the following elements:

- Quantitative characterization of fruit, nut, and kernel traits (Atangana et al. 2001)
- Analysis of chemical or nutritional (Thiong'o et al. 2002) and physical properties of the products (Leakey, Greenwell, et al. 2005) for food or industrial use
- Development of market-oriented "ideotypes" (Leakey and Page 2006)
- Predictive tests for mature tree characteristics (e.g., for timber; Ladipo et al 1991a,b)
- Extensive and long-term field trials to test provenances, progenies, and clones

Each of these techniques allows the identification of the rare individuals in a population with elite characteristics to be separated from the rest for vegetative propagation to create a cultivar. This process can be achieved within

Table 5.1. A sample of west african trees, shrubs, and liane species appropriate for growth in multistrata agroforests and for domestication

Species	*Use*	*Common names*	*Mature height (m)*
Anthocleista schweinfurthii	2	Ayinda	15–20
Antrocaryon micraster	1	Aprokuma/onzabili	40–50
Baillonella toxisperma	3,4	Maobi	45–55
Calamus ssp.	4	Rattan	35–45
Canarium schweinfurthii	3,4	Aiele/Africa canarium/incense tree	45–55
Chrysophyllum albidum	1	Star apple	30–40
Cola acuminate	1	Kola nut	15–25
Cola lepidota	1	Monkey kola	10–20
Cola nitida	1	Kola nut	20–30
Coula edulis	1	Coula nut/African walnut	25–35
Dacryodes edulis	1	African plum/Safoutier	15–25
Entandrophragma spp.	4	Sapele/tiama/utile/sipo	50–60
Garcinia kola	1,2,4	Bitter kola	20–30
Gnetum africanum	1	Eru	0–10
Irvingia gabonensis	1	Bush mango/andok	20–30
Khaya spp	3,4	African mahogany	50–60
Lovoa trichiloides	4	Bibola/African walnut	40–50
Milicia excelsa	4	Iroko/mvule/odum	45–55
Nauclea diderichii	4	Opepe/kusia/bilinga	35–45
Pentacethra macrophylla	1	Oil bean tree/Mubala/Ebe	20–30
Raphia hookeri, other spp.	4	Raphia palm	5–15
Ricinodendron heudelotti	1	Groundnut tree/nyangsang/essessang	40–50
Terminalia ivorensis	4	Framire/Idigbo	45–55
Terminalia superba	4	Frake/afara/limba	45–55
Tetrapleura tetraptera	1	Prekese/Akpa	20–30
Treculia africana	1	African breadfruit/eroup	20–30
Trichoscypha arborea	1	Anaku	15–25
Triplochitn schleroxylon	1,4	Ayous/obeche/wawa	55–65
Vernonia amydalina	1	Bitter leaf	0–10
Xylopia aethiopica	1	Spice tree	15–25

From Leakey 1996b
Uses: 1 = Foods: fruits, nuts, vegetables
2 = Medicinal products
3 = Extractives and fiber
4 = Timber, cane, etc.

a rigorous tree improvement program conducted by a university or research institute on a field station, or it can be implemented by farmers on their farms in communities working in partnership with researchers and nongovernmental organizations (NGOs) (Leakey, Tchoundjeu, et al. 2005a). In the latter case, the researchers typically play a mentoring role, providing the NGOs with training

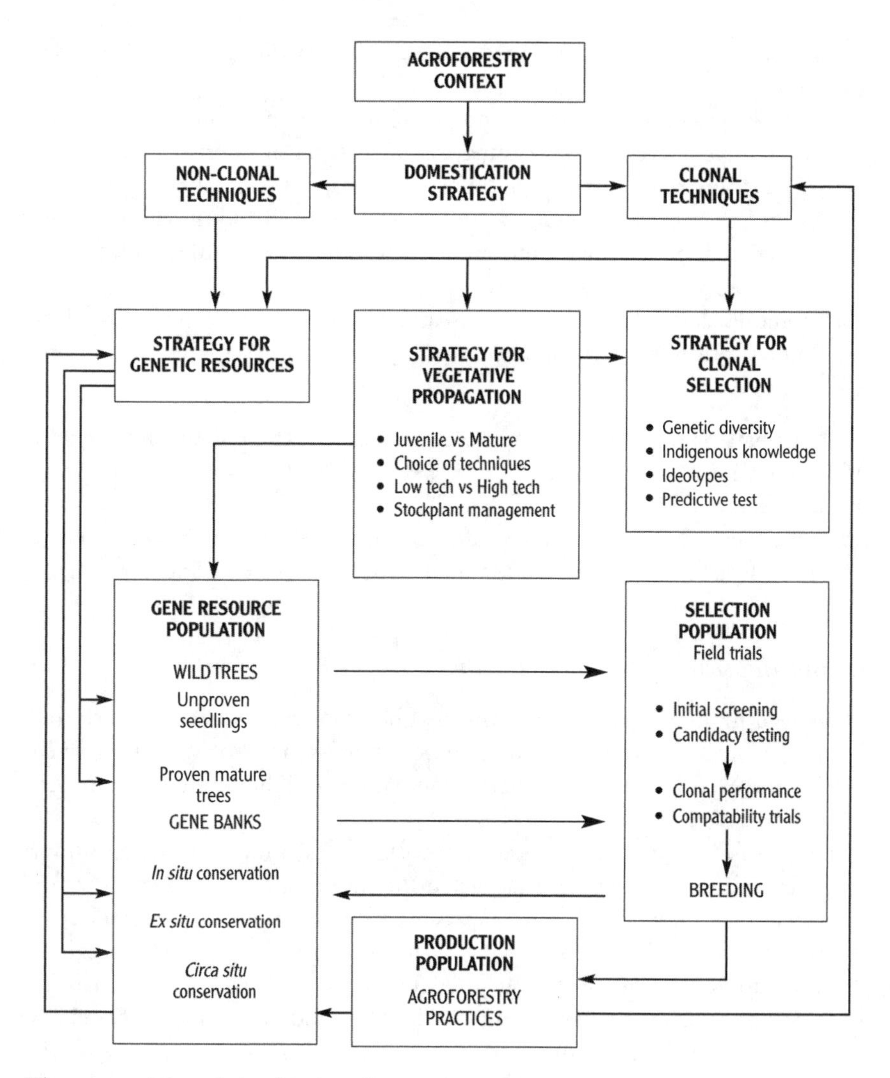

Figure 5.2. The relationship between strategies for domestication, vegetative propagation, and clonal selection

and technical support. This is the participatory domestication approach (Leakey et al. 2003), which has a number of advantages over the more common formal tree improvement program. For example, it:

1. Builds on and enhances the local social and cultural traditions
2. Promotes small-scale local processing and entrepreneurial activity at the community level
3. Benefits and empowers local people by giving them "rights" to the

germplasm they have developed based on their traditional and indigenous knowledge

4. Benefits and empowers local women, who are often involved in the labor of intensive harvesting, processing, and marketing of forest fruits and who play a strong role in managing home gardens and village nurseries
5. Meets the immediate needs of poor people for food and nutritional security, because indigenous fruits are rich in minerals, vitamins, protein, oils, and carbohydrates
6. Promotes the early adoption of research results, which is important when domestication programs are supported by donor projects with short or uncertain funding

Participatory domestication is being implemented in West and Central Africa for a range of fruit and nut tree species, as well as some medicinal tree species (Leakey, Tchoundjeu, et al. 2005; Tchoundjeu et al. 2006). The fruit tree *Sclerocarya birrea* has been domesticated in South Africa (Holtzhausen et al. 1990) and Israel (Mizrahi and Nerd 1996) using the research station approach.

Domestication of Nontree Crops

The recent interest in the domestication of novel crops is not restricted to tree species, and a wide range of new herbaceous crops are also being developed (Smartt and Haq 1997). This set of species requires a slightly different approach to domestication.

Many of these are leaf, fruit, and root vegetables, which have been minor crops for a long time but never attracted the investment to make them more significant as domestic food sources or as market commodities. Nevertheless, in recent years interest has been growing in many indigenous vegetables as candidates for domestication, leading to calls for greater research inputs (Guarino 1997; Schippers 2000; Schippers and Budd 1997; Sunderland et al. 1999a). Proponents argue that the traditional herbaceous food plants are often more nutritious than the major crops (Mnzana 1997) and that they are important for both the livelihoods of rural people and the diversification of farming systems (Bhag Mal et al. 1997). The domestication of these species is accompanied by the need to conserve the genetic resources of these species (Putter 1994). It is, however, also recognized that circa situ conservation could be one of the effective ways to provide protection of the resource (Eyzaguirre 1997; Nkefor et al. 1999).

Although not strictly herbaceous, one traditional African vegetable receiving considerable attention is eru (*Gnetum africanum*). This shade-tolerant species is being heavily exploited in Cameroon for both domestic and export markets. Until recently, when vegetative propagation techniques were developed (Shiembo et al. 1996), it was virtually impossible to cultivate because seed germination is very difficult. Now domestication initiatives are in progress at

Limbe Botanic Gardens and at International Centre for Research in Agroforestry (ICRAF, now World Agroforestry Centre) in Cameroon (Ndam et al. 2001) and ongoing studies in the Congo (Mialoundama 1993). These are examining the light requirements and types of support structures (living and nonliving) as well as starting some genetic selection so that eru can become a component of multistrata systems. The consequence of these efforts is that farmers are starting to cultivate the species. There is an urgent need for other new shade-tolerant crops for multistrata cultivation systems so that the ground flora can be composed of useful and economically important crops, which thrive and provide an early return on investment.

Another group of plants with great potential as shade-tolerant understory crops in complex farming systems consists of the very numerous herbaceous medicinal plants. Traditionally, this is a very valuable group of plants with deep sociocultural importance. However, much more attention has been paid to their ethnobotany, medicinal properties, and markets than to their domestication. Nevertheless, the very considerable potential for domestication and cultivation has been recognized (Mander 1998; Mander et al. 1996) and in a few cases implemented (e.g., Shah and Kalakoti 1996).

Maintaining Genetic Diversity

The domestication process must include an appropriate strategy to conserve a substantial proportion of the genetic variability for future use and to ensure that the genetic resource is used wisely. Whether based on clonal or nonclonal (i.e., seed) approaches, a sound domestication strategy includes the following:

- Establishing a gene bank (ex situ conservation)
- Wise utilization of the genetic resource in cultivation (circa situ conservation)
- Protecting some wild populations (in situ conservation) (see Fig. 5.2).

Clonal approaches require processes of vegetative propagation and clonal selection. A domestication strategy results in the formation of three populations: the genetic resource population, the selection and breeding population, and the production population. This strategy should ensure that the genetic resource in the original wild population is maintained in selection programs and, subsequently, through breeding to broaden the genetic base of the cultivars in the production and genetic resource populations.

Commercialization of Agroforestry Tree Products

The success of a domestication program is dependent on the availability and expansion potential of markets for the products, potentially with more than one target market for a single species. It is therefore important that agroforestry tree domestication programs are implemented in parallel with studies on postharvest storage and processing as an integral element of developing new crops for agricultural diversification (Leakey, Tchoundjeu, et al. 2005).

Many of the AFTPs from tropical trees are already marketed locally and regionally and a few are marketed internationally. For most AFTPs an international market is not a priority, and the promotion of international markets could lead to situations where large companies would see opportunities to be producers. This may result in these larger producers undermining the small-scale producers, who are the key stakeholders of agroforestry tree domestication. This, in turn, could undermine the opportunity for agroforestry to meet some of the Millennium Development Goals (Garrity 2004; Fig. 5.3). It is clear therefore that commercialization has to be implemented sensitively. Although it is necessary to achieve the objectives of agricultural diversification (Leakey and Izac 1996), it can also potentially promote large-scale monocultures that are detrimental to the potential livelihood and environmental benefits of agroforestry/ecoagriculture. Consequently, it is best if the commercialization of AFTPs is focused on the development of local markets, builds on local knowledge, and supports local traditions and culture (Shackleton et al. 2003). This approach also promotes social equity because women are often key players in local marketing.

A recent study of the "winners and losers" from different commercialization models (Sullivan and O'Regan 2003) has found that there are different qualities among producers, markets, and the natural resource that determine who or what are the so-called winners and losers (Table 5.2). The conclusion from this study was that positive AFTP commercialization outcomes can be maximized if the importance of community involvement is appreciated by external players and if the communities themselves work together and use their own strengths to manage and use their resources effectively. Furthermore, this study concluded that to improve the livelihood benefits from commercializing AFTPs it is important to improve:

1. The quality and yield of the products through:
    - Domestication and the dissemination of germplasm
    - Enhancing the efficiency of postharvest technology (extraction, processing, storage, etc.)
2. The marketing and commercialization processes by:
    - Diversifying markets for existing and new products
    - Investing in marketing initiatives and campaigns promoting:
        - ⋆ Supply contracts with equitable distribution of benefits
        - ⋆ Opportunities in national and international cuisine that build on indigenous knowledge and cultural heritage
        - ⋆ Improved sensory perceptions (taste, aroma, etc.)
        - ⋆ Market chain investments

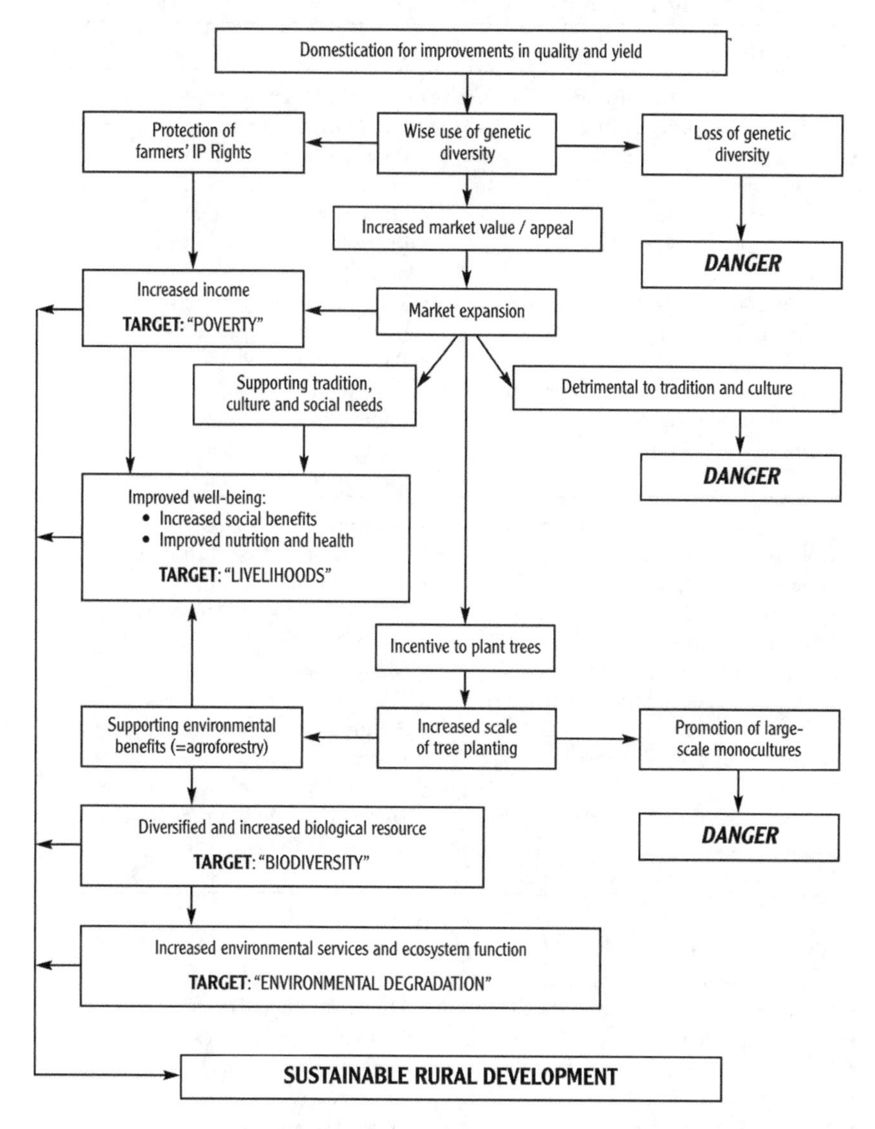

Figure 5.3. Potential impacts on the sustainability of domesticating agroforestry trees

- ⋆ Trading partnerships in local businesses with plans for sustainability (including exit strategies)
- ⋆ Commercialization pathways that recognize the role of women
- ⋆ Health and nutritional benefits

To focus domestication on the needs of the markets those involved in the selection process must work closely with the companies processing and market-

Table 5.2. The relationship between strategies for domestication, vegetative propagation, and clonal selection

Winner qualities	*Loser qualities*
In individuals, households, and enterprises	
• Individuals organized as a group • Well informed about markets • Good access to transport • Coordinated production • Small "input cost:revenue received" ratio • Consistently good-quality products • Skilled in bargaining • Well networked with good partnerships • Easy and equitable access to resource • Fits with other livelihood strategies and sociocultural norms	• Poorly organized group structure • Poorly informed of markets • Poor access to transport • Uncoordinated production • Large "input cost:revenue received" ratio • Variable-quality products • Unskilled in bargaining • Poorly networked • Uncertain and restricted access to resource • Competes with other livelihood strategies and sociocultural norms
In product marketing	
• Commercial opportunities • Diversity of end markets • Diversity of end products • Positive marketing image • Unique characteristics of product • Raw product quality well matched to market • Many buyers of raw materials and products • Many sellers of raw materials and products • Buyers aware of product or brand	• Undeveloped/poor market interest • Limited markets • Fad or single-niche products • No or negative marketing image • Many other substitutes • Raw product requires processing • A monopsony—only one buyer of raw materials • A monopoly—only one seller • Buyers ignorant of product or brand
In the tree resource	
• Abundant resource • Plant part used is readily renewable • Harvesting does not destroy the plant • Easily propagated • Genetically diverse with potential for domestication • Multiple uses for products • High yield of high-quality product • Valuable product • Consistent and reliable yield from year to year • Already cultivated within farming system • Already being domesticated by local farmers • Fast growing	• Rare resource • Slow replacement of harvested product • Destructive and damaging harvesting • Difficult to propagate • Genetically uniform or little potential for selection • Narrow-use options • Low yielding and/or poor-quality product • Low-value product • Inconsistent and unpredictable production • Wild resource that is difficult to cultivate • Totally wild resource • Slow growing

Table 5.2. (*Continued*)

Winner qualities	*Loser qualities*
• Short time to production of product • Compatible with agroforestry land uses • Hardy • Widely distributed	• Long time to production • Competitive with crops; labor intensive, etc. • Sensitive to adverse environmental conditions • Only locally distributed

From Shackleton et al. 2003; Leakey et al., 2005a

ing the products so that market-oriented traits are selected (Leakey 1999b). Clement et al (2004) have drawn attention to both the need to ensure a functional production-to-consumption chain, and the importance of retaining focus on smallholder farmers as the clients of the overall program.

Sustainable Agriculture: Closing the Loop

Can ecological stability be a practical reality in agriculture? The problem here is that over the long term ecosystems are always in a state of flux. Nevertheless, relative stability should arise if species from all trophic levels are present at a density that ensures the functioning of their food chains, life cycles, and other necessary elements. This is difficult to achieve. At a plot level, populations of soil bacteria can be controlled by populations of nematodes (Neher and Barbercheck 1999), but populations of monkeys can only be controlled by populations of top predators (e.g., leopards) at a landscape level. Will farmers be willing to have functioning populations of top predators? Can ecoagriculture provide enough of these ecological services to result in environmentally sustainable farming? If so, then ecoagriculture can close the sustainability loop.

To investigate these processes experimentally is very difficult. However, a multidisciplinary study is under way in Bahia, Brazil, to determine the level of diversification that is required to achieve sustainable cocoa production. This study, using a Nelder fan design, is investigating the changes in microclimate, biological diversity above and below ground, cocoa production, and so forth, in time and space, when cocoa is grown at 40 to 2000 plants ha-1, under the shade of 1 or 8 upper-story tree species and 1 or 16 understory tree species at the constant density of 40 and 80 plants per ha, respectively.

Another aspect of the foregoing debate relates to the relative merits or demerits of indigenous versus exotic crops. In the short term, at least, exotic crops can have fewer pest and disease problems, perhaps because of the absence of the pest or disease organism; or the vector of the disease; or some organism important in the life cycle or food chain of the organism. However, in the long term this absence can either be filled by the introduction of the missing organism, or be replaced by a local species that previously did not fulfill the function. Sometimes this situation can have devastating effects because the predators or parasites that normally control the population growth of the pest or disease may also be absent in the environment. It is therefore clear that the introduction of exotics can pose risks for the successful establishment of ecological equilibrium. The domestication of indigenous plants as new crops can therefore minimize the risks posed by alien species (Ewel et al. 1999).

Domestication for Wildlife Conservation

The benefits of tree domestication for people are hopefully clear, but perhaps beyond the provision of a food source, the benefits to wildlife are less clear. The following case studies illustrate examples of other ways in which tree domestication could promote wildlife conservation.

Prunus africana

For the reasons already discussed, it is not likely that a tree species will be domesticated when there is a clash between the conservation and production interests. However, conflicts of interest are not inevitable. For example, the overexploitation of *Prunus africana* for the medicinal properties of the bark (Cunningham and Mbenkum 1993) can pose serious threats to the endemic birds and monkeys in its afromontane habitat. The current domestication of *P. africana* as an agroforestry tree, selected for its pharmaceutical properties, may have considerable benefits for wildlife in densely populated highlands of Africa, and is, for example, supported by Birdlife International (Nurse et al. 1994). It is unlikely that the wildlife benefiting from the fruits arising from the domestication of this tree for its bark properties will have negative impacts on its cultivation. However, the situation would be different with trees cultivated for the production of fruits, as in this case the restoration of populations of frugivorous birds and monkeys would be detrimental to fruit production.

Pausinystalia johimbe

P. johimbe is a second example of an agroforestry tree with possible wildlife conservation benefits (Sunderland et al. 1999b), which is being domesticated for its potential to generate income for smallholder farmers in West Africa (Tchoundjeu et al. 1999). This is another tree with medicinal products located in the bark. In this case, the products are a cardiac stimulant and johimbine, a potent, clinically proven aphrodisiac. The cultivation and domestication of this tree and marketing of its products could perhaps undermine the markets for products from such threatened species as the rhino and tiger, whose aphrodisiac properties are of dubious merit. Any threats this might impose through the expansion of human populations could be addressed by the addition of extracts from the thunder vine (*Tripterygium wilfordii*), which has reversible effects on slowing sperm motility.

Indigenous Fruits and Nuts in Papua New Guinea

In the Pacific Islands, there have been several attempts to commercialize indigenous nut crops, such as the galip nut (*Canarium indicum*), pili nut (*Canarium ovatum*), tropical almond (*Terminalia catappa*), and Tahitian chestnut (*Inocarpus fagifer*). Although there was great interest in the commercialization of these nuts, these initiatives have not all been a success because the work on postharvest processing and marketing was sometimes done in isolation from efforts to develop a reliable supply of good-quality nuts. In the Solomon Islands, for example, production of galip nut has been significantly increased from 3 metric tons in 1989 to almost 100 metric tons in 1992 and included over 1000 farmers (Evans 1994). Subsequently, the demand outstripped supply, leading to a decline in the commercial activity. This illustrates the importance of running domestication and commercialization in parallel. In 2002, a new project was started to domesticate cut nuts (*Barringtonia procera*) and Tahitian chestnut in the Solomon Islands (Pauku 2005), while a similar project is being planned for Galip nuts in the East New Britain Province of Papua New Guinea (PNG). Besides these nuts there is high demand for Pacific Island nut oils for body lotions, pharmaceutical products, and high-quality soap. In addition, fruits such as the Pacific lychee (*Pometia pinnata*) have also been successfully bottled for sale in local tourist markets.

The PNG Constitution recognizes traditional community ownership over 97% of the land area. Less than 2.8% of PNG has formal protection area status. Only just over 1% is state-owned national parks, the rest being traditionally owned and protected under the Fauna (Protection and Control) Act. Because of the high proportion of land under community ownership, PNG has officially recognized an integrated conservation and development (ICAD) approach to wildlife conservation. It is hoped that PNG can exemplify the potential for ecoagriculture, based on the domestication of new indigenous tree crops in concert with providing future habitat and protection for a number of threatened animals and retention of natural vegetation despite growing pressures on the forest from agriculture.

Conclusions

The vision of agroforestry presented here is as "an integrated landuse that, through the capture of intraspecific diversity and the diversification of species on the farm, combines increases in productivity and income generation with environmental rehabilitation and the creation of biodiverse agroecosystems" (Leakey 1999a, 141). Tree domestication plays a role in creating incentives for farmers to plant trees within their farming systems and to develop mature agro-ecosystems, which increase the biodiversity present on their land. This has value for the maintenance of wildlife populations in areas of high primary productivity, which otherwise would be lacking in wild flora and fauna. However, it is essential to also understand the importance of this wildlife in agroecosystem function, and consequently in sustainable land use and rural development. It is this improvement in agroecosystem function by the development of new and valuable crops that can address the series of complex interactions within the cycle of land degradation (see Fig. 5.1) and improve both the livelihoods of farmers as well as support the biodiversity and the restoration of the environment (Fig. 5.3).

References

Abbiw, D. 1990. *Useful Plants of Ghana: West African Uses of Wild and Cultivated Plants*. Intermediate Technology Publications, Royal Botanic Gardens, Kew, UK.

Atangana, A.R., Z. Tchoundjeu, J.-M. Fondoun, E. Asaah, M. Ndoumbe, R.R.B. Leakey. 2001. Domestication of *Irvingia gabonensis*: phenotypic variation in fruit and kernels in two populations from Cameroon. *Agroforestry Systems* 53:55–64.

Bhag Mal, R.S. Paroda, S. Kochhar. 1997. Underutilized crops and their implications in farming systems in India. In: *Domestication, Production and Utilization of New Crops*, ed. J. Smartt and N. Haq. International Centre for Underutilized Crops, University of Southampton, UK: 30–45.

Clement, C.R., J.C. Weber, J. van Leeuwen, C.A. Domian, D.M. Cole, L.A.A. Lopez, H. Argüello. 2004. Why extensive research and development did not promote use of peach palm fruit in Latin America. *Agroforestry Systems* 61:195–206.

Cunningham, A.B. and F.T. Mbenkum. (1993). *Sustainability of Harvesting* Prunus africana *Bark in Cameroon: A Medicinal Plant in International Trade*. Report of WWF/UNESCO/Kew People and Plants Programme, Worldwide Fund for Nature (WWF), Godalming, Surrey, England.

Evans, B.R. 1994. *Marketing Galip Nut (*Canarium *spp.) in Kandrian and Gloucester Districts, West New Britain, Papua New Guinea*, PDM, Canberra, Australia, 100pp.

Ewel, J.J., D.J. O'Dowd, Bergelson, J., Daeler, C.C., D'Antonio, C.M., Diego Gomez, L., Gordon, D.R., Hobbs, R.J., Holt, A., Hopper, K.R., Hughes, C.E., LaHart, M., Leakey, R.R.B., Lee, W., Loope, L.L., Lorence, D., Louda, S., Lugo, A.E., McEvoy,

P.B., Richardson, D.M. and Vitousek, P.M. 1999. Deliberate introductions of species: research needs. *BioScience* 49:619–630.

Eyzaguirre, P. 1997. Conservation through increased use: complementary approaches to conserving Africa's traditional vegetables. In: *Traditional African Vegetables*, ed. L. Guarino. International Plant Genetic Resources Institute, Rome: 16–19.

Garrity, D. 2004. World agroforestry and the achievement of the Millennium Development Goals. *Agroforestry Systems* 61:5–17.

Guarino, L. 1997. *Traditional African Vegetables*. International Plant Genetic Resources Institute, Rome.

Holtzhausen, L.C., E. Swart, and R. van Rensburg. 1990. Propagation of the marula (*Sclerocarya birrea* subsp. *caffra*). *Acta Horticulturae* 275:323–334.

ICRAF (International Centre for Research in Agroforestry). 1997. *ICRAF Medium-Term Plan 1998–2000*. ICRAF, Nairobi, Kenya.

Irvine, F.R. 1961. *Woody Plants of Ghana with Special Reference to Their Uses*. Oxford University Press, London.

Ladipo, D.O., R.R.B. Leakey, J. Grace. 1991a. Clonal variation in a four-year-old plantation of *Triplochiton scleroxylon* K. Schum. and its relation to the Predictive Test for Branching Habit. *Silvae Genetica* 40:130–135.

Ladipo, D.O., R.R.B. Leakey, J. Grace. 1991b. Clonal variation in apical dominance of *Triplochiton scleroxylon* K. Schum. in response to decapitation. *Silvae Genetica* 40:135–140.

Leakey, R.R.B. 1996. Definition of agroforestry revisited. *Agroforestry Today*, 8(1):5–7.

Leakey, R.R.B. 1999a. Agroforestry for biodiversity in farming systems, In: *Biodiversity in Agroecosystems*, eds. W.W. Collins and C.O. Qualset. CRC Press, New York: 127–145.

Leakey, R.R.B. 1999b. Potential for novel food products from agroforestry trees. *Food Chemistry* 64:1–14.

Leakey, R.R.B. 2001a. Win:win landuse strategies for Africa: building on experience with agroforests in Asia and Latin America. *International Forestry Review* 3:1–10.

Leakey, R.R.B. 2001b. Win:win landuse strategies for Africa: capturing economic and environmental benefits with multistrata agroforests. *International Forestry Review* 3:11–18.

Leakey, R.R.B., P. Greenwell, M.N. Hall, A.R. Atangana, C. Usoro, P.O. Anegbeh, J.-M. Fondoun, and Z. Tchoundjeu. 2005. Domestication of *Irvingia gabonensis*, IV: tree-to-tree variation in food-thickening properties and in fat and protein contents of dika nut. *Food Chemistry* 90:365–378.

Leakey, R.R.B. and A.-M.N. Izac. 1996. Linkages between domestication and commercialization of non-timber forest products: implications for agroforestry. In: *Domestication and Commercialization of Non-timber Forest Products for Agroforestry*, ed. R.R.B. Leakey, A.B. Temu, M. Melnyk, and P. Vantomme. FAO, Rome: 1–7.

Leakey, R.R.B. and A.C. Newton. 1994a. Domestication of "Cinderella" species as a start of a woody-plant revolution. In: *Tropical Trees: Potential for Domestication and the Rebuilding of Forest Resources*, ed. R.R.B. Leakey and A.C. Newton. HMSO, London: 3–5.

Leakey, R.R.B. and A.C. Newton. 1994b. *Domestication of Tropical Trees for Timber and Non-timber Products*. MAB Digest 17. UNESCO, Paris.

Leakey, R.R.B. and T. Page. 2006. The "ideotype concept" and its application to the selection of "AFTP" cultivars. *Forests, Trees and Livelihoods* 16:5–16.

Leakey, R.R.B., K. Schreckenberg, and Z. Tchoundjeu. 2003. The participatory domestication of West African indigenous fruits. *International Forestry Review* 5:338–347.

Leakey, R.R.B. and A.J. Simons. 2000. When does vegetative propagation provide a viable alternative to propagation by seed in forestry and agroforestry in the tropics and sub-tropics? In: *Problem of Forestry in Tropical and Sub-tropical Countries: The Procurement of Forestry Seed—the Example of Kenya*, ed. H. Wolf and J. Arbrecht. Ulmer Verlag, Stuttgart, Germany: 67–81.

Leakey, R.R.B., Z. Tchoundjeu, K. Schreckenberg, S. Shackleton, and C. Shackleton. 2005. Agroforestry tree products (AFTPs): targeting poverty reduction and enhanced livelihoods. *International Journal of Agricultural Sustainability* 3:1–23.

Mander, M. 1998. *Marketing of Indigenous Medicinal Plants in South Africa: A Case Study in Kwazulu-Natal*. FAO, Rome.

Mander, M., J. Mander, and C. Breen. 1996. Promoting the cultivation of indigenous plants for markets: experiences from KwaZulu Natal, South Africa. In: *Domestication and Commercialization of Non-timber Forest Products in Agroforestry Systems*, ed. R.R.B. Leakey, A.B. Temu, M. Melnyk, and P. Vantomme. FAO, Rome: 104–109.

Mialoundama, F. 1993. Nutritional and socio-economic value of *Gnetum* leaves in Central African forest. In: *Tropical Forests, People and Food: Biocultural Interactions and Applications to Development*, ed. C.M. Hladik, A. Hladik, O.F. Linares, H. Pagezy, A. Semple and M. Hadley. Parthenon Publishing Group, Carnforth, UK: 177–182.

Mizrahi, Y. and A. Nerd. 1996. New crops as a possible solution for the troubled Israeli export market. In: *Progress in New Crops: Proceedings of the Third National New Crops Symposium*, ed. J. Janick. American Society of Horticultural Science Press, Alexandria USA: 37–45.

Mnzana, A. 1997. Comparing nutritional values of exotic and indigenous vegetables. In: *African Indigenous Vegetables*, ed. R.R. Schippers, and L. Budd. IPGRI, Rome, and Natural Resources Institute, Chatham, UK: 70–75.

Nair, P.K. 1989. Agroforestry defined. In: *Agroforestry Systems in the Tropics*, ed. P.K.R. Nair. Kluwer Academic Publishers, Dordrecht, Netherlands: 13–18.

Ndam N., J.P. Nkefor, and P.C. Blackmore. 2001. Domestication of *Gnetum africanum* and *G. buchholzianum* (Gnetaceae), over-exploited wild forest vegetables of the central African region. *Systematics and Geography of Plants* 71:739–745.

Neher, D.A. and M.E. Barbercheck. 1999. Diversity and function of soil mesofauna. In: *Biodiversity in Agroecosystems*, ed W.W. Collins and C.O. Qualset. CRC Press, New York: 27–47.

Nkefor, J.P., N. Ndam, P.C. Blackmore, and T.C.H. Sunderland. 1999. The conservation through cultivation programme at the Limbe Botanic Garden: achievements and benefits. In: *Non-wood Forest Products of Central Africa: Current Research Issues and Prospects for Conservation and Development*, ed. T.C.H. Sunderland, L.E. Clark, and P. Vantomme. FAO, Rome: 79–86.

Nurse, M.C., C.R. MacKay, J.B. Young, and C. Asanga. 1994. *Biodiversity Conservation through Community Forestry in the Montane Forests of Cameroon*. Birdlife International 21st World Conference, Rosenheim, Germany.

Pauku, R.L. 2005. Domestication of indigenous nuts for agroforestry in the Solomon Islands. Ph.D. Thesis, James Cook University, Cairns, Queensland, Australia, 381pp.

Putter, A. 1994. *Safeguarding the Genetic Basis of Africa's Traditional Crops*. CTA, the Netherlands and IPGRI, Rome.

Sanchez, P.A. 1995. Science in agroforestry. *Agroforestry Systems* 30:5–55.

Sanchez, P.A., R.J. Buresh, and R.R.B. Leakey. 1997. Trees, Soils and Food Security. *Philosophical Transactions of the Royal Society, Lond. B* 352:949–961.

Schippers, R.R. 2000. *African Indigenous Vegetables: An Overview of the Cultivated Species*. Natural Resources Institute, Chatham, UK/ACP-EU Technical Centre for Agricultural and Rural Cooperation, Wageningen, The Netherlands.

Schippers, R.R. and L. Budd. 1997. *African Indigenous Vegetables*. IPGRI, Rome, Italy and Natural Resources Institute, Chatham, UK.

Shackleton, S.E., R.P. Wynberg, C.A. Sullivan, C.M. Shackleton, R.R.B. Leakey, M. Mander, T. McHardy, S. den Adel, A. Botelle, P. du Plessis, C. Lombard, S.A. Laird, A.B. Cunningham, A. Combrinck, and D.P. O'Regan. 2003. Marula commercialization for sustainable and equitable livelihoods: synthesis of a southern African case study. In: *Winners and Losers in Forest Product Commercialization*. Final Technical Report to DFID Forestry Research Programme (R7795), Centre for Ecology and Hydrology, Wallingford, UK.

Shah, V. and B.S. Kalakoti. 1996. Development of *Coleus forskohlii* as a medicinal crop. In: *Domestication and Commercialization of Non-timber Forest Products in Agroforestry Systems*, ed. R.R.B. Leakey, A.B. Temu, M. Melnyk and P. Vantomme. FAO, Rome: 212–217.

Shiembo, P.N., A.C. Newton, and R.R.B. Leakey. 1996. Vegetative propagation of *Gnetum africanum* Welw., a leafy vegetable from West Africa. *Journal of Horticultural Science* 71(1):149–155.

Simons, A.J. and R.R.B. Leakey. 2004. Tree domestication in tropical agroforestry. *Agroforestry Systems* 61:167–181.

Smartt, J. and N. Haq. 1997. *Domestication, Production and Utilization of New Crops*. International Centre for Underutilized Crops, University of Southampton, UK.

Sullivan, C.A. and D.P. O'Regan. 2003. *Winners and Losers in Forest Product Commercialization*. Final Report to DFID Forestry Research Programme (R7795), Centre for Ecology and Hydrology, Wallingford, UK. www.ceh-wallingford.ac.uk/research/winners/literature.html.

Sunderland T.C.H., L.E. Clark, and P. Vantomme. 1999a. *Non-wood Forest Products of Central Africa: Current Research Issues and Prospects for Conservation and Development*. FAO, Rome.

Sunderland, T.C.H., M.-L. Ngo-Mpeck, Z. Tchoundjeu, and A. Akoa. 1999b. The ecology and sustainability of *Pausinystalia johimbe*: An over-exploited medicinal plant from the forests of Central Africa. In: *Non-wood Forest Products of Central Africa: Current Research Issues and Prospects for Conservation and Development*, ed. T.C.H. Sunderland, L.E. Clark, and P. Vantomme. FAO, Rome: 67–78.

Tchoundjeu, Z., E. Asaah, P. Anegbeh, A. Degrande, P. Mbile, C. Facheux, A. Tsobeng, A. Atangana, and M.-L. Ngo-Mpeck, 2006. AFTPs: Putting participatory domestication into practice in West and Central Africa. *Forests, Trees and Livelihoods* 16:53–69.

Tchoundjeu, Z., B. Duguma, M.L. Tiencheu, and M-L. Ngo-Mpeck. 1999. The domestication of indigenous agroforestry trees: ICRAF's strategy in the humid tropics of

West and Central Africa. In: T.C.H. Sunderland, L.E. Clark, and P. Vantomme, *Non-Wood Forest Products of Central Africa: Current Research Issues and Prospects for Conservation and Development*, ed. T.C.H. Sunderland, L.E. Clark, and P. Vantomme. FAO, Rome: 161–170.

Thiong'o, M.K., S. Kingori, and H. Jaenicke. 2002. The taste of the wild: variation in the nutritional quality of marula fruits and opportunities for domestication. *Acta Horticulturae* 575:237–244.

Tribe, D. 1994. *Feeding and Greening the World: The Role of International Agricultural Research*. CAB International, Wallingford, UK.

Young, A. 1997. *Agroforestry for Soil Management*. CAB International, New York.

Chapter 6

Tropical Agroforestry

Götz Schroth and Maria do Socorro Souza da Mota

Ecoagriculture and agroforestry are related, but distinct, concepts. Agroforestry is defined through its *methods* (the use of trees on farms and in agricultural landscapes), whereas ecoagriculture is defined through its *objectives* (the integration of biodiversity conservation, livelihoods, and productivity in agricultural landscapes). Not all types of agroforestry are species-rich or particularly beneficial for biodiversity and therefore qualify as ecoagriculture (e.g., two-species systems of an intensively managed tree crop with an exotic shade tree species), nor do species-rich land-use systems necessarily contain trees (e.g., certain grasslands). However, in many cases, land-use systems or arrangements that combine productivity, improved livelihoods, and the conservation of wild biodiversity in the same landscape (i.e., meet the ecoagriculture criteria) also qualify as agroforestry. Agroforestry has, therefore, a key role to play in ecoagriculture strategies.

Compared with earlier-recognized benefits of agroforestry practices (e.g., contributions to soil fertility and livelihoods), biodiversity conservation has only recently attracted attention as a potential benefit of agroforestry. Therefore, it is particularly important to develop a conceptual framework for analyzing agroforestry's effect on biodiversity in order to focus research efforts and practical application toward key issues. Following Schroth et al. (2004a), we distinguish conceptually three different ways that agroforestry can benefit biodiversity in tropical landscapes: by reducing deforestation or other conversion of natural habitat into cultivated land; by improving the habitat quality of cultivated areas; and by making the matrix between natural habitat areas more benign for wild species and buffering their limits against adverse influences.

Agroforestry as a Means to Reduce the Conversion of Natural Habitat

Any land-use practice that reduces the conversion of natural habitat into cultivated land can be assumed to have a beneficial effect on biodiversity conservation. It has often been argued that sustainable and productive agroforestry systems reduce the conversion of forest into agricultural land because farmers no longer need to search for fertile soil to replace exhausted agricultural land. Several authors, including Angelsen and Kaimowitz (2004), have shown that this assumption is not justified in such a general form. However, under some conditions the adoption of agroforestry practices can have a forest-conserving effect, for example, where agroforestry practices require increased labor inputs per unit cultivated area compared with previous land-use practices (e.g., slash-and-burn agriculture or pasture) and labor availability is limited, because then no free labor would be available to further expand the cultivated area.

This may be the case for a traditional community living in a sustainable development reserve or an extractive reserve, where immigration is legally restricted, if it adopts tree crop agroforestry as an alternative to more extensive slash-and-burn agriculture. For example, in the Tapajós region of the Brazilian Amazon, two principal land-use practices used by traditional farmers are cassava growing in slash-and-burn agriculture and rubber agroforestry (Schroth et al. 2003). Rubber agroforestry is a (semi-) permanent land-use practice that does not require rotation between areas and very rarely replanting. Therefore, promoting this practice (which was basically abandoned during part of the 1990s due to low rubber prices) through higher product prices and technical inputs (Schroth et al. 2004c) could absorb labor from the cassava sector, and in addition make people's lives easier by replacing the hard work in the slash-and-burn plots by tapping in shaded rubber groves. Of course, the substitution of slash-and-burn agriculture by rubber agroforestry would not be complete because (1) cassava is not only the farmers' principal traded commodity but also their basic food; (2) rubber is only tapped in the rainy season and early in the morning, leaving time for other activities such as cassava cropping; and (3) new rubber groves are generally established in slash-and-burn plots, usually with cassava (Schroth et al. 2003).

In most cases, the adoption of agroforestry practices as such will not be sufficient to limit deforestation. If capital, labor, and market outlets are not limiting, new agroforestry technologies could even boost deforestation rates if they make forest conversion more profitable (Angelsen and Kaimowitz 2004). Relying on direct forest-conserving effects of agroforestry introduction is thus risky. However, agroforestry has a substantial potential to develop synergies with direct measures of forest protection. Where access to new land is restricted, for example, by a park boundary or management plan of an established protected area, agroforestry practices that are profitable and do not require regular shifts in cultivated areas may help the population comply with legal restrictions to further deforestation. Also, agroforestry practices may provide forest products that would otherwise have to be extracted from natural forest in a potentially unsustainable manner. For example, Murniati et al. (2001) showed that in the surroundings of the Kerinci-Seblat National Park in Sumatra, farmers who owned complex agroforests were less dependent on forest resources and also had less labor available to exploit them than farmers who only possessed rice fields.

Where legislation obliges farmers to maintain farm areas under forest, agroforestry practices can help them comply with such legislation. For example, in the Brazilian Amazon region land holders are obliged to maintain 80% of their farm areas under natural forest cover, and agroforestry practices can help to reduce the cost of maintaining this "legal reserve." In the Manaus region, the extractive use and management of fruit producing tucumã palms (*Astrocaryum tucuma*) that occur at relatively high density in secondary forests, fallows, and cultivated (i.e., disturbed) areas in the central Amazon was shown to be profitable and a useful supplement to normal agricultural activities while helping farmers to make better use of their "legal reserves" (Schroth et al. 2004d) (Fig. 6.1). This land-use practice is located at the transition between agroforestry and extractivism, but the agroforestry character was prominent in this case through the focus on the management and in situ improvement of the spontaneous palm population (Schroth et al. 2004d).

Another legal requirement in the Brazilian Amazon is to keep the banks of rivers and water sources (as well as steep slopes and hilltops) under permanent forest cover ("areas of permanent preservation"). Especially in regions with a pronounced dry season, this regulation can present a hardship to land users because these riparian areas are particularly productive for certain crops. In the Tapajós region, a locally developed agroforestry practice consists of the underplanting and/or encouragement of fruit-producing palm species such as açai (*Euterpe oleracea*) and bacaba (*Oenocarpus distichum*) in such riparian forests (Fig. 6.2). Although present legislation allows uses of riparian forests only under certain conditions, this traditional practice is much preferable to the conversion of these areas into agricultural fields and reduces the opportunity cost of maintaining these riparian forests for the land users.

Agroforestry practices may also help focus farmers' activities on secondary instead of primary forest areas, thereby reducing the environmental impact of farming practices. For example, tree crops such as cocoa (*Theobroma cacao*) can be planted in tree fallows and old tree crop plantations after clearing the undergrowth without burning (Fig. 6.3). Such practices can be further strengthened by creating more valuable secondary forests through the conservation of forest tree regeneration and planting of valuable tree seedlings into fields during the cropping phase.

In summary, although agroforestry should not usually be expected to reduce

Figure 6.1. The extractive use and management of spontaneous tucumã palms, in the central Amazon (from Schroth et al. 2004d, with permission)

forest clearing by itself, it has a considerable potential to develop synergies with direct forest conservation measures and other environmental regulation.

Agroforestry as a Means to Improve the Habitat Value of Cultivated Areas

Agroforestry systems present higher habitat value for many species of plants and animals than simpler, treeless, and more intensively managed land uses. This is a result of the permanent or temporary tree cover, the structural complexity

Figure 6.2. Enrichment of riparian forests through palm planting in the Tapajós region

(including vertical stratification), more stable and protected conditions, and often extensive management of at least part of the system (for example, the tree crowns). Especially in areas where little primary habitat remains, agroforestry areas can play a crucial role in conserving part of the original biodiversity in human-dominated landscapes (Michon and de Foresta 1999).

When considering the diversity of agroforestry or other agricultural systems, it is important to distinguish between planned diversity of agricultural or otherwise useful species, and wild biodiversity. Agricultural or agroforestry areas such as home gardens can be rich in species and varieties of cultivated crops and trees and contribute substantially to the conservation of agrobiodiversity but at the same time be intensively managed and contain few wild species of plants

Figure 6.3. Young cacao plants under old rubber trees, Brazilian Amazon

(which are weeded out) and animals (which may find unsuitable habitat conditions and may be hunted). Another agroforestry area may contain a single planted crop species, such as rubber, coffee, or cocoa trees, but through extensive management or periodic abandonment at times of low product prices may provide niches for many wild species of plants and animals.

Figure 6.4 illustrates conceptually the relationship between planned diversity and wild biodiversity for different multistrata agroforestry systems (i.e., systems based on differently sized trees and tree crops). It suggests that extensive management (i.e., relatively little input of labor and capital per unit of cropped area) is a key factor for the presence of wild biodiversity in agroforestry systems because it allows niches of "benign neglect" to develop where wild species can get established. However, due to the spatiotemporal complexity of agroforestry systems, even a relatively intensively managed system may contain extensively managed compartments. For example, a shaded coffee or cocoa plantation may be fairly intensively managed in the under- and midstory but still offer habitat for wild species of birds and insects in the tree canopies. Agroforests based on canopy trees such as rubber or damar even offer extensively managed habitat in the under- and midstory (Schroth et al. 2004b), silvopastoral systems in fence-

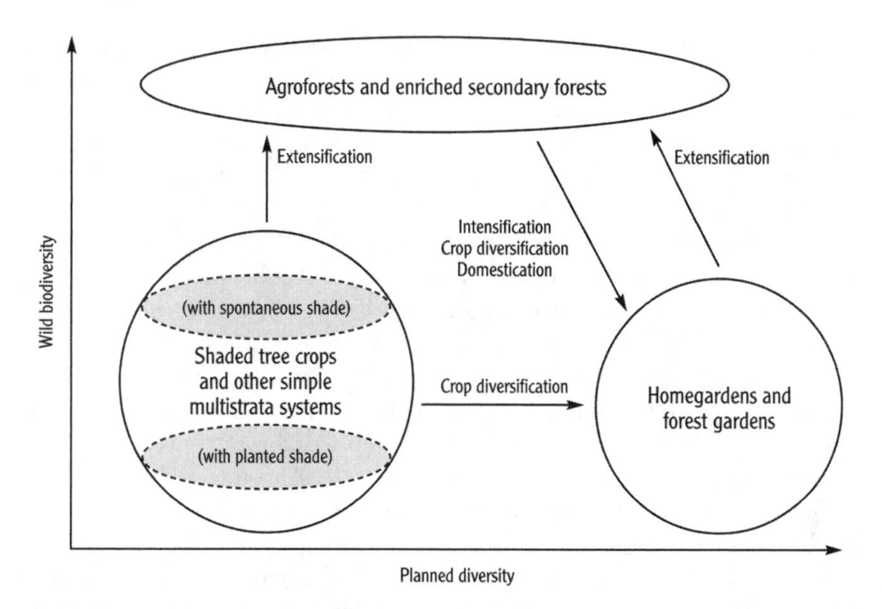

Figure 6.4. Relative position of different multistrata agroforestry practices on two conceptual gradients of planned diversity and wild (or unplanned) biodiversity, and processes through which these practices may convert into one another. *Intensification* and *extensification* refer to the input of labor and external resources (fertilizers, pesticides) per unit of planted area (from Schroth et al. 2004b)

rows or windbreaks composed of trees (Harvey et al. 2004), and fallow systems during the fallow phase (Finegan and Nasi 2004).

For these different types of agroforestry systems, research has demonstrated the presence of wild plant and animal species that would not be present, or not at the same density, in simpler agricultural systems and monocultures (see reviews cited earlier). Occasionally, agroforestry systems may even host rare or endangered species, such as rare monkey, bird, and cat species in cocoa agroforests in Brazil and Costa Rica (Alves 1990; Harvey et al. 2004), and even rhinos and tigers in damar agroforests in Sumatra (Michon and de Foresta 1999). It is, however, important to keep in mind that agroforestry areas are not primary habitat and cannot fully replace it in its functions for the maintenance of native biodiversity. For example, the underplanting of thinned forest with shade-tolerant tree crops such as cocoa and coffee has been shown to impoverish the fauna and flora of the under- and midstory, as would be expected. Certain sensitive, forest-dependent groups such as understory birds tend to be underrepresented or missing from such systems (Alves 1990). Even forestlike systems such as rubber, damar, and durian agroforests in Sumatra are depleted of certain sensitive forest bird species (Thiollay 1995). It is also important to remember that wild species that are present in agroforestry areas may be only temporary visitors and still require native habitat for their long-term survival. Nevertheless, the secondary habitat provided by agroforestry areas may be of critical importance for the conservation of many species in regions that are largely depleted of natural habitat.

Pressure to intensify land use is a threat to both planned and wild biodiversity of agroforestry systems. To increase the yields of modern varieties of tree crops such as coffee and cocoa and improve their response to fertilizer inputs, traditional, multispecies shade canopies have in many places been removed or replaced by monospecific, often exotic, shade trees over the past decades (Perfecto et al. 1996; Rice and Greenberg 2000). In lowland Sumatra, extensively managed "jungle rubber" groves are being replaced by intensively managed rubber or oil palm plantations (Williams et al. 2001). Also, throughout the tropics, increased population pressure obliges farmers to switch from traditional, long tree fallows to short fallows.

Because of the high density of fruit trees and moderate levels of disturbance, agroforestry landscapes may present higher density and productivity of certain wildlife species than undisturbed forest (Wilkie and Lee 2004). Hunting in agroforestry systems and landscapes is common ("garden hunting") and may have the positive effect of deflecting hunting pressure from primary habitat. However, research is necessary on the question to what extent the presence of fruit trees in agroforests increases the productivity of landscapes for wildlife by increasing food sources, and to what extent it only attracts wildlife into sink areas where it is susceptible to hunting. In many places hunting pressure is clearly excessive and risks annihilating the biodiversity benefits of agroforestry. To conserve high levels of wild biodiversity in agroforestry landscapes, changes in consumption preferences and attitudes toward wildlife are certainly necessary in some regions (Wilkie and Lee 2004). In regions where hunting is little practiced by the local population, such as parts of Sumatra, in contrast, high density of wildlife such as wild pigs and monkeys in agroforestry areas may present a significant problem for farmers, for example, through the destruction of tree regeneration (Williams et al. 2001).

Agroforestry as a Means to Create a Benign Matrix for Habitat Fragments

A potential key role of agroforestry in biodiversity conservation in human-dominated landscapes is the creation of a matrix that ensures connectivity between areas of natural habitat, such as forest remnants, and to buffer such areas from microclimatic and other influences of adjacent cultivated land. In combination, these two mechanisms may reduce the negative impacts of fragmentation on natural ecosystems and populations of plants and animals and increase the conservation value of natural habitat islands. Some studies have directly quantified such landscape-scale benefits of agroforestry, whereas others provide indirect evidence for them.

For example, the importance of shaded coffee plantations in Central America as stopovers and overwintering sites of migratory birds is well established (Perfecto et al. 1996; Rice and Greenberg 2004), as well as the use of windbreaks and other tree vegetation in pasture landscapes as corridors and habitat for birds and a variety of other wildlife (Harvey et al. 2004; Rice and Greenberg 2004). Complex agroforests have been shown to harbor, at least temporarily, large and sometimes endangered wildlife, as mentioned earlier. Remnant trees in fields, pastures, and small forest islands within agricultural land, earlier considered "living dead" for lack of regeneration, have recently been shown to exchange pollen over larger distances than previously assumed and thereby ensure genetic connectivity between forest remnants (Boshier 2004). Indirect evidence exists that tall-statured agroforests may also have a potential to buffer forest edges from wind and other disturbances, thereby reducing edge-related mortality of forest trees (Mesquita et al. 1999; Gascon et al. 2000). Furthermore, valuable tree vegetation provides an incentive to farmers to control the use of fire in agricultural and pasture land and prevent the passage of wildfires, thereby also benefiting natural vegetation.

Based on such evidence, some pioneer projects have started to use agroforestry specifically to buffer and connect forest fragments. For example, a project in the largely deforested state of São Paulo, Brazil, uses agroforestry buffers and stepping stones to make a pasture-dominated matrix less hostile for the last existing forest remnants (Cullen Jr. et al. 2004), whereas another initiative in southern Bahia, Brazil, promotes the maintenance of traditional shaded cocoa plantations and legally required on-farm forest reserves to ensure connectivity between remnants of the Atlantic rainforest (Center for Applied Biodiversity Science 2000). Efforts in Queensland, Australia, aim at establishing ecological corridors through farmland between protected areas (Tucker et al. 2004).

With climate change and associated habitat shifts for many species and populations becoming reality, the importance of matrix management in landscape-scale conservation strategies will increase further. As Hannah (2004) summarizes, "(t)he potential roles for agroforestry (in this situation) include providing transitional habitat for species whose ranges are shifting, supplemental habitat for species in decline, food sources outside protected areas, microclimatic conditions appropriate for many species, and flexible options for future regeneration of natural forest." Furthermore, diversified agroforestry systems may also be less susceptible to exceptional weather conditions, such as very dry or wet years, than monocultures, and may help farmers to adapt to a changing climate.

Although the more systematic use of agroforestry in landscape-scale conservation strategies is a fascinating perspective, there are potential problems and trade-offs that need to be considered. Farmers who plant trees along the forest boundary, thereby contributing to the conservation of a particular forest and its species, may benefit from the increased presence of pollinators (Ricketts et al. 2004) and predators of pest species, but also suffer harvest losses from incursions of wildlife such as monkeys, elephants, and wild pigs into their crop fields (Naughton-Treves and Salafsky 2004). Some crops are more suitable for planting at the forest margin than others. For example, rubber trees attract a lot of wildlife through their abundant seed crop but are not susceptible to wildlife damage (the seeds themselves, though edible in principle, are not usually used) and so are well suited for transitional areas, but fruit trees are much more likely to be damaged by wild animal species and require closer supervision. Cocoa can be severely damaged by wildlife such as monkeys or rodents (Wood and Lass 1985), and this needs to be considered when attempting to use shaded cocoa groves in conservation corridor strategies. Timber trees and relatively unpalatable crop species such as coffee and tea are options for sites that are most exposed to wildlife incursions.

Although the proximity to natural habitat may in some cases pose problems to agroforestry areas, the contrary can also be the case if invasive, exotic tree species are used in agroforestry areas that subsequently invade natural habitat. Although the weediness of species such as *Leucaena leucocephala* was already recognized as a problem by Indonesian coffee planters in the early 20th century, the problem of invasiveness has not received much attention among agroforesters until recently (Richardson et al. 2004). Humid tropical regions have so many potentially useful tree species that the use of potentially invasive exotics should be avoidable. In dry areas, however, there is often a trade-off between outstanding growth characteristics and invasiveness (e.g., *Prosopis* spp. in Ethiopia, Kenya, South Africa, India, and other places where this genus has been introduced), and decisions are more complex. Where possible, agroforesters should work with local tree species that may also have more uses for local people (Richardson et al. 2004).

The Changing Role of Agroforestry with Changing Land Use Intensity

A strategy of improving livelihoods through increased agricultural production and conserving wild biodiversity requires a vision of a "desirable landscape" that allows the simultaneous achievement of both objectives. Within this landscape vision, agroforestry will in most cases play a role, but this role will differ from one case to the next.

In a region where farmland is adjacent to still expansive areas of natural habitat, such as many parts of the Amazon, the most important role of agroforestry in a conservation strategy will not be to serve as habitat for wild plants and animals, but rather to make efficient use of areas already under agriculture, thereby making restrictions to the expansion of agricultural land acceptable. Where markets for tree products exist or can be created, the planting or retention of trees on agricultural land will increase its value to land users and may reduce incentives to move further to the forest frontier. It will also help to reduce the use of fire. The use of species that thrive in disturbed areas, such as tucumã palms (*Astrocaryum tucuma*) on uplands (see Fig. 6.1) or açai palms (*Euterpe oleracea*) on riparian sites in the Amazon (see Fig. 6.2), can be an important supplement to food crop agriculture and may help to concentrate the labor force on already cleared land. However, any profitable activity may also make resources available for further expansion of agriculture, so measures of direct protection of natural habitat, either through legal measures or through agreements with land holders (e.g., conservation incentive agreements) are important.

As natural habitat becomes increasingly fragmented and its total area reduced, agroforestry areas will increasingly play a role of secondary habitat and beneficial matrix land use in a landscape conservation strategy, without losing their role in supporting the sustainable and profitable use of agricultural land. Many such areas of long-standing agricultural use are dominated by slash-and-burn agriculture and pasture, the latter especially in Latin America. Many measures are possible to improve both habitat value and connectivity of such landscapes through agroforestry. The introduction or (easier) retention of spontaneous trees in pastures, complemented with live fences and windbreaks, may increase not only the habitat value and permeability of pasture areas for many wild species but also agricultural productivity (Harvey et al. 2004; Rice and Greenberg 2004).

Although such measures do not require a major change of economic practices, this is different when tree crop agriculture is proposed as an alternative to slash-and-burn production of food crops to increase cash income and agricultural output per unit land area and reduce the use of fire (see Tomich et al., chapter 18, this volume). However, in many regions traditions of tree cropping exist, for example, in home gardens or traditional agroforests of rubber or cocoa, and can be built upon. Reliability of markets and prices are likely to be a dominating factor in any effort to promote tree crop agriculture. For example, in the Tapajós region of the Brazilian Amazon, farmers with a tradition of rubber agroforestry were frustrated when Brazilian rubber markets stopped functioning in the early 1990s.

In fragmented landscapes, the spatial arrangement of land-use types relative to primary habitat islands becomes of primary importance for conservation. In some tropical regions, agricultural land merges gradually into forest, with the most intensively managed land in valleys and close to homesteads, and extensively managed agroforests forming the transition to forest land (e.g., Murniati et al. 2001, in Sumatra). Such gradual transitions are desirable because they reduce edge effects and help to keep fire away from the forest boundary. Managed secondary forests, riparian strips, and even remnant trees in agricultural land will all make a contribution to conservation in such mosaic landscapes by providing habitat and increasing landscape connectivity.

In densely populated regions that have long been under intensive agricultural use and where most primary habitat has been converted, agroforestry areas may provide secondary habitat and refuge for those elements of the original biodiversity that tolerate a certain level of disturbance. This is the case, for example, in parts of lowland Sumatra, where natural forest areas have been strongly reduced and jungle rubber systems are the principal forested areas remaining in the landscape (Michon and de Foresta 1999; Thiollay 1995). With the ongoing conversion of these traditional land-use systems into intensively managed rubber and oil palm plantations, much of this remaining biodiversity is likely to be lost from the landscape. Similarly, Augusseau et al. (2006) predict the disappearance of forests and fallows from their study area in southern Burkina Faso due to increasing land-use pressure, leaving trees in crop fields (the traditional parklands) as the principal tree vegetation in the landscape.

Scaling Up Agroforestry and Ecoagriculture

Scaling up agroforestry and its strategic use in both the improvement of livelihoods and the conservation of wild biodiversity in the tropics will require measures at the scientific, institutional, and political levels. This section focuses on the question of what type of research, on what issues, is needed to identify and overcome constraints to the adoption of ecoagriculture practices on a broader scale.

As this review has shown, much research still needs to be done to quantify the specific benefits of certain agroforestry practices and their arrangements with other land uses in land-use mosaics on wild biodiversity. But more important, much evidence on the potential of agroforestry to improve landscape-scale conservation has accumulated over the past one or two decades and is now available. Rather than waiting for more quantitative information to be produced by further basic research, scientists (and the politicians who fund them) should now address directly the constraints that farmers and institutions face in adopting agroforestry and other conservation-friendly land-use practices as a means of improving their livelihoods and conserving wild biodiversity. Much valuable and scientifically rigorous information will be produced by monitoring the impact and success of land-use innovations in practice, and this will help to further refine the approaches in a continuous feedback process.

This cannot be achieved through traditional on-station research with a static three-year work plan and corresponding budget prepared in advance and perhaps remotely controlled from a university institute or committee somewhere in a temperate country. Rather, scientists need to engage in partnerships with local farmers and the institutions that influence their land-use decisions, such as extension agencies, local government agencies, industries that can provide market outlets for agroforestry products, and the like. Their research programs and approaches need to be flexible, including the constant reappraisal of farmers' land-use opportunities and priorities and the constraints they face in implementing them. These may range from complex and even unclear environmental legislation, through difficulties in organizing production and marketing, to technical constraints at all steps from planting and managing trees to providing quality products that meet both consumer demand and legal requirements. This obviously requires decentralized, flexible organization of research institutions and their funding. It may also require that scientists' performance and success be measured in terms of the impact that is achieved or facilitated through their research (e.g., land-use improvements adopted by farmers, legislative improvements adopted by governments, etc.), rather than in terms of publications produced and lectures delivered, which are ultimately measures of activity and not of achievement.

Nowadays, with the importance rightly given by development agencies to the role of institutions, both governmental and nongovernmental, in the development process, biological-technical land-use research is often given low priority. However, such research addresses very real constraints faced by tropical farmers. Biological-technical problems such as low productivity in traditional agroforestry systems, low and variable product quality in many fruits and extractive products, high pest and disease pressure in cocoa agroforests, and uncertainty about sustainable harvest levels with many extractive products, are very real constraints that require more research. Development projects that address institutional and marketing issues but do not recognize such biological-technical constraints will have little impact in the field. Nor will projects that help farmers to solve biological-technical constraints without addressing the organizational, marketing, and perhaps even legal impediments to capitalizing on these innovations. Clearly, research that aims at improving farmers' livelihoods and conserving biodiversity needs to be interdisciplinary and designed to address issues along the whole chain from sustainable production, through the development of legal frameworks for new land-use activities and capturing subsidies where these are needed and available, to marketing of the product.

The intensification of traditional agroforestry practices that have developed under conditions of relative land abundance and scarcity of labor, such as rubber and cocoa agroforests, is necessary for their conservation but is at the same time a threat to their biodiversity. It is therefore a particular challenge for agroforestry science to devise ways to increase the profitability of such complex agroforestry systems without losing too much of their complexity and wild biodiversity. For example, the introduction of more productive, disease-resistant plant breeds into traditional agroforests may encourage the thinning of shade canopies because these new varieties may require higher levels of light and nutrients to realize their productive potential. Also, the high value of improved (and perhaps purchased) seedlings may preclude the extensive management that is a precondition for the development of some traditional agroforests. However, highly productive individuals of tree crops such as rubber (Schroth et al. 2004c), that are at the same time adapted to extensive agroforest conditions and pressure from endemic diseases, have been found in traditional agroforests and could provide a basis for new plant breeds specifically developed for low-input, extensive agroforests. Clearly, compromises are required. In some cases, this may mean combining intensification on the field level with the conservation of natural habitat on field boundaries, riparian buffer strips, and on-farm reserves.

Many trees in traditional agroforestry systems (e.g., rustic coffee and cocoa groves) are remnants of previous vegetation, which will eventually die out, so practices need to be developed and encouraged to actively regenerate and manage such tree vegetation. An important strategy for this is to encourage diversification through the increased use and commercialization of timber and non-timber products from such complex agroforestry systems, thereby creating additional sources of revenue and reducing the susceptibility of the farmers to fluctuations of commodity prices. Participatory selection of tree germplasm and the development of methods to regenerate trees within established agroforests or at the time of their establishment or renovation are important areas of research. However, developing the markets for these products and, in the case of timber, overcoming legal restrictions to commercialization as a precondition for sustainable management, will in many cases be the key issues to be addressed.

Apart from these biological-technical issues, a large amount of work remains to be done on the question of how land-use mosaics can be designed and planned together with local people so that they "work" from both a livelihoods and a conservation perspective. Where a legal framework exists, as in the buffer zone of a protected area, and the means to implement it are available, zoning may be a way to encourage agroforestry practices in biologically sensitive parts of a landscape, for example, near forest boundaries or in critical corridor areas. However, in many cases zoning will not be effective for lack of enforcement mechanisms. In such situations, growing markets for ecosystem services, including the conservation of biodiversity, may offer ways to encourage certain land uses and discourage others. Where biodiversity-friendly agroforestry practices are less profitable than alternative practices (e.g., because of lower yields), where agroforestry practices need higher initial investments than alternative practices to get started (e.g., for tree planting), or where the biodiversity conserved through the use of agroforestry negatively affects crop harvests (e.g., wildlife damage), temporary or even permanent subsidies may be needed to make such land uses attractive for farmers. But transactions costs may be high unless farmers develop a self-interest in adopting conservation-friendly land-use practices on a large scale. This reinforces the need for work to increase the productivity and profitability of such environmentally friendly land-use practices without losing their conservation benefits. Some countries such as Brazil, have legislation that obliges wood-consuming industries to replant annually a number of trees proportional to their consumption or pay someone to do it on their behalf; this can offer opportunities for farmers to meet the investment costs of tree planting and the establishment of agroforestry systems. In other cases, industries may be interested in promoting sustainable production methods by giving farmers access to preferential markets or by funding programs that promote agroforestry production methods. A huge amount of applied research and development work must be done to make such schemes work for both local farmers and wild biodiversity.

Conclusions

Agroforestry has an important role to play in ecoagriculture strategies. Traditional agroforestry practices such as extensively managed, species-rich tree crop systems or long fallows with extractive uses are particularly valuable in this sense. However, simpler practices such as trees, windbreaks, and live fences in farmland that fit easily into modern production methods are also valuable and should not be overlooked. Innovative farmers throughout the tropical world have developed ecoagriculture practices such as tree crop agroforests, enriched riparian forests, and no-burn systems of food crop production that should be conserved and may have potential to be more widely implemented. In some cases, the more systematic exchange of experience between different tropical regions may help farmers to choose technical innovations from a larger pool of experiences and solve land-use problems. This seems, for example, to be the case with rubber agroforestry in Indonesia and the Amazon, where technical innovations developed locally many decades ago could be applicable elsewhere. Scientists have an important role to play by helping farmers to make traditional and innovative land-use solutions work under present-day conditions, provided that scientific institutions and funding agencies live up to the challenge and adapt to the new requirements. Key challenges are to make land-use practices work for both livelihoods and biodiversity on the plot scale, and to develop methods to integrate them on landscape and regional scales.

References

Alves, M.C. 1990. The Role of Cacao Plantations in the Conservation of the Atlantic Forest of Southern Bahia, Brazil. Master Thesis, University of Florida, Gainesville.

Angelsen, A. and D. Kaimowitz. 2004. Is agroforestry likely to reduce deforestation? In: *Agroforestry and Biodiversity Conservation in Tropical Landscapes*, ed. G. Schroth, G.A.B Fonseca, C.A. Harvey, C. Gascon, H.L. Vasconcelos, A.-M.N Izac. Island Press, Washington, DC: 87–106.

Augusseau, X., P. Nikiéma, E. Torquebiau. 2006. Tree biodiversity, land dynamics and farmers' strategies on the agricultural frontier of south-western Burkina Faso. *Biodiversity Conservation* 15:613–630.

Boshier, D.H. 2004. Agroforestry systems: important components in conserving the genetic viability of native tropical tree species? In: *Agroforestry and Biodiversity Conservation in Tropical Landscapes*, ed. G. Schroth, G.A.B Fonseca, C.A. Harvey, C. Gascon, H.L. Vasconcelos, A.-M.N Izac. Island Press, Washington, DC: 290– 313.

Center for Applied Biodiversity Science. 2000. *Designing Sustainable Landscapes: The Brazilian Atlantic Forest.* Center for Applied Biodiversity Science at Conservation International and Institute for Social and Environmental Studies of Southern Bahia, Washington, DC.

Cullen, L. Jr., J.F. Lima, T.P. Beltrame. 2004. Agroforestry buffer zones and stepping stones: tools for the conservation of fragmented landscapes in the Brazilian Atlantic Forest. In: *Agroforestry and Biodiversity Conservation in Tropical Landscapes*, ed. G. Schroth, G.A.B Fonseca, C.A. Harvey, C. Gascon, H.L. Vasconcelos, A.-M.N Izac. Island Press, Washington, DC: 415–430.

Finegan, B., R. Nasi. 2004. The biodiversity and conservation potential of shifting cultivation landscapes. In: *Agroforestry and Biodiversity Conservation in Tropical Landscapes*, ed. G. Schroth, G.A.B Fonseca, C.A. Harvey, C. Gascon, H.L. Vasconcelos, A.-M.N Izac. Island Press, Washington, DC: 153–197.

Gascon, C., G.B. Williamson, G.A.B. da Fonseca. 2000. Receding forest edges and vanishing reserves. *Science* 288:1356–1358.

Hannah, L. 2004. Agroforestry and climate change—integrated conservation strategies. In: *Agroforestry and Biodiversity Conservation in Tropical Landscapes*, ed. G. Schroth, G.A.B Fonseca, C.A. Harvey, C. Gascon, H.L. Vasconcelos, A.-M.N Izac. Island Press, Washington, DC: 473–486.

Harvey, C.A., N.I.J Tucker, A. Estrada. 2004. Live fences, isolated trees, and windbreaks: tools for conserving biodiversity in fragmented tropical landscapes. In: *Agroforestry and Biodiversity Conservation in Tropical Landscapes*, ed. G. Schroth, G.A.B Fonseca, C.A. Harvey, C. Gascon, H.L. Vasconcelos, A.-M.N Izac. Island Press, Washington, DC: 261–289.

Mesquita, R.C.G., P. Delamônica, W.F. Laurance. 1999. Effect of surrounding vegetation on edge-related tree mortality in Amazonian forest fragments. *Biological Conservation* 91:129–134.

Michon, G., H. de Foresta. 1999. Agro-forests: incorporating a forest vision in agroforestry. In: *Agroforestry in Sustainable Agricultural Systems*, ed. L.E. Buck, J.P. Lassoie, E.C.M. Fernandes. Lewis Publishers, Boca Raton, FL: 381–406.

Murniati, D.P. Garrity, A.N. Gintings. 2001. The contribution of agroforestry systems to reducing farmers' dependence on the resources of adjacent national parks: a case study from Sumatra, Indonesia. *Agroforestry Systems* 52:171–184.

Naughton-Treves, L., N. Salafsky. 2004. Wildlife conservation in agroforestry buffer zones: opportunities and conflict. In: *Agroforestry and Biodiversity Conservation in Tropical Landscapes*, ed. G. Schroth, G.A.B Fonseca, C.A. Harvey, C. Gascon, H.L.Vasconcelos, A.-M.N Izac. Island Press, Washington, DC: 319–345.

Perfecto, I., R.A. Rice, R. Greenberg, M.E. van der Voort. 1996. Shade coffee: a disappearing refuge for biodiversity. *BioScience* 46:598–608.

Rice, R.A., R. Greenberg. 2000. Cacao cultivation and the conservation of biological diversity. *Ambio* 29:167–173.

Rice, R.A., R. Greenberg. 2004. Silvopastoral systems: ecological and socioeconomic benefits and migratory bird conservation. In: *Agroforestry and Biodiversity Conservation in Tropical Landscapes*, ed. G. Schroth, G.A.B Fonseca, C.A. Harvey, C. Gascon, H.L. Vasconcelos, A.-M.N Izac. Island Press, Washington, DC: 453– 472.

Richardson, D.M., P. Binggeli, G. Schroth. 2004. Invasive agroforestry trees: problems and solutions. In: *Agroforestry and Biodiversity Conservation in Tropical Landscapes*, ed. G. Schroth, G.A.B Fonseca, C.A. Harvey, C. Gascon, H.L. Vasconcelos, A.-M.N Izac. Island Press, Washington, DC: 371–396.

Ricketts, T.H., G.C. Daily, P.R. Ehrlich, C.G. Michener. 2004. Economic value of tropical forest to coffee production. *Proceedings of the National Academy of Sciences, USA* 101:12579–12582.

Schroth, G., P. Coutinho, V.H.F. Moraes, A.K.M Albernaz. 2003. Rubber agroforests at the Tapajós river, Brazilian Amazon: environmentally benign land use systems in an old forest frontier region. *Agriculture, Ecosystems and Environment* 97:151–165.

Schroth, G., M.S.S. da Mota, R. Lopes, A.F. de Freitas. 2004d. Extractive use, management and in-situ domestication of a weedy palm, *Astrocaryum tucuma*, in the central Amazon. *Forest Ecology and Management* 202:161–179.

Schroth, G., G.A.B Fonseca, C.A. Harvey, C. Gascon, H.L. Vasconcelos, A.-M.N Izac, A. Angelsen, B. Finegan, D. Kaimowitz, U. Krauss, S.G. Laurance, W.F. Laurance, R. Nasi, L. Naughton-Treves, E. Niesten, D.M. Richardson, E. Somarriba, N.I.J. Tucker, G. Vincent, D.S. Wilkie. 2004a. Conclusion: agroforestry and biodiversity conservation in tropical landscapes. In: *Agroforestry and Biodiversity Conservation in Tropical Landscapes*, ed. G. Schroth, G.A.B Fonseca, C.A. Harvey, C. Gascon, H.L.Vasconcelos, A.-M.N Izac. Island Press, Washington, DC: 487–502.

Schroth, G., C.A. Harvey, G.Vincent. 2004b. Complex agroforests: their structure, diversity, and potential role in landscape conservation. In: *Agroforestry and Biodiversity Conservation in Tropical Landscapes*, ed. G. Schroth, G.A.B Fonseca, C.A. Harvey, C. Gascon, H.L.Vasconcelos, A.-M.N Izac. Island Press, Washington, DC: 227–260.

Schroth, G., V.H.F Moraes, M.S.S. da Mota. 2004c. Increasing the profitability of traditional, planted rubber agroforests at the Tapajós river, Brazilian Amazon. *Agriculture, Ecosystems and Environment* 102:319–339.

Thiollay, J.-M. 1995. The role of traditional agroforests in the conservation of rain forest bird diversity in Sumatra. *Conservation Biology* 9:335–353.

Tucker, N.I.J., G. Wardell-Johnson. C.P. Catterall, J. Kanowski. 2004. Agroforestry and biodiversity: improving conservation outcomes in tropical northeastern Australia. In:

Agroforestry and Biodiversity Conservation in Tropical Landscapes, ed. G. Schroth, G.A.B Fonseca, C.A. Harvey, C. Gascon, H.L.Vasconcelos, A.-M.N Izac. Island Press, Washington, DC: 431–452.

Wilkie, D.S. and R.J. Lee. 2004. Hunting in agroforestry systems and landscapes: conservation implications in West-Central Africa and Southeast Asia. In: *Agroforestry and Biodiversity Conservation in Tropical Landscapes*, ed. G. Schroth, G.A.B Fonseca, C.A. Harvey, C. Gascon, H.L. Vasconcelos, A.-M.N Izac. Island Press, Washington, DC: 346–370.

Williams, S.E., M. van Noordwijk, E. Penot, J.R. Healey, F.L. Sinclair, G. Wibawa. 2001. On-farm evaluation of the establishment of clonal rubber in multistrata agroforests in Jambi, Indonesia. *Agroforestry Systems* 53:227–237.

Wood, G.A.R., R.A. Lass. 1985. *Cocoa*. Blackwell Scientific, Oxford.

Chapter 7

Livestock Systems

Constance L. Neely and Richard Hatfield

Introduction

Livestock is becoming agriculture's most important subsector worldwide (ILRI 2006a, Jutzi 2006). Partly fuelling this trend is the "livestock revolution," resulting from an unprecedented demand for meat by consumers—an increase of 70 million metric tons consumed between the 1970s and 1990s in developing countries alone. This holds profound social, environmental, and economic implications (Delgado et al. 1999; FAO LEAD 2006). Although this revolution is manifesting itself in different ways globally, it poses a key question—is it destined to aggravate a worsening environmental situation? or can it be worked to the advantage of both people and the natural resource base supporting them? This chapter briefly considers aspects of livestock production as they relate to livelihoods and the conservation of biological diversity, and presents a set of indicative examples in which these two objectives are achieved simultaneously. The chapter also provides examples that demonstrate good practices, particularly from ruminant-dominated systems, from around the world. The final sections highlight some cross-cutting themes of community-based approaches, genetic diversity management and livestock value chains, and then discuss policy priorities to promote ecoagriculture innovations in livestock systems.

Context: Livestock, Landscapes, and Livelihoods

Currently, land used for grazing livestock makes up 26% of the world's land area. Cattle numbers globally are approximately 1.34 billion animals and are

Table 7.1. Systems of the interplay between livestock and conservation

	Extensive systems	*Intensive systems*
Commercial systems	Ranching	Feed lots and large-scale crop–livestock systems
Traditional systems	Mobile pastoralism	Smallholder crop–livestock systems

considered to affect more ecoregions of significant biodiversity than any other agricultural commodity (Clay 2004). This will be exacerbated if, as predicted, the unprecedented increase in demand for animal-based food products continues.

In considering the interplay of livestock and conservation, production systems can be categorized along two dimensions: commercial versus traditional or subsistence-dominated systems; and intensive versus extensive systems (Table 7.1). Each contains particular implications for production, livelihoods, and environmental impact.

Extensive Systems—Core Biodiversity Issues

Extensive systems, including commercial ranching and pastoralism, tend to affect biodiversity negatively through two major means: land conversion "off-site" and degradation "on-site," where *degradation* can be defined as decline in core biological processes (water and mineral cycles, energy flow, succession) and accompanying biodiversity. Of the 13 million ha annual loss of forests (FAO 2005), land cleared for livestock is approximately 1.5 million ha per year (de Haan et al. 2001). Deforestation along with the conversion of natural habitats and fragmentation has resulted in significant losses of biodiversity. Meanwhile, existing livestock systems exhibit decreased soil water retention and stream flow, increased sediment loads, and decreased plant biomass and diversity. Three drivers of these trends include inappropriate management, burgeoning market demand, and local poverty.

Intensive System—Core Biodiversity Issues

The chief concerns of intensive operations, whether individual scale or smallholder, relate to the high levels of required inputs and the effect of by-product disposal. These include contamination and/or eutrophication of watering points and water bodies by introduction of nutrients, sediment loading, metals, pathogenic bacteria, hormones, and antibiotics through surface and subsurface flow; changes in nutrient cycles (pigs and poultry alone produce 6.9 million tons of nitrogen annually); and disease transmission. Further, large quantities of water are used during animal production and processing. Livestock, particularly ruminants, are considered to contribute 5 to 10% of the greenhouse gases carbon dioxide and methane (e.g., Castel 2005; Harvey et al. 2005; Osofsky et al. 2005; Goodland and Pimentel 2000; World Bank Group 1998). Using the concept of the ecological footprint, a method to evaluate the area of the earth's surface dedicated to human consumption, beef production in the United States accounts for 79% of the land area required to produce all the country's food, with other animal-based food making up an additional 13% of the total food footprint (Palmer 1998).

Livestock-Based Livelihoods

Researchers, development practitioners, and policymakers are struggling to achieve the targets of the Millennium Development Goals addressing hunger, poverty, and the environment. Livestock are a major strategy in this regard, contributing to the livelihoods of 70% of the world's rural poor. Estimates indicate that more than 675 million rural poor, particularly the landless and women, are dependent on livestock for food and income (WRI 2005). Livestock serve to raise productivity on farms by providing nutrient inputs, transportation, draft, and energy. Further, livestock enhance household nutrition; are an important source of financial capital, savings, and hyperinflation insurance; and can be sold, traded, or used as collateral (Ashley et al. 1999; IFAD 2004; Thornton et al. 2002; WRI 2005). Additionally, these benefits have a ripple effect in that they can extend to the non-owning poorest community members through gifts or sales at reduced prices (Shackleton et al. 1999).

How the poor stand to gain from burgeoning global demand for livestock products is unclear; however, the livestock revolution clearly represents significant potential for their participation in one of the few sectors in which they hold a comparative advantage (Delgado et al. 1999). The importance of livestock-based livelihoods to society as a whole and to the poor in particular, and the growing realization that productivity and conservation must be mutually reinforcing, points to the potential beneficial payoffs of investment in ecoagriculture initiatives.

Innovations in Extensive Livestock Grazing Systems

Grasslands evolved with enormous herds of wild ungulates (Frank 2004, Frank et al.1998), and yet the belief that there are too many livestock on the land is pervasive. Research has shown that standard concepts of carrying capacity are less applicable in dynamic environments such as the semiarid to arid rangelands used by pastoralists. Research has also shown that systems where herd size fluctuates according to availability of natural resources and livelihood needs are more sustainable than constant stocking rates (e.g., Behnke and Abel 1996; Savory 1999). Under proper management, livestock have been shown to be beneficial to, if not required for, enhancing rangeland production, biodiversity, and subsequent moisture availability. Innovative approaches to achieving both livelihood and biodiversity aims include grazing for habitat management, cooperative corridors, adaptations of traditional pastoralism, comanagement of livestock and wildlife, disease and predator management, and game ranching.

Management and Planned Grazing Approach

The Holistic Management decision-making framework recognizes that grazing pressures can be timed to maximize plant productivity and overall biodiversity (Savory 1999; Voisin 1988). Taking into account that time, rather than simply numbers of animals, needs to be managed, the major benefits of planned grazing together with herd management in dryland areas include enhanced (1) nutrient (and moisture) provision by livestock through dung and urine, (2) microbial decomposition, (3) infiltration as herd hoof action breaks the soil surface crusts, and (4) soil–seed contact for germination of dormant seeds. Livestock are herded and moved so that they are able to provide these benefits. Careful planning and monitoring of timing ensure plants are not overgrazed and increased animal density enhances herd hoof action, both of which are key to producing the benefits listed above. Land cover can be dramatically increased, addressing a major cause of drought. Experience in Zimbabwe (Figure 7.1) has demonstrated that increased soil cover increases water infiltration and subsequent stream and water table recharge and availability. This approach is practiced mainly on private lands in Southern Africa, North America, and Australia, and is now generating significant interest in communal systems in sub-Saharan Africa: for example, Maasai communities in Kenya in collaboration with World Vision Kenya, and Hwange communities in Zimbabwe.

The main impacts have been increased awareness, biodiversity, and income. Stinner et al. (1997) followed 25 farmers and ranchers in the United States on landholdings ranging from 7 to 90,000 ha. Prior to practicing Holistic Management, 9% considered biodiversity as important in their operations versus 95% afterward. For example, from 1984 to 1992, one 4800 ha ranch achieved a 300% increase in perennial species, 50% tighter plant spacing, and enhanced productivity in terms of beef from 66 kg/ha to 171 kg/ha. In Australia, other practitioners have been able to increase stock numbers while increasing ground

Figure 7.1. Using livestock to heal the land in Zimbabwe. Photos a (grazing area) and b (waterway) represent the desertified Hwange communal lands in Zimbabwe at the end of the rainy season. Photos c (grazing area) and d (waterway) depict the neighboring Holistic Management learning site for the community using the Holistic decision making framework (Savory, 1999) and greatly increased livestock to reverse desertification. These pictures were taken on the same day as a and b at the end of the rainy season. With increased soil cover, there is increased infiltration and thus more water available to recharge streams, bore holes, and springs. An added value of increased biomass and water availability has been increased wildlife (photos by Allan Savory).

cover, enhancing species diversity, improving water quality, building bird diversity for pest control, and setting aside rangeland for timber corridors allowing wildlife movement (Forge and Reid 2006).

Livestock Grazing for Habitat Protection

Gordon and Duncan (1998) noted the impact of large herbivores and reported a loss of species in European wetlands when livestock were removed. The

English landscape has some 4000 sites of special scientific interest, 2000 of which are of international importance. These are grassland, heathland, wood pasture, floodplain, and coastal marshes—each of which is recognized as requiring livestock grazing to conserve the associated wildlife habitats, including prevention of scrub encroachment and removing plant material without cutting and burning, thus enhancing species mobility (English Nature 2005).

Cooperative Corridors

Within the Landscape Development, Biodiversity and Cooperative Livestock System (LACOPE) project, research teams from seven European countries are contrasting four pastoral systems, ranging from reindeer husbandry in Norway to sheep grazing in Spain and Poland and cattle grazing in the Swiss Alps, to understand how large-scale grazing systems can lend themselves to biodiversity improvements and habitat creation for target species of the EUROPA NATURA 2000 System. This project stems from a perception that large-scale nature conservation efforts have not been successful. Early results have raised the importance of institutional mechanisms and the legal setting as well as the current socioeconomic and technical trend (Gueydon and Röder 2003).

Adapting Traditional Pastoralism

Pastoralism depends heavily on mobility for its productivity. Decreased mobility raises the question of whether pastoralism should seek to reenhance mobility or focus instead on adapting to modern conditions. In Kenya, for example, many communal lands exist under a cooperative land title (group ranches). Two mutually exclusive dynamics are widely evident: continued movement between group ranches under traditional reciprocity; and modern investment into other land-use types, for example, wildlife conservancies and conversion of key resource "patches" to crops. However, gains from new uses remain private to individual group ranches, who are also able to graze neighboring lands. The latter are, therefore, in effect subsidizing the new initiatives of the former. Such dynamics can be anticipated over time to generate a move toward transaction-based exchanges rather than general reciprocity.

Comanagement of Livestock and Wildlife

Wildlife is a common option on rangelands, particularly where land is being managed for high plant biodiversity, which can, in turn, support wildlife assemblages rather than individual species. However, the issue that draws the most debate—and presents real management issues—is the degree to which livestock and wildlife are compatible.

Hard Boundaries: The Limitations

Historically livestock and wildlife have been seen as incompatible, and this is still the dominant worldview, demanding "hard" boundaries separating the two in order to protect wildlife. This rationale extends beyond protected areas and is common within the growing number of community "wildlife conservancies," particularly in Eastern and Southern Africa. This approach, however, removes resources upon which community members, typically poorer households, depend for their main livelihood, fueling community resentment and ultimately undermining support for wildlife conservation. In addition, wildlife do not remain within the hard boundaries, thereby creating management challenges; and wildlife populations do not appear to thrive under this model, except in situations where conservancies are large, which is often socially and politically challenging.

Permeable Boundaries and Grass Banking: The Opportunities

Out of necessity, variations are being tested, partly based on traditional management systems and partly informed by recent experience in Western contexts. Three examples are deferred grazing, grass-banking, and seed-banking. All involve the element of controlling plant recovery time. In an African context, deferred grazing typically involves dividing land into three to four grazing blocks, each used once during the year. Grass banking refers to either an annual dry-season reserve or an emergency (four- to seven-year drought cycle) reserve. In East Africa, the term means annual reserve and is therefore akin to deferred grazing. In other settings, for example, the Malpai Borderlands of Texas/New Mexico where private ranchers have "pooled" grazing resources, it refers to a four- to five-year emergency reserve. Under seed banking, plant consumption is held off until seeding stage, some of which may be harvested meanwhile. These measures have implications for wildlife and biodiversity. One variation is for grass banks to double as dry-season grazing and wildlife refuges (often the de facto situation in African community conservancies). A further variation is that such grass banks not be fixed in location but rotate, as a means of promoting periodic "harvesting" and regeneration. These models are being explored with communities by such organizations as the African Wildlife Foundation in the Kilimanjaro region of Kenya and Tanzania.

No Boundaries

A newer view is that grazing systems should return to a situation where boundaries between livestock and wildlife are erased, and where the whole range is brought into production. In such a system no particular, fixed reserve is necessarily designated. Such approaches are being implemented by those who believe in the positive effects of animal feeding and/or disturbance, for example, Holistic Management. Under such a system, the maximum available land area is used, and dry-season stress is countered by sufficiently increased plant productivity, biomass, and diversity across the whole area.

Such efforts are at the center of efforts in Syria around the Talila Reserve east of Palmyra, where the government and Food and Agriculture Organization of the United Nations (FAO) have used a participatory decision-making approach to reintroduce gazelles and oryx such that Bedouins have access to grazing in the reserve on a seasonal basis, improved grazing lands outside of the reserve, and increased wildlife (Williamson 2006; Mirreh 2005).

Managing Risk in Livestock–Wildlife Systems: Disease Transmission and Predator Management

Wildlife are known disease reservoirs and sometimes vectors. However, as a review of current knowledge on the wildlife–livestock disease interface (Kock 2005) concludes, "the interface . . . will need to be better understood by all the stakeholders. More research is needed, as are new philosophies and attitudes, and new approaches to livelihoods and resource use." Traditional African management focused on managing livestock–wildlife interaction in time and space, a practice still pursued today but made more challenging by decreased mobility. Such practices suggest the possible usefulness of intensified rotation for managing interaction in a smaller space. Organizations such as VETAID are assisting pastoralists in managing disease. An additional risk factor is predators. Two well-documented approaches to managing predators are predator-proof corrals and guard dogs. Several initiatives are under way in Eastern and Southern Africa, for example, that improve traditional night corrals to be effective against lions, leopards, and hyenas. These include the Laikipia Predator Project in Kenya (WCS 2006), and Holistic Management in Zimbabwe. The kraals in Zimbabwe are woven mats of twigs, portable by ox cart. The use of livestock guardian dogs has a 1000-year history throughout Europe, with breeds such as Anatolean shepherd dogs and Spanish mastiffs, which specifically guard against wolves, bears, and foxes.

Game Ranching

Game ranching strictly refers to raising wildlife for consumptive use. It is now widely accepted that game ranching, despite being perceived as conferring ecological advantages to the land, cannot outcompete livestock ranching. This is primarily due to the favorable management characteristics of domesticated animals (for example, handling) and productivity gains bred into livestock over millennia. However, game ranching can be a favorable complementary activity to livestock, whether used consumptively or not, particularly based on evidence from Zimbabwe (Jansen et al. 1992).

Innovations in Intensive Small- and Large-Holder Farming Systems

Innovative approaches for biodiversity conservation in intensive livestock systems include silvopastoral systems, integrated crop–livestock systems, integrated fisheries in protected areas, beekeeping, bushmeat domestication, integrating livestock into no-tillage systems, and innovations in industrial agriculture.

Silvopastoral Systems

Silvopastoral systems that incorporate livestock and trees can increase complexity and biodiversity benefits (Dagang and Nair 2003). These systems support higher floral and faunal biodiversity while providing habitat, food, and refuge for wildlife, increasing alternative livelihood inputs of tree products, and providing grazing and shade for livestock (Harvey et al. 2005). Such systems are considered to be an important way to buffer negative effects of land-use conversion (Dagang and Nair 2003) and can serve as corridors between protected areas. In Tanzania, the Hifadhi Ardhi Shinyanga Soil Conservation Initiative (HASHI), implemented between 1986 and 1998, resulted in the revival of a traditional silvopastoral practice called Ngitili, which has since produced 70,000 ha of restored woodland. The practice, once used for fodder production, has now been extended to bring social benefits while conserving biodiversity, restoring land and river edges, providing grazing areas, and providing fuel, timber, and thatch. Regeneration of indigenous tree species and reduction in wildfires has resulted in household income diversification such as bee keeping (UNDP 2006).

Integrated Crop–Livestock Farming Systems

Livestock play an important role in agroecological and conservation agricultural systems to enhance or restore agricultural biodiversity on-site. Contributions include recycling of nutrients and additions to manures that increase plant biomass and soil biota. Parris (2002) notes that farms based on multiple crops and livestock with natural pasture are richer in biodiversity than monocultural farms. In Asia (ILRI 2006a) mixed farming systems have been shown to provide 90% of milk, 77% of ruminant meat, 47% of pork and poultry, and 31% of eggs while promoting links between system components—land, crops, water, livestock, and biodiversity.

Cut and Carry. Cut and carry systems have become popular as a means of raising livestock in pens with inputs carried from small plots of intensively grown fodder trees/shrubs and grasses (Ramos 2000). Used by smallholders with few animals, the system provides a means of protecting other areas of the landscape from destruction by grazing animals.

Keeping Elephants, Lions, Livestock, and Maize. In Zimbabwe, one landowner has increased his maize yield while living with elephants. After multiple raids on his crops, he built a deep ditch around the field to keep the elephants out and built a small lion-proof kraal on the field that could be rotated during the nongrowing season. Community cattle were kept in the kraal at night to capture dung and urine, after which maize was planted using no-tillage techniques. Maize yields from this field were 15 times those of his traditionally treated fields. The rest of the community is keen to mimic this effort to preserve wildlife and integrate conservation agriculture, livestock, and agroforestry (Lybian Sibanda, pers. comm.).

Integrated Fisheries in Protected Areas. Utilizing indigenous knowledge and collaborative management in Laos, communities set up 68 Fish Conservation Zones (FCZs) along the Mekong River between 1993 and 1998 that resulted in increased fish stocks of 50 types of fish. Local livelihoods rely on fish from the Mekong River and wildlife from the protected areas. Communities adjacent to the Xe Pian National Biodiversity Conservation Area established FCZs outside the protected area and found that with increased fishing income the terrestrial wildlife in the protected area has been enhanced (CIP-UPWARD 2003).

Bees for Biodiversity. In Tanzania, which has the fourth-highest loss of forest in Africa, beekeeping has become a livelihood strategy that raises awareness on the value of forest preservation. Heifer Tanzania has introduced the Stingless Bee Project. Marketed as Tabora Honey destined for European markets, partners have been able to create livelihood opportunities on the forest edges near Kilimanjaro to increase income and environmental awareness. This has been done through the formation of 43 farmer groups targeting the socially disadvantaged and the establishment of 700 hives (Erwin Kinsey, Godfried Narh, and Evod Banzi, pers. comm.).

Grasscutters in Ghana—Changing Our Approach to Bushmeat. The grasscutter (*Thyronomys swinderianus*) is a rodent found in the savanna and forests of Ghana and is considered a delicacy. In order to collect them in the wild, hunters have historically set poison baits or brushfires to drive out animals. These practices are thought to be the main cause of bushfires and are responsible for much forest conversion. While the grasscutter is wild, it can be bred in captivity and a collaboration between Heifer Ghana and the Ministry of Agriculture promotes grasscutter raising as small businesses that improve the environment, health, and sustainable wildlife utilization. Some 500 people have been trained in grasscutter production and the Grasscutters Farmers Association has now become a major source of safe and healthy meat for the general public. This is an important element in the Ghana Forestry and Wildlife Policy and the national poverty reduction strategy to promote grasscutter raising on the periphery of game reserves (Heloo 2006).

INTEGRATING LIVESTOCK INTO NO-TILLAGE SYSTEMS. In Brazil, which hosts some of the largest areas of conservation tillage, producers and researchers are exploring ways to integrate overseeded pasture into no-tillage systems to boost overall productivity, retain ground cover, provide wildlife corridors, offset the degradation of the natural grassland systems that are host to 50 species per meter, and reduce wildlife losses (Carvalho and Batello 2006; FAO 2006a).

Innovations in Industrial Agriculture

Overconcentration of genetically uniform animals, land-use conversions for concentrated feed production, and nutrient imbalances associated with waste are key areas where biodiversity conservation is most threatened in large-scale intensive, industrial agriculture. Currently, conversion of Brazilian grasslands to produce soybean meal for livestock production in China, India, and other countries is receiving media attention. However, those soybean fields are in demand for soya oil, although the coming demand for bioenergy may dwarf that which is attributed to livestock feed requirements (Andrew Speedy, pers. comm.).

Naylor et al. (2005) points out that crop and livestock systems must undergo a recoupling, be it physical or through pricing and policy mechanisms, to address social and environmental costs. Technological examples include feed efficiency enzymes or the use of buffer strips for waste capture, requiring policies related to land and water pricing and damage limitation. In European countries and the United States, technology, quality control, regulations, and environmental conservation incentive assistance are mechanisms used for discouraging environmental damage.

Modern technologies associated with pig and poultry production in particular, animals in which feed conversion is more efficient, are serving to reduce the pressures on rangelands and forests (FAO LEAD 2006).

The Netherlands is experimenting with the concept of tradable quotas that incorporate nutrient accounting systems for hog production. The FAO LEAD (2006) group is working in Asia and Latin America to evaluate policies and technologies based on the idea of "polluter pays, provider gets." Consumers are being viewed as a means to shift the paradigm (Naylor et al. 2005), and eco-labeling can enhance consumer awareness of ecofriendly management practices. An example is Quality Assured Beef, a European ecolabel that encourages working with natural processes while including grazing requirements and limited grain consumption (Clay 2004). Innovative production practices to move away from grain production to perennial species, such as pastured poultry and pork, have been demonstrated by smaller-scale livestock producers (Eco-Friendly Foods, 2006). These practices can have landscape implications when large commercial operations draw their products from cooperatives of innovative small-scale producers.

Cross-Cutting Themes

Achieving ecoagriculture systems with livestock will require more than technological and management innovations. Other key issues cutting across production systems include the value of community-based approaches, the importance of managing livestock and grassland diversity, and developing livestock value chains to enhance biodiversity conservation.

The Value of Community-Based Approaches

Low and highly variable rainfall creates risky environments for households. People's knowledge is therefore a valuable resource for managing risky environments, contrasting with the narrower understanding that underlies introduced technologies, many of which have failed (Mortimore, 2006). As Barrow (1996) points out "the lesson to be learnt . . . is that lack of attention to social concerns, values, rules, and regulations can be a root cause of much biodiversity loss." For example, the extent of a herd owner's social network is directly related to the family's chances of long-term survival; complex tenure and usufruct rights to key resources have evolved that do not conform to generalised forms of property rights, such as the Turkana *Ekwar* system in northern Kenya where individuals own individual trees along important rivers; or groups in Shinyanga, Tanzania, that have tenure rights over trees under the *Ngitiri* conservation system (Barrow 1996). Successful resource management in drylands depends on social controls, knowledge, and rights over a wide array of resources.

LINKING LANDCARE AND LIVESTOCK. To advance ecoagricultural initiatives, it is important to consider the institutional structures that can promote livestock production systems and biological diversity conservation. Multistakeholder or local community efforts can raise awareness and promote collective implementation of productive agricultural systems and conservation efforts at a landscape scale. These are often termed community-based natural resources management (CBNRM) efforts. An example is that of Landcare—an approach that connects land managers in a landscape approach for collective conservation action. The Landcare approach, originating in Australia, has spread to many developed and developing countries, with an emphasis on awareness raising on conservation, community participation, and sustainable livelihood strategies that also improve land and water management and biodiversity (chapter 16).

Bringing the Actors Together—an Example from the Kenya Livestock Working Group. The role of livestock keepers in sustainable agriculture and rural development was identified as a priority subject area within the Sustainable Agriculture and Rural Development (SARD) Initiative. Launched as a global effort at the 2005 International Farming Systems Association Global Learning Opportunity (IFSA GLO), the Global Livestock Working Group identified a mandate that includes connecting the many and diverse actors working on livestock and wildlife-based issues; sharing good practices to address issues affecting livestock keepers; and advocating fair livestock development opportunities among the poor through educational efforts and policy influence. In Kenya, the group consists of Samburu and Masai communities, government, community-based organizations, Kenya civil society organizations, business and industry, the research community, international nongovernmental organizations (NGOs), and the FAO to build a sustainable livestock value chain that starts with the protection and enhancement of the environment and ends with safe and quality food for consumers.

Livestock and Grassland Genetic Diversity and Its Biodiversity Impacts

Genetic diversity relates not only livestock but also grassland ecosystems that cover 40% of the surface of the earth, which must be recognized for their support of flora, livestock, wildlife, and human populations and as having been the foundation of various food grain crops and pharmaceuticals.

Livestock Genetic Diversity. Within livestock systems, the diversity of livestock breeds plays its own unique and important role in the maintenance of grassland ecosystems. In some areas of the world, especially those in harsh environments, human settlement may not have been possible without well-adapted livestock breeds (Scherf et al. 2006). Breeding being an essential part of the management, use, and development of animal genetic resources, emphasis must be placed on the diversity of breeds, because changes in breeds bring changes in the food and value chain (Hoffman and Scherf 2005). The Raika pastoralists of the Thur Desert in Rajasthan, India, place great importance on their rich variety of livestock breeds, sheep in particular, and the community pays close attention while selecting breeding stock, making sure that the young can survive the challenging climate and produce the best goods. The Raikas have been known to memorize the pedigrees of the animals for seven to eight generations. More recently, however, they face threats from privatization and decreasing pasture lands, along with drought, thus forcing the sale of livestock and the mixing of different breeds. Other pastoral groups' strong adherence to their culture and livestock values has contributed to the present animal diversity. For example, the Maasai communities in East Africa have resisted cross-breeding programs and thus contributed significantly to the high livestock genetic pool in the region (Kohler-Rollefson 1993).

GRASSLAND BIODIVERSITY. Half of the centers for plant diversity (CPDs) include grassland habitat and coincide with high grassland diversity where conservation practices are protecting the grassland species (White et al. 2000). In a seven-year experiment in the United States and United Kingdom, Tilman and others (2001) found that plant diversity and niche complementarity had progressively stronger effects on ecosystem functioning, noting that higher-diversity plots outperformed monocultures in productivity and carbon storage.

Livestock Value Chains to Enhance Biodiversity Conservation

Specialization is particularly risky in drier environments, and diversification of enterprises is the key means of achieving security. Two current global trends hold significant implications for livestock and biodiversity conservation: growth in urban populations and growth in disposable income. Although both increase market possibilities, they also offer opportunities for both good practice and conservation, as well as their opposites.

BRANDING FOR NATURE. In response to the emergence of multiple new market segments, interest in the formation of small "cooperatives" to improve access to markets has been growing (e.g., IUCN 2005). These associations not only improve continuity of supply (thereby allowing better access to higher-value contracts) and bargaining power, but also introduce quality-control standards. Standards can then be translated into higher-value "brands," capitalizing on facets such as better animal treatment, social causes, and conservation value. Recent "franchise" designs, such as the FAO initiative in the Maldives, are a specific example, whereas others are being proposed elsewhere, including East Africa (Ron Kopicki, pers. comm.)

WILDCARE DAIRY PRODUCTS, LTD. Dairy farmers in the United Kingdom have benefited from a brand that delivers a promise to consumers of benefiting wildlife and farmers. The Agri-Trade Direct Ltd., a farmers' buying consortium, found a way to add value to their commodity, and a new company emerged, Wildcare Dairy Products, Ltd. The White and Wild brand, now on sale at 500 supermarkets across the country, provides a premium for farmers to implement a number of commitments, including putting in place measures that enhance wildlife, such as dedicating 10% of their land to wildlife (Wildcare 2006).

OL PEJETA BEEF. Kenya's Ol Pejeta Trust, managing 40,000 ha of prime African savanna, markets beef in East African capital cities under the "free-range" banner. At the same time, the Trust is piloting the purchase and processing of Samburu nomadic pastoralist livestock in the region as a means of enhancing pastoralist livelihoods and development, as well as a means of providing income that increases the ability of pastoralists to tolerate important wildlife populations that use community lands. Community lands are an integral part of the Laikipia region's wildlife populations and strong tourism sector growth, with the majority of wildlife habitating private individual and community lands, rather than government-protected areas.

ENHANCING BIODIVERSITY AT THE ABATTOIR. In Kenya, the Livestock Self-Help Stakeholders Association (LISSA) has developed innovations that allow for more hygienic and environmentally friendly processing. The aim is for better products and lower costs. This has been done by building a biogas plant at Bahati abattoir to convert waste into gas that can then be used for lighting and heating water for cleaning. This practice has been coupled with treatment ponds to discourage offsite contamination and the planting of trees around the processing plant to prevent soil erosion and encourage wildlife. The sludge from the biogas plant is also composted and sold to local farmers. Additionally, the abattoir managers and workers are encouraging environmental awareness and conservation by promoting tree planting among pastoralists who bring cattle and sheep for slaughter (Kibue 2005).

Policies for Enhancing Biodiversity Conservation in Livestock-Based Systems

Across the globe, common characteristics are repeatedly identified in connection with the general failure of interventions targeting extensive livestock production systems. These also have repercussions, sometimes positive but commonly negative, for conservation. Efforts by conservation organizations often experience the same failures. Although bilateral donors and governments have invested significant funds, results have been limited due to four main factors: supply-driven solutions by external players rather than local demand-driven solutions; theoretical rather than practical applications; central rather than locally based implementation; and excessive focus on the production aspect of systems, with particular neglect of markets (e.g., European Commission 2003).

Suggested Policy Approach: "What" versus "How"

Hutton et al. (2005) note that sustainable use of ecosystems by communities may be the only viable political and economic strategy for broad landscape conservation in many countries. From a policy perspective they also note that the perceived lack of effectiveness of people-oriented approaches may well be real in many instances, but that this is usually more to do with the way programs are implemented rather than the concept (i.e., the "how" rather than the "what"). In particular relation to pastoralism, lack of sustainability has as much, if not more, to do with weak common property institutions than biology. At the same time, the common property literature shows the conditions necessary for effective management institutions (Hutton et al 2005).

Informed Decision Making and Democratized Expertise

Mearns (1996) proposes two specific cross-cutting policy needs of "overriding importance." The first is informed decision making, where new thinking and knowledge about range ecology necessitate an urgent need for training of a new generation of range managers that does not aim to produce a rigid scientific assessment appropriate for all producers for all time. The second is democratized expertise, which calls for new forms of professionalism among decision makers and those responsible for implementing policies, programs, and projects, based on a participatory approach. Although much is now known about facilitating a genuinely participatory development process, this requires operational application. This resonates with a statement from a Maasai participant in a recent workshop: "Is it possible to get researchers and institutions interested and involved in our pastoralist reality?"

Benefit Sharing among Local Users: Facilitating Collective Action

The property rights literature highlights the fact that the key to maximizing benefits between multiple users of a resource or service is the development of mechanisms that minimize players' transactions costs associated with arriving at the best solutions (e.g., the Coase theorem). Typically, the most appropriate mechanism involves a forum of some kind where trust can be built and negotiated solutions achieved at manageable cost. Notable examples include conflict resolution initiatives (e.g., Karamajong women's groups, *alokita* in Uganda, IISD 2005), payment for ecosystem services mechanisms worldwide (e.g., IUCN 2006), and comanagement between local and other resource claimants (e.g., Bolivian protected area comanagement).

Enhancing Rights

Some countries have sought to implement a pastoral code that would systematize pastoral land-use rights within a system of legal protection. The Sahelian countries of Mauritania, Niger, and Mali have each established a Code Pastorale. This Code seeks to regulate traditional forms of common access to rangeland resources while also taking into account modern legislative measures to protect individual and group-specific land rights, for example, stipulating negotiated settlements while conserving options for mobility and guaranteeing wetland access (IISD 2005).

Building on New Global Initiatives

Two recent initiatives illustrate promising directions for global action. The FAO Globally Important Agricultural Heritage Sites (GIAHS) initiative (FAO 2006c) focuses on identifying and promoting landscapes that are rich in biological diversity and have evolved from the dynamic adaptation of a people and their environment, and the needs and aspirations for sustainable development. A number of the GIAHSs incorporate livestock, such as those associated with Ladakh on the Tibetan plateau in India, the Maasai of East Africa, and reindeer herding in Siberia—an ancient practice dating back 5000 years. Once herded in groups of 3000, numbers have been diminished by one-third, affecting not only the herder communities but also the unique contribution the reindeer add to the taiga and tundra biome.

World Initiative on Sustainable Pastoralism

A second notable initiative is the World Initiative on Sustainable Pastoralism (WISP), which aims to build momentum for greater recognition of the need for sustainable pastoral development by bringing together pastoralists and knowledge on pastoralism to dispel myths undermining pastoralists. WISP works to ensure that appropriate policies and support systems are established for the self-evolution of pastoralists toward an economically, socially, and ecologically sustainable livelihood system (IISD 2005).

Conclusions

From the few examples presented here it is clear that innovations exist for livestock systems in ecoagriculture landscapes, yet the topic needs much greater attention. Against the backdrop of the looming livestock revolution is a great opportunity to enhance our research and implementation efforts to ensure that biodiversity conservation is well integrated into livestock production and processing practices. These need to be linked to the policy dimension.

The livestock sector is subject to powerful global trends such as urbanization, land degradation, and growing consumer demand, which increase opportunities as well as challenges for biodiversity conservation at a landscape scale. Consequently, the link between livestock and land needs to be strengthened through inter alia awareness raising, institutional reforms, upscaling community-based strategies, technological advances, and increased practice-to-policy linkages, including evidence submitted for global negotiations (Convention on Biological Diversity, Convention to Combat Desertification, Commission on Sustainable Development, etc.). It is imperative that ecoagriculture practices and policies aimed at livestock-related food security and biological diversity be included among the highest-priority strategies to meet the targets of the Millennium Development Goals and the multilateral environmental agreements.

References

Ashley, S., S. Hoden, and P. Bazeley. 1999. *Livestock in Poverty Focused Development.* Livestock in Development, Crewkerne, UK.

Barrow, E. 1996. *Who Gains Who Loses? Biodiversity in Savannah Systems.* African Wildlife Foundation Community Conservation Discussion Paper No. 3. Nairobi, Kenya.

Behnke, R. and N. Abel. 1996. Intensification or overstocking: when are there too many animals? *World Animal Review* 87:4–9.

Carvalho, P. and C. Batello. 2006. Access to land, livestock production and ecosystems conservation in the Brazilian Campos biome: the natural grasslands dilemma. *Agriculture, Ecosystems and Environment.* Submitted.

Castel, V. 2005. Effect of grazing system on water quality and availability. Paper presented on developing a shared vision for livestock production. A workshop of CGIAR Challenge Programme on Water & Food (CPWF), Kampala, Uganda, 5–8 September 2005.

CIP-UPWARD. 2003. *Conservation and Sustainable Use of Agricultural Biodiversity.* International Potato Center, Los Banos, Philippines.

Clay, J. 2004. *World Agriculture and Environment.* Island Press, Washington, DC.

Dagang, A.B.K. and P.K.R. Nair. 2003. Silvopastoral research and adoption in Central America: recent findings and recommendations for future directions. *Agroforestry Systems* 58:149–155.

De Haan, C., T. Schillhorn van Veen, B. Brandenburg, J. Gauthier, F. le Gall, R. Mearns, and M. Simeon. 2001. *Livestock development. Implications for Rural Poverty, the Environment and Global Food Security*. Directions in Development. The World Bank, Washington, DC.

Delgado C., M. Rosegrant, H. Steinfeld, S. Ehui, and C. Courbois. 1999. *Livestock to 2020: The Next Food Revolution*. Food, Agriculture, and Environment Discussion Paper 28. International Food Policy Research Institute, Washington, DC.

Eco-Friendly Foods. 2006. http://www.ecofriendly.com/index.html.

English Nature. 2005. *The Importance of Livestock Grazing for Wildlife Conservation*. http://www.english-nature.org.uk/.

European Commission. 2003. *Livestock and Livestock Products Production and Marketing System in Kenya*. 12 October 2003.

FAO. 2005. *Global Forest Resources Assessment*. http://www.fao.org/forestry/foris/webview/forestry2/index.jsp?siteId=101&sitetreeId=16807&langId=1&geoId=0.

FAO. 2006a. *Crop-Livestock Systems in Conservation Agriculture: The Brazilian Experience*. Food and Agriculture Organization of the United Nations, Rome.

FAO. 2006b. *Livestock, Environment, and Development Center*. LEAD Virtual Centre. Food and Agriculture Organization of the United Nations, Rome.

FAO. 2006c. Globally Important Agricultural Heritage Sites. http://www.fao.org/AG/agL/agll/giahs/default.stm.

FAO LEAD. 2006. *Livestock's Long Shadow: Environmental Issues and Options*. Livestock, Environment and Development Center, Food and Agriculture Organization of the United Nations, Rome.

Forge K. and N. Reid. 2006. *Wool Production and Biodiversity. A Holistic Solution for Fine Wool and Healthy Profits at "Lana."* Land, Water and Wool: Shaping the Future, Australian Government, Land and Water Australia. http://www.landwaterwool.gov.au.

Frank, D.A. 2004. The Interactive effects of grazing ungulates and aboveground production on grassland diversity. *Oecologia* 143(4):629–34.

Frank, D.A., S.J. McNaughton, and B.F. Tracy. 1998. The ecology of the Earth's grazing ecosystems. *BioScience* 48(7):513–521.

Goodland, R. and D. Pimentel. 2000 Environmental sustainability and integrity in natural resources systems. In: *Ecological Integrity*, ed. D. Pimentel, L. Westra, and R. Noss. Island Press, Washington, DC.

Gordon, I and P. Duncan. 1998. Pastures new for conservation. *New Scientist* 117(1604):54–59.

Gueydon, A., and N. Röder. 2003. Institutional settings in co-operative pastoral systems in Europe: first results from the Lacope Research Project. Presented at The Commons in Transition: Property on Natural Resources in Central and Eastern Europe and the Former Soviet Union, a Regional Conference of the International Association for the Study of Common Property, Prague, 11–13 April 2003. http://dlc.dlib.indiana.edu/archive/00001060/.

Harvey, C.A., F.L. Sinclair, J. Sáenz, M. Ibrahim, C. Villanueva, R. Gómez, M. López, J. Montero, A. Medina, D. Sánchez, S. Vílchez, and B. Hernández. 2005. *Opportunities for Conserving Biodiversity within Agricultural Landscapes in Central America: Lessons from the FRAGMENT Project*. Henry A. Wallace/CATIE Inter-American Scientific Conference Series. CATIE, Turrialba, Costa Rica.

Heloo, John. 2006. *Grasscutters in Ghana*. Presentation at the Heifer International Advocacy Event, Little Rock, Arkansas.

Hoffman, I., and B. Scherf. 2005. *Management of Farm Animal Genetic Diversity: Opportunities and Challenges*. FAO, WAAP Book of the Year 2005.

Hutton, J., W. Adams, and J. Murombedzi. 2005. *Back to Barriers? Changing Narratives in Biodiversity Conservation*. Forum for Development Studies, No. 2. Tromsø, Norway

IFAD (International Fund for Agricultural Development). 2004. *Livestock Services and the Poor: A Global Initiative Collecting, Coordinating and Sharing Experiences*. Rome, Italy.

IISD (International Institute for Sustainable Development). 2005. *Herding on the Brink: Towards a Global Survey of Pastoral Communities and Conflict*, Occasional Working Paper from the IUCN Commission on Environmental, Economic and Social Policy. Gland, Switzerland.

ILRI (Internal Livestock Research Institute). 2006a. *Pastoralism: The Surest Way Out of Poverty in East African Drylands*. http://www.ilri.cgiar.org/ILRIPubAware/Uploaded%20Files/2006711123340.NR_EV_060629_002_Pastoralism%20counters%20Poverty.pdf.

ILRI (International Livestock Research Institute). 2006b. *Africa's High Returns: G8 Investment Pays Off*. ILRI Livestock Brief, Nairobi.

IUCN (World Conservation Union). 2005. *Sustainable Financing of Protected Areas: A Global Review of the Challenges and Options*. Best Practice Protected Area Guidelines Series No. 13. IUCN, Gland, Switzerland.

IUCN (World Conservation Union). 2006. World Initiative on Sustainable Pastoralism. http://www.iucn.org/wisp/.

Jansen, D., I. Bond, B. Child. 1992. *Cattle, Wildlife, Both, or Neither: Results of a Financial and Economic Survey of Commercial Ranches in Southern Zimbabwe*. World Wide Fund for Nature (WWF) Multispecies Project No. 27, Harare, Zimbabwe.

Jutzi, S. 2005. *Global Challenges and Opportunities Facing Production and Use of Livestock and Livestock Products*. http://www.engormix.com/e_articles_view.asp?art=101.

Kibue, M. 2005 Challenges in the development of a functioning livestock marketing chain in Kenya: a best practice case study. In: *Farming Systems and Poverty: Making a Difference—Proceedings of the 18th International Symposium of the International Farming Systems Association: A Global Learning Opportunity*, 31 October–3 November 2005, Rome, ed. J. Dixon, C. Neely, C. Lightfoot, M. Avila, D. Baker, C. Holding, and C. King. International Farming Systems Association, Rome.

Kock, R. 2005. *What Is This Infamous "Wildlife/Livestock Disease Interface?" A Review of Current Knowledge for the African Continent*. AU-IBAR, Nairobi, Kenya.

Kohler-Rollefson, I. 1993. Traditional pastoralists as guardians of biological diversity. *Indigenous Knowledge and Development Monitor* 1:14-16.

Mearns, R. 1996. When livestock are good for the environment: benefit-sharing of environmental goods and services. Special Paper for the World Bank/FAO Workshop Balancing Livestock and the Environment, 27–28 September 1996, Washington, DC.

Mirreh, M. M. 2005. Range rehabilitation and biodiversity conservation in the Syrian steppe. Paper presented at the workshop Synergies between the Three UN Conventions: UNCCD, UNCBD, and UNFCC, 10–12 January 2005, organized by ACSAD, Abu Dhabi.

Mortimore, M. 2006. Managing Agricultural Transition in African Drylands. *LEISA,* 22:32–34.

Naylor, R., H. Steinfeld, W. Falcon, J. Galloway, V. Smil, E. Bradford, J. Alder, and H. Mooney. 2005. Agriculture: losing the links between livestock and land. *Science* 310(5754):1621–1622.

Osofsky, S. A., S. Cleaveland, W.B. Karesh, M.D. Kock, P.J. Nyhus, L. Starr, and A. Yang (eds). 2005. *Conservation and Development Interventions at the Wildlife/Livestock Interface: Implications for Wildlife, Livestock and Human Health.* IUCN, Gland, Switzerland, and Cambridge, UK.

Palmer, A.R. 1998. Evaluating ecological footprints. *Electronic Green Journal* Special Issue 9(December).

Parris, K. 2002. *OECD Observer.* Sustainable agriculture depends on biodiversity. http://www.oecdobserver.org/news/fullstory.php/aid/755/Sustainable_agriculture_depends_on_biodiversity.html.

Ramos, G. 2000. *A Primer, Securing the Future: By Promoting the Adoption of Sustainable Agroforestry Technologies.* University of the Philippines at Los Banos, Laguna. http://www.pcarrd.dost.gov.ph/cin/AFIN/afin%20FAQ's.htm.

Savory, A. 1999. *Holistic Management: A New Framework for Decision Making.* Island Press, Washington, DC.

Scherf, B., B. Rischkowsky, I. Hoffmann, M. Wieczorek, A. Montironi, R. Cardellino. 2006. *Livestock Genetic Diversity in Dry Rangelands.* United Nations Educational, Scientific, and Cultural Organization (UNESCO), Paris.

Shackleton, C., S. Shackleton, T. Netshiluvhi, F. Mathabela, and C. Philri. 1999. The direct use value of goods and services attributed to cattle and goats in the Sand River Catchment, Bushbuckridge. Unpublished report No. ENV-P-C 99003. Council for Scientific and Industrial Research (CSIR), Pretoria.

Stinner, D., B. Stinner, and E. Martsolf. 1997. Biodiversity as an organizing principle in agroecosystem management: case studies of holistic resource management practitioners in the USA. *Agriculture, Ecosystems and Environment* 62:199–213.

Thornton, P., R. Kruska, H. Henninger, P. Kristjanson, R. Reid, F. Atieno, A. Odero, and T. Ndegwa. 2002. *Mapping Poverty and Livestock in the Developing World.* International Livestock Research Institute (ILRI), Nairobi.

Tilman, D., P.B. Reich, J. Knops, D. Wedin, T. Mielke, and C. Lehman. 2001. Diversity and productivity in a long-term grassland experiment. *Science* 294(5543):843–845. http://www.sciencemag.org/cgi/content/abstract/294/5543/843.

UNDP (United Nations Development Program). 2006. Report 2 (of 5): Greening the Desert, Tanzania. http://www.tve.org/ho/doc.cfm?aid=887.

Voisin, A. 1988. *Grassland Productivity.* Island Press, Covelo, CA.

WCS (Wildlife Conservation Society). 2006. *Wildlife Conservation Society, Laikipia Predator Project Highlights.* WCS, New York.

White, R.P., S. Murray, and M. Rohweder. 2000. *Grasslands Ecosystems: Pilot Analysis of Global Ecosystems.* World Resources Institute, Washington, DC. http://www.wri.org/wr2000.

Wildcare. 2006. *Branding to Make a Difference: The Wildcare Dairy.* http://www.effp.org.uk/StellentEFFPLIVE/groups/public/documents/case_studies/wildcaredairyproduc_ia431ff321.hcsp.

Williamson, D. 2006. *Range Improvement and Development of a Wildlife Reserve in the Syrian Steppe*. Food and Agriculture Organization of the United Nations (FAO), Rome. http://www.fao.org/docrep/004/y2795e/y2795e06a.htm.

World Bank Group. 1998. *Meat Processing and Rendering: Pollution Prevention and Abatement Handbook*. http://ifcln1.ifc.org/ifcext/environ.nsf/AttachmentsByTitle/gui_meat_WB/$FILE/meat_PPAH.pdf.

WRI (World Resources Institute). 2005. *World Resources 2005: The Wealth of the Poor: Managing Ecosystems to Fight Poverty*. Chapter 2: Ecosystems and the livelihoods of the poor. United Nations Development Programme, United Nations Environment Programme, The World Bank, World Resources Institute.

Part II

Biodiversity and Ecosystem Management in Ecoagriculture Landscapes

Overview

The development of ecoagriculture as a concept has been accompanied by a growing appreciation in the scientific literature of the importance of biodiversity sustained within agricultural landscapes, as well as the importance of biodiversity in sustaining productive agricultural ecosystems. The Millennium Ecosystem Assessment (MA 2005) devoted numerous chapters to the topic, at scales ranging from genes to landscapes. The seven chapters in this section synthesize new knowledge and practice about how biodiversity and ecosystem services can be conserved in diverse types of ecoagriculture landscapes while achieving livelihood benefits.

The chapter by **Celia A. Harvey** provides an overview of how the structure, composition, and management of agricultural landscapes influence their ability to conserve biodiversity and identifies key principles that can be used to guide conservation and restoration efforts in farming regions. The chapter also highlights key research questions and describes strategies for protecting and restoring native habitat within the agricultural matrix: retaining elements that enhance landscape connectivity, ensuring heterogeneity at both the field and landscape level, and reducing the intensity of land management.

Barbara Gemmill-Herren, Connal Eardley, John Mburu, Wanja Kinuthia, and Dino Martins focus on wild pollinators, which are essential to the reproduction of many crops, especially fruits and vegetables, as well as many wild plants and the wild fauna dependent on them. They present evidence that maintaining a full suite of pollinators will increase agricultural productivity, and argue that maintaining pollinators requires protection of their habitats, in forests, hedgerows, living fences, and other non-domesticated elements in the agricultural landscape.

Hans Herren, Johann Baumgärtner, and Gianni Gilioli argue that conventional analysis and research methods are inadequate to address problems of complex systems, such as ecoagriculture landscapes. They describe the evolution of thinking about agricultural ecosystems from "integrated pest management" to "adaptive management" (i.e., a systematic, cyclical process for continually improving management procedures). They illustrate the approach with an application to pest management in Ethiopia.

Meine van Noordwijk, Fahmuddin Agus, Bruno Verbist, Kurniatun Hairiah, and Tom P. Tomich consider lessons from managing ecoagriculture for both biodiversity conservation and watershed services. At the landscape scale, the spatial organization of tree and forest landscape elements can provide filters for overland flow of water and sediments as well as corridors for forest biota, connecting areas with more specific conservation functions. At plot and regional scales, the relationship is more variable because watershed functions not only depend on plot-level land use but also on the spatial organization of trees in a landscape, infiltration, dry-season flow, and other factors.

Jan Sendzimir and Zsuzsanna Flachner then examine the role of natural disturbance in maintaining landscape biodiversity, using the example of flooding in the Hungarian reach of the Tisza River Basin. In the preindustrial period, regular flooding sustained landscape complexity and many ecosystem services. For example, floods maintained hydrological connections that sustained fish nurseries and migration, stored water for use during droughts, and distributed and mixed fallen fruit in novel combinations that stimulated agrobiodiversity. Because regional decline appears driven by interrelated crises of ecology, economy, and society, the authors question the conversion of the Tisza River Basin from a fruit/nut/fishery polyculture to a wheat or corn monoculture and suggest that production patterns adapted to intermediate natural disturbances be studied as models for future ecoagriculture landscapes.

David Molden, Rebecca Tharme, Iskandar Abdullaev, and Ranjitha Puskur examine the challenge of conserving biodiversity in irrigated agricultural systems. They identify opportunities for positive change, including increasing water productivity (growing more food with less water); national and international policies, including trade; and many water management designs and practices that support diverse landscapes, crops, and connectivity for plant and animal movement to maintain biodiversity. They argue that widespread implementation of these innovations will require that they be integrated into institutions, incentive structures, and education.

The feasibility of managing production and conservation activities at a landscape scale depends on the ability of stakeholders working in the landscape to assess and agree on the baseline conditions and to jointly monitor progress. Although direct and qualitative assessment by land managers remains central to monitoring efforts, a variety of new remote-sensing methods have also been developed to quantify landscape changes. **Aaron Dushku, Sandra Brown, Tim Pearson, David Shoch, and Bill Howley** describe and illustrate the application of some of these methods to monitor land cover, biomass and carbon stocks, and proxy indicators for biodiversity or habitat quality in different parts of the landscape and over time.

Overall, these chapters demonstrate that agricultural landscapes of many different types have the potential to sustain biodiversity and ecosystem services. Successful approaches require effective collaboration between farmers and those responsible for landscape-scale biodiversity conservation and protection of ecosystem services. The importance of spatial patterns and temporal dynamics in determining outcomes suggests that landscape-specific, adaptive management approaches will be needed, but these site-specific strategies can be informed by studying and understanding general principles of ecoagriculture.

References

Millenium Ecosystem Assessment (MA). 2005. *Ecosystems and Human Well-Being: Synthesis.* Island Press, Washington, DC.

Chapter 8

Designing Agricultural Landscapes for Biodiversity Conservation

Celia A. Harvey

Introduction

Achieving conservation goals within agricultural landscapes is challenging because each landscape has its own unique set of biophysical, agricultural, and socioeconomic characteristics. These characteristics collectively present both opportunities and challenges for biodiversity conservation. In addition to the obvious variation in biophysical conditions (soils, vegetation type, topography, etc.), agricultural landscapes vary widely in the history and types of land uses, the management practices used, and the spatial arrangement of land uses. Whereas some landscapes are highly homogenized, consisting of a single crop (e.g., banana, wheat, or corn) extending uninterrupted as far as the eye can see, other landscapes are complex mosaics of small cultivated areas, interspersed with pastures, hedges, riparian forests, and patches of native habitat (e.g., smallholder agroforestry systems). In addition, agricultural landscapes vary greatly in the degree to which native vegetation has been lost or fragmented, and in the degree to which they have been affected by human settlements. The particular social, cultural, and economic contexts of different agricultural landscapes add additional layers of complexity, which can either facilitate or constrain the implementation of conservation initiatives.

If agricultural landscapes are to be included within large-scale conservation efforts, it will be important to understand what landscape characteristics are compatible with biodiversity conservation and how these landscapes can be designed and managed to achieve conservation goals. This chapter explores the relationships between landscape characteristics and biodiversity conservation and creates a framework of general principles that can be used to guide conservation activities and set future research priorities within rural landscapes worldwide. The chapter first provides an overview of how landscape structure, composition, and management influence the patterns of terrestrial animal and plant diversity, then identifies the key principles for conserving and restoring biodiversity in agricultural landscapes that are emerging from the scientific literature. The final section highlights the remaining uncertainties and research that must be addressed if agricultural landscapes are to play significant roles in future conservation efforts.

Landscape characteristics and biodiversity

Despite the great variety of agricultural landscape types, all landscapes can be characterized in terms of four basic characteristics that are critical to biodiversity conservation: landscape composition (i.e., the types of habitats or land uses present); landscape structure (i.e., the way in which different land uses are spatially arranged in the landscape); landscape management; and regional context (Table 8.1). (A glossary of common terms used to describe landscapes is provided in Box 8.1 for readers unfamiliar with landscape ecology.)

Landscape composition

Agricultural landscapes are generally composed of a mixture of agricultural and nonagricultural land uses, which together create a larger mosaic. Depending on its land-use history, a landscape may include remnant or relict areas of native vegetation (such as forests, riparian forests, native grasslands, prairies, wetlands, or natural meadows), as well as other nonagricultural areas such as border strips, windbreaks, hedges, live fences, or forest plantations, which are established and maintained by farmers for productive reasons. Patches of native vegetation tend to have the greatest conservation potential because they are more structurally and floristically complex, cover a greater range of environmental gradients and microhabitats, and provide higher-quality habitat and resources for a wider array of species than agricultural lands. Consequently, studies that have compared animal and plant diversity across different land-use types within agricultural landscapes in a diverse range of geographic settings, including Central America (Daily 2001; Daily et al. 2003; Harvey et al. 2006; Perfecto et al. 2003) Southeast Asia (Lawton et al. 1998; Schulze et al. 2004), North America (Blann 2006), and Australia (Fischer et al. 2004, 2005), have invariably found higher species richness and abundance of native species in the native habitats. However, the types of agricultural land uses are also important in determining the landscape's conservation potential, because some agricultural lands may provide

Table 8.1. Key landscape characteristics that can influence the patterns of plant and animal diversity within agricultural landscapes

Landscape composition	*Landscape structure*	*Landscape management*	*Regional context*
Types of land uses present (both agricultural and nonagricultural)	Patch sizes and shapes	Crop management practices (e.g., soil preparation, crop rotations, use of fallows, soil management practices, harvesting methods, etc.)	Native ecosystem type, native plant and animal assemblages, and biophysical characteristics
Floristic and structural composition of different land use types	Spatial arrangement of agricultural and nonagricultural land uses	Pasture and animal management (e.g., stocking densities and grazing regimes, use of agrochemicals)	Location in broader context, especially relative to protected areas
Land use constituting the agricultural matrix	Location of patches of native habitat relative to one another (distances, arrange-) ment, and configuration	Control of invasive, aggressive, or overabundant plant or animal species (including domestic species)	Temporal dynamics of land-use change in the region
Proportion of landscape under native vegetation	Degree of connectivity of native habitat within the agricultural landscape	Degradation of remaining patches of native vegetation through extraction of firewood, timber, and other products	History of agriculture in the region (agricultural frontier or ancient agricultural region?)
	Landscape heterogeneity	Temporal dynamics of land-use change within landscape (and particularly patterns of conversion and fragmentation of native habitat)	
	Permeability of the agricultural matrix		
	Abundance and types of edge habitats and borders		

Table 8.1. (*Continued*)

Landscape composition	*Landscape structure*	*Landscape management*	*Regional context*
		Human settlements and associated infrastructure (roads, power lines, etc.)	

complementary or supplementary habitat, shelter, foraging sites, and food resources for certain plant and animal species (e.g., Fuller et al. 2004; Wilson et al. 2005). In general, agricultural land uses that include a greater variety of plant species have a more complex structure, are subject to less intensive management practices (e.g., polycultures, agroforestry systems), and conserve greater animal and plant diversity than those that are floristically and structurally uniform or intensively managed (Benton et al. 2003; Freemark 2005; Schroth et al. 2004). One of the best-known examples of this is the high diversity of plants, birds, bats, other mammals, and insects in complex, multistrata coffee agroforestry systems compared to industrialized coffee plantations that lack a shade canopy (Moguel and Toledo 1999; Somarriba et al. 2004).

The composition of land uses is also important at the landscape scale. Four broad types of landscapes are often differentiated on the basis of the availability of native habitat at the landscape scale: intact landscapes (where most of the landscape is still under the original native habitat); variegated landscapes (where 60 to 90% of the native habitat remains); fragmented landscapes that retain only 10 to 60% of the native habitat; and relict landscapes in which remnants of native habitat represent less than 10% (Hobbs 2002; McInytre and Hobbs 1999; see Figure 8.1 for examples of these landscape types). Most agricultural landscapes fall into the fragmented and relictual landscape categories, though in some cases where agriculture has only recently entered an area (e.g., shifting cultivation landscapes on forest frontiers) or has affected only part of the landscape, the landscape may be classified as variegated or even intact. It is generally assumed that the intact and variegated landscapes are better able to conserve biodiversity over the long term than either fragmented or relictual landscapes, though empirical data are lacking.

Similarly, landscapes that encompass a diverse array of land-use types (both agricultural and nonagricultural) are found to conserve more biodiversity than highly homogenized landscapes consisting of a single crop, because the various land uses collectively provide a broader array of habitat and resource types and

Box 8.1. Glossary of agricultural and landscape ecology terms used in this chapter

Agricultural intensification. The use of practices associated with increasing agricultural yields per unit of land, including enhanced use of purchased inputs (pesticides, chemical fertilizers, manufactured seeds, and machinery) per unit land, the use of high-yield crop varieties, irrigation, and mechanization (Matson et al. 1997)

Connectivity. The degree to which a landscape facilitates or impedes movement of organisms among resource patches (Forman 1995); The degree to which patches or landscapes are linked by the flow of organisms through intervening patch types, possibly via habitat corridors or stepping stones (Fremark et al. 2002)

Core area or habitat. Center portions of large, undisturbed tracts of land occupied by interior species (as opposed to edge species)

Corridor. A linear strip of a habitat that differs from the adjacent land on both sides, connecting otherwise isolated, large remnant habitat patches (Forman 1995)

Edge. The portion of a habitat near its perimeter, where influences of the surrounding land uses prevent development of interior or core-area environmental conditions (Forman 1995)

Landscape. A large, heterogeneous land area (e.g., multiple square miles or several thousand hectares) consisting of a cluster of interacting ecosystems repeated in a similar fashion

Landscape composition. The types of different habitats or elements present within the landscape (Forman 1995)

Landscape structure. The spatial relationships among the different elements (patches, corridors, matrix) forming the landscape (Forman 1995)

Matrix. The background ecosystem or land-use type in a mosaic, characterized by extensive cover, high connectivity, and/or major control over landscape functioning (Forman 1995). In agricultural landscapes, the matrix usually consists of agricultural land such as crop fields or pastures.

Metapopulations. A network of semi-isolated populations with some level of regular or intermittent migration and gene flow among them, in which individual populations may be extinct but then be recolonized from other subpopulations (Meffe and Carroll 1997)

Mosaic. A pattern of patches, linear corridors, and matrix in a landscape (Forman 1995)

Patch. A relatively homogeneous type of habitat use that is spatially segregated from other similar habitat and differs from its surroundings (Forman 1995)

Figure 8.1. (a) Intact landscape, aerial view on Siberian Tiaaga; (b) variegated landscape, Monteverde Cloud Forest, San Louis Valley, Costa Rica; (c) fragmented landscape, Mahantango Creek Watershed, Pennsylvania, USA; (d) relictual landscape, rice terraces in Nepal

may support different suites of species (Forman 1995). Landscape heterogeneity, not only in terms of landscape composition but also in terms of landscape structure, is emerging as a key determinant of biodiversity patterns in agricultural landscapes worldwide (Benton et al. 2003; Tews et al. 2004).

Landscape Structure: Patches, Edges, and Matrices

The size of a given landscape habitat is not the only factor influencing its conservation value; patch quality is also a key determining factor, with patches that are more floristically and structurally intact having greater conservation value than those that have been highly modified or degraded (Kennedy et al. 2003).

Patch shape also influences the distribution of animal and plant species within agricultural landscapes, primarily through associated edge effects (Forman 1995; Harris 1988). The borders of native habitat fragments in agricultural matrices experience altered microclimatic effects (such as changes in solar radiation, wind, and humidity), greater exposure to predators, agricultural inputs and human impacts, and altered disturbance regimes (Laurence et al. 1997). These so-called edge effects can cause important changes in species composition and are often considered detrimental to the original suite of native species while benefiting generalist and disturbance-related species. When other factors are held constant, patches with more compact shapes suffer fewer edge effects and conserve a greater proportion of native species than patches that are irregularly shaped and more exposed to the agricultural matrix (Forman 1995; Wiens 2002, 2005). However, the extent to which edge conditions penetrate a fragment is also a function of patch size (which determines the amount of core area within the patch) and patch context (Ewers and Didham 2006), with habitat patches that are surrounded by structurally similar land uses being less likely to suffer deleterious edge effects than those surrounded by highly contrasting land uses.

The location of a given native habitat patch within the surrounding landscape can also mitigate or confound the effects of fragment size (Ewers and Didham 2006; Gascon et al. 1999). The types of land uses surrounding a given patch and the degree to which the patch is connected with similar habitat patches influence the abundance and species richness of the assemblage within the patch, as well as ecological processes such as seed dispersal, seed predation, predator–prey interactions, and the invasion of nonnative species (Blann 2006). Native habitat patches that are located near other similar habitat patches or that are surrounded by a greater proportion of native habitat tend to be used by more animal and plant species and have higher chances of maintaining populations over the long term than those that are highly isolated (Wiens 2005).

Landscape-Level Structural Attributes

One of the key structural characteristics that determines patterns of biodiversity within agricultural landscapes is the arrangement of native habitats within the landscape. The location of different habitats patches relative to one another largely determines the movement of individuals and genes across the landscape (Forman 1995; Wiens 2005). For example, a given species may move between forest patches within an agricultural landscape if these fragments are located close to each other, but may entirely avoid dispersal or movement if these forest fragments are located farther apart or if they are separated by a hostile matrix that the species is either physically or behaviorally incapable of crossing. Fragments that are closer or more connected to source populations have higher species richness than those that are more isolated (Hilty et al. 2006; Wiens 2005).

A related feature of landscape structure that is important in determining biodiversity patterns is the degree of landscape connectivity. Landscapes are considered to be functionally connected for a given species if this species can easily move from one patch of native habitat to another, either using corridors or stepping stones or moving across the agricultural matrix (Forman 1995). Connectivity is a critical issue in agricultural landscapes because individual patches of remnant habitat are rarely large enough to support self-sustaining populations; instead, the persistence of populations often depends on the ability of individuals to forage across the landscape, to colonize disjoint habitat patches, and to interact with individuals in other parts of the landscape (e.g., Pardini et al. 2005). For many species, landscape connectivity is also critical for seasonal migration and dispersal (Bennett 1999). Landscape connectivity is a function of both the distribution and types of land uses and the particular requirements and behaviors of individual plant and animal species (Bennett 1999; Hilty et al. 2006). In general, landscapes with high degrees of connectivity (i.e., those that have a high proportion of the land under native habitat, short distances between remnant patches, and networks of corridors or stepping stones that facilitate animal dispersal) appear to have the highest probabilities of conserving species populations and assemblages over the long term (Bennett 1999).

The abundance, distribution, and types of edges and boundaries will also influence biodiversity patterns by influencing animal movement (Wiens 2005). Certain types of boundaries, such as roads, developed areas, and power lines, can serve as complete barriers to some animal species and isolate populations that were previously connected. Similarly, landscapes that have high edge to area ratios or edges between sharply contrasting land uses (e.g., forest patches bordering an open field) may restrict or inhibit animal movement to a greater degree than landscapes that are less fragmented, conserve more interior habitat, and have "softer" edges (Ewers and Didham 2006; Wiens 2005).

A final structural characteristic that is critical in determining biodiversity patterns is landscape heterogeneity. *Landscape heterogeneity* refers to the diversity of land uses (including their structural and floristic diversity) as well as the spatial diversity of land uses within the landscape. Heterogeneous landscapes are characterized by having not only a diverse array of land uses but also a complex mosaic of patches of different shapes and sizes arrayed in a non uniform fashion. Because heterogeneous landscapes provide a wide variety of niches and resources for plant and animal species (Gascon et al. 1999), they hold much greater conservation value than landscapes that are spatially, floristically, and structurally uniform (McNeely and Scherr 2003; Schroth et al. 2004).

Landscape Management

Within a given agricultural landscape, farmers apply a multitude of management practices to both agricultural and nonagricultural lands. These practices can vary in type, intensity, and frequency, as well as in their spatial and temporal application. In landscapes dominated by agricultural crops, the most common practices involve land preparation, fertilization, agrochemical application, burning, crop rotations, soil management practices, and harvesting methods; whereas in landscapes dedicated to animal production, the prevalent practices include site preparation (using fire or tillage), weeding, application of fertilizers and other agrochemicals, and animal management (stocking densities, rotation patterns, etc.). In forested habitats, common practices include thinning, weeding, pruning, and replanting, as well as hunting of wildlife and extraction of timber, firewood, and other products. Collectively, these landscape management practices change the distribution and quality of available habitat and resources, inflict mortality on plant and animal species, alter patterns of landscape connectivity or heterogeneity, degrade the remaining terrestrial and aquatic habitats, disrupt ecosystem processes, and change patterns of fires, predators, pests, and other disturbances (Harvey, Alpizar, et al. 2005a; Tscharntke et al. 2005).

Another critical facet of landscape management is the degree to which the landscape has already been affected by human settlements and their associated infrastructure (roads, paths, railroads, drainage ditches, dams, power lines, fences, etc.). Housing developments and roads have a powerful effect on the patterns of land use, reducing the size of natural areas and separating them into smaller parcels, creating noise and air pollution, and serving as physical barriers (and sources of mortality) to animals (Trombulak and Frissell 2000). In landscapes that have already been heavily modified by human activity, the long-term prognosis for biodiversity is likely to be poor, unless active interventions are undertaken.

Regional Location

The location of agricultural landscapes in the broader regional context is also important. Agricultural landscapes that are located in the buffer zones of national parks or other kinds of protected areas are likely to have a more intact flora and fauna than landscapes that are surrounded by a sea of agriculture because the large areas of natural vegetation serve as important sources of colonists and propagules (Schroth et al. 2004). The regional context also determines the larger-scale dynamics of plant and animal communities. Landscapes that occur in active agricultural frontiers are likely to have highly dynamic plant and animal communities that undergo frequent changes in abundance, species richness, and composition as additional habitat in the surrounding area is destroyed, fragmented, and converted to agriculture. In contrast, in agricultural regions that have been cultivated for many centuries (as is the case in most of Europe), the species assemblages occurring within agricultural landscapes may be relatively more stable and consist primarily of those species that have adapted to the new landscape characteristics (Sutherland 2004). This history of land use has important implications for which species can be conserved within different landscapes: in newly deforested areas where native habitat is still available, it is often possible to conserve many of the original forest species, whereas in landscapes that have been modified for many centuries, conservation activities are often directed toward species typical of open and modified habitats associated with agriculture (Sutherland 2004).

Guiding Principles for Conserving and Restoring Biodiversity in Agricultural Landscapes

Although our understanding of the patterns and dynamics of biodiversity in agricultural landscapes is incomplete, it is already evident that there several broad principles that can be used to guide conservation initiatives in agricultural regions (Table 8.2). These principles build on the relationships between landscape characteristics and biodiversity already discussed, as well as several key scientific papers that have proposed landscape-level criteria for maintaining biodiversity and ecosystem functions (e.g., Blann 2006; Dale et al. 2000; Fischer et al. 2005; Freemark 2005; Kennedy et al. 2003; Taberelli and Gascon 2005):

1. *Maintain large areas of protected native vegetation within the region, to serve as sources of species, individuals, and genes.* The establishment and protection of national parks, biological reserves, community-conserved areas, and other types of protected areas continues to be of utmost importance for biodiversity conservation and should form the cornerstone for regional, national, and international conservation planning (Margules and Pressey 2000; Rodrigues et al. 2004). However, at the regional level, the maintenance of populations, species, and ecological assemblages will depend on integrating the management of protected areas with the land-use dynamics of the adjacent agricultural areas (DeFries et al. 2005).
2. *Conserve the remaining areas of native habitat within the agricultural landscape, giving priority to patches that are large, intact, and ecologically important.* The conservation of remnant native habitat is probably the single most important factor determining the conservation potential of a given agricultural landscape. Consequently, special emphasis should be placed not only on conserving the greatest amount of native habitat possible but also on protecting the largest patches and those that have the highest quality. However, even tiny patches of native vegetation, such as remnant trees occurring in pastures or tiny strips of uncultivated meadows along field borders, can provide critical supplementary habitats and resources for some animals and plants and merit consideration in conservation planning (e.g., Fischer and

Lindenmayer 2002; Harvey, Villanueva, et al. 2005, Harvey et al. 2006). Special attention should also be paid to the protection of rare landscape elements (such as wetlands, creeks, riparian areas, mountain zones, old-growth forests, and native grasslands that are often replaced or degraded by agriculture) and "keystone structures" (such as tree hollows) that may provide critical habitat for rare and endangered species (Fischer et al. 2005; Kennedy et al. 2003).

3. *Prevent the further destruction, fragmentation, or degradation of native vegetation within the agricultural landscape.* In addition to ensuring that native habitat remains in the landscape, it is critical to ensure that these habitats are not subdivided into smaller fragments, isolated from adjacent fragments of a similar type, or degraded by human and agricultural activities. Actions that may help ensure the long-term integrity of the remaining native remnants include long-term land-use planning and zoning (including restrictions on the conversion of land to agriculture or residential use), careful positioning of roads, paths, and other infrastructure, and the adoption of best management approaches across the agricultural matrix. However, the priority actions for a given landscape will need to be tailored to the specific nature and intensity of threats.
4. *Maintain landscape connectivity, at multiple spatial scales, for as wide a group of plant and animal species as possible.* To achieve the greatest conservation benefits, land-use planners should strive to maintain connectivity both among the native habitat patches occurring within the agricultural landscape as well as between these patches and larger protected areas in the surrounding region. At both spatial scales, connectivity can be enhanced by conserving, restoring, or enlarging remaining patches of native vegetation, carefully managing the agricultural matrix to make it more permeable to animal movement, and retaining (or reestablishing) landscape elements that promote connectivity. Gallery forests and riparian areas are natural landscape linkages and warrant special attention. However, other linear features such as greenways, live fences, hedges, and windbreaks can also contribute to landscape connectivity by serving as corridors or stepping stones for some species (Bennett 1999; Forman and Baudry 1984; Harvey, Villanueva, et al. 2005; Hilty et al. 2006).
5. *Actively manage the agricultural landscape to maintain heterogeneity at both the patch and the landscape level.* Landscape heterogeneity can be achieved by diversifying patch types, including patches of different sizes and shapes, retaining landscape elements such as hedges, live fences, windbreaks, remnant trees, or patches of noncrop habitat that interrupt the monotony of agricultural land, including noncultivated borders and noncropped areas both within and across fields, avoiding large areas of monocropping, and incorporating floristically and structurally complex vegetation types.

6. *Use appropriate best-management practices to make agricultural systems more compatible with biodiversity conservation.* A wide range of best-management practices has been proposed for different agricultural systems, including coffee, cacao, citrus, sugarcane, and cattle production, among others (see Clay 2004 for an overview). These best-management practices seek to make agricultural systems more environmentally friendly and compatible with biodiversity conservation by reducing agrochemical use, preventing the depletion of soil and water resources, reducing habitat conversion, and optimizing the provision of habitats and resources for native species within agricultural systems. Although best-management practices are generally designed for the plot or field level, application at broader spatial scales will be needed to achieve the desired conservation benefits. In addition, vigorous monitoring of the impacts of these best-management practices is needed to ensure that they provide the desired biodiversity outcomes.
7. *Identify and address threats to the conservation of native habitats and biodiversity within the agricultural landscape.* An analysis of both current and potential threats to biodiversity will be critical for long-term success. In agricultural landscapes where there is still abundant and contiguous native habitat, the key threats to conservation are likely to be habitat loss, fragmentation, and conversion, whereas in landscapes that have already been heavily influenced by agriculture, the principal threats may include loss of connectivity, landscape homogenization, spread of nonnative species, unsustainable harvesting of wildlife or plant species, and inappropriate management practices, among others. Once threats have been clearly identified, preemptive measures must be taken to reduce or mitigate their potential negative impacts.
8. *Restore areas of native habitat on degraded portions of the agricultural landscape.* In landscapes that have been highly modified and degraded by agriculture, the restoration of habitat and landscape connectivity should be a priority. Habitats can be restored by letting areas regenerate naturally, enriching existing habitat patches with native species, or actively revegetating with native plant species (Lamb et al. 2005). In some cases, rapidly growing, nonnative plantations can be established to catalyze the natural regeneration of native species and facilitate the restoration process (Parrotta 1992), but care has to be taken that these nonnative species do not become invasive. Depending on their location within the landscape, restored areas can serve as new habitats, as buffer areas to existing habitat fragments, or as corridors or stepping stones that enhance landscape connectivity (Lambeck and Hobbs 2002).
9. *Take marginal lands out of production and allow them to revert to native vegetation.* Most farming landscapes have some marginal areas that are inappropriate

for agricultural production, for example, because of steep slopes, degraded soils, or low fertility. If these marginal lands are taken out of production and converted to natural vegetation through natural regeneration or reforestation, the benefits for biodiversity conservation and the provision of ecosystem services could be considerable (Dale et al. 2000).

10. *Apply specific conservation strategies for species or communities that are of particular conservation concern.* No matter how carefully agricultural landscapes are designed and managed, they are unlikely to be able to maintain all of the species that were originally present. In particular, species that require large areas of natural habitat, specialize on certain habitats or specific resources, are highly sensitive to fragmentation and habitat loss, or are unable to move across agricultural lands are likely to suffer population declines and eventually be lost from the landscape, unless additional measures are taken (Blann 2006; Fischer et al. 2005). If conserving these species within the agricultural landscape is a high priority, case-specific recovery plans may be necessary.

Table 8.2. A summary of ten guiding principles for conserving biodiversity within agricultural landscapes

Principles for conserving biodiversity in agricultural landscapes
1. Maintain large areas of protected native vegetation within the region to serve as sources of species, individuals, and genes
2. If possible, maintain (or reestablish) connectivity between native habitats within agricultural landscapes with large contiguous areas of native vegetation
3. Conserve the remaining areas of native habitat within the agricultural landscape, giving priority to patches that are large, intact, and ecologically important
4. Prevent the further destruction, fragmentation, or degradation of native habitat patches within the agricultural landscape
5. Maintain landscape connectivity at multiple scales for as wide a group of plant and animal species as possible
6. Actively manage the landscape to maintain heterogeneity at both the patch and the landscape level
7. Use best management practices to make agricultural systems more compatible with biodiversity conservation
8. Identify and address threats to the conservation of native habitats and biodiversity
9. Restore areas of native habitat in degraded portions of the agricultural landscape
10. Take marginal lands out of production and allow them to revert to native vegetation
11. Apply specific conservation strategies for species or communities that are of particular conservation concern

Based on Blann 2006; Fischer et al. 2005; Freemark et al. 2005; Harvey et al. 2005; Harvey et al. 2006; Taberelli and Gascon 2005.

Important Considerations in Applying Guiding Principles

Although the foregoing principles are broadly applicable to a wide range of agricultural landscapes, the particular importance or relevance of different principles will be landscape specific and will require an understanding of the unique circumstances of individual landscapes. Classifying landscapes based on the amount of habitat remaining (e.g., the fragmented, variegated, and intact landscape categories discussed earlier) may be a useful starting point for identifying priority actions (Hobbs 2002; McInytre and Hobbs 1999). Broadly speaking, conservation efforts in intact and variegated landscapes should focus on conserving the existing fragments of native vegetation and preventing their degradation, fragmentation, or loss, whereas in fragmented and relictual habitats emphasis must be placed on restoring habitats and conserving landscape connectivity (Table 8.3). Once these landscape-level priorities have been identified, a more detailed analysis of the particular biophysical and socioeconomic context combined with an analysis of both the opportunities and the threats to biodiversity conservation can help customize a broad-scale and long-term

Table 8.3. Priority actions for agricultural landscapes with varying degrees of habitat loss and fragmentation

Landscape type	*Intact (> 90% intact)*	*Variegated (60–90% habitat remaining)*	*Fragmented (10–60% habitat remaining)*	*Relictual (< 10% habitat remaining)*
Conserve	Existing natural habitat (= matrix)	Existing natural habitat (= matrix)	Fragments of natural habitat that are in good condition	NA (nothing to maintain)
Enhance	NA	Buffer areas and connecting areas	Quality of remaining fragments of natural habitat	NA
Reconstruct or restore	NA	NA	Buffer areas, corridors, and stepping stones	New habitat, buffer areas, corridors, and stepping stones
Manage	NA	NA	Agricultural matrix	Agricultural matrix

Based on Hobbs 2002 and McInytre and Hobbs 2001
NA, not applicable

conservation strategy that achieves specific conservation priorities (Dale et al. 2000; Kennedy et al. 2003; Mattison and Norris 2005). Because agricultural landscapes are highly dynamic entities with frequent changes in the types of land uses, the sizes of cultivated areas, and the management regimes applied, it is critical that these conservation strategies be flexible, adaptive, and able to respond quickly to change.

Uncertainties and Research Needs

Although our understanding of the patterns of biodiversity within agricultural landscapes has grown enormously in recent years, many gaps remain in our knowledge of the complex relationships between landscape characteristics and biodiversity patterns. Addressing these could help us refine the principles for biodiversity conservation and better tailor them to specific landscape conditions. In addition, although there is broad consensus on the general principles that should guide conservation efforts within agricultural landscapes, there is still much uncertainty about how to effectively apply these principles in different contexts to achieve the desired conservation outcomes.

To address these knowledge gaps, two types of research are required: (1) additional research aimed at better understanding the complex patterns and dynamics of species and assemblages within existing agricultural landscapes, and (2) empirical evaluations of the effectiveness of conservation interventions across a broad suite of agricultural landscapes in different geographic settings.

Enhancing Our Understanding of Biodiversity in Agricultural Landscapes

Research aimed at better understanding the ability of different agricultural landscapes to support biodiversity should focus on several key areas. First, it is important to improve our understanding of how different taxa use different parts of the landscape as habitat, foraging sites, or travel paths, and to what degree they depend on the different types of habitats within the landscape. Detailed studies of species that are sensitive to habitat loss and fragmentation would enable us to identify which landscape elements and potential travel pathways need to be conserved or restored (Hilty et al. 2006).

More information is also needed about the long-term dynamics of both populations and assemblages within agricultural landscapes. Demographic studies will be particularly useful for understanding whether individual populations persist, increase, or decline over time. Similarly, long-term studies of plant and animal assemblages are needed to determine whether assemblages remain constant over time, or whether they lose species of conservation interest and exhibit lagged responses to landscape change (Blann 2006; Tscharnke et al. 2005).

Carefully planned comparisons of the biodiversity in landscapes that vary in the amount of native habitat, the degree of landscape connectivity, the level of heterogeneity, and the intensity of management practices would also be useful for determining the relative importance of these different landscape characteristics (Ewers and Didham 2006; Fahrig 2001; Kennedy et al. 2003; Taberelli et al. 2004). If planned correctly, these studies could also help elucidate the potential synergisms and interactions that may occur between different landscape characteristics.

Studies should also be designed to identify and examine landscape thresholds in the amount of native habitat remaining within the landscape, the maximum distance between isolated habitat patches, and/or the connectivity of native habitats below which abrupt changes occur in the persistence of populations or in the composition of ecological assemblages. A number of recent papers have examined potential thresholds in agricultural landscapes, but the results to date are highly taxa and site specific (e.g., Grashof-Bokdam and van Langevelde 2004; Lindenmayer and Luck 2005; Lindenmayer et al. 2005; Radford et al. 2005). However, if generic thresholds can be identified for different keystone species or species of conservation concern in different types of landscapes and ecological regions, this would be a major contribution to conservation and landscape planning.

Evaluating the Impact of Conservation Interventions

Although it is important that these more basic questions about relationships between landscape characteristics and biodiversity conservation be addressed, there is also an urgent need to apply, evaluate, and modify the guiding principles that have already been identified. Conservation interventions based on these principles should be widely applied across different spatial scales (farms, communities, landscapes, and regions) and across landscapes spanning the wide range of agricultural land uses and biophysical and socioeconomic contexts. By systematically documenting the impacts of these interventions on biodiversity conservation as well as on agricultural production and farmer livelihoods it should be possible to elucidate what success can (or cannot) be expected, what approaches are most effective, what resources and conditions are required (financial, manpower, legal, socioeconomic, etc.), and what major challenges for large-scale implementation lie ahead. It is clear that there are both tradeoffs and synergies between designing and managing agricultural landscapes for conservation with production goals, but additional work is needed to develop decision-support tools that can quantify and evaluate trade-offs and explore alternative land-use scenarios (Brown 2005; McNeely and Scherr 2003). Tools are also required for prioritizing areas for conservation and/or restoration within agricultural landscapes, exploring the potential impacts of different landscape configurations, cropping patterns or management schemes, and monitoring and evaluating potential impacts of both existing and new policy mechanisms that promote conservation within agricultural landscapes (Dudley et al. 2005; Grice et al. 2004; Hobbs 2002). By combining strategic research with the systematic evaluation of different conservation interventions within agricultural landscapes, it should be possible to develop detailed guidelines and action plans that are effective, compatible with existing farming systems, and tailored to the biological and socioeconomic realities of different agricultural landscapes.

References

Bennett, A.F. 1999. *Linkages in the Landscape: The Role of Corridors and Connectivity in Wildlife Conservation.* World Conservation Union (IUCN), Gland, Switzerland.

Benton, T.G., J.A. Vickery, and J.D. Wilson. 2003. Farmland biodiversity: is habitat heterogeneity the key? *Trends in Ecology and Evolution* 18:182–188.

Blann, K. 2006. *Habitat in Agricultural Landscapes: How Much Is Enough?* A state-of-the-science literature review. Defenders of Wildlife, Washington, DC.

Brown, K. 2005. Addressing trade-offs in forest landscape restoration. In S. Mansourian, D. Vallauri and N. Dudley (eds). *Forest Restoration in Landscapes: Beyond Planting Trees.* Springer, New York: 59–62.

Clay, J. 2004. *World Agriculture and the Environment.* Island Press, Washington, DC.

Daily, G.C. 2001. Ecological forecasts. *Nature* 411:245.

Daily, G.C., G. Ceballos, J. Pacheco, G. Suzan, and A. Sanchez-Azofeifa. 2003. Countryside biogeography of neotropical mammals: conservation opportunities in agricultural landscapes of Costa Rica. *Conservation Biology* 17(6):1814–1826.

Dale, V., S. Brown, R. Haeuber, N. Hobbs, N. Huntly, R. Naiman, W. Riesbsame, M. Turner, and T. Valone. 2000. Ecological Society of America report: ecological principles and guidelines for managing the use of land. *Ecological Applications* 10:639–670.

DeFries, R., A. Hansen, A.C. Newton, and M.C. Hansen. 2005. Increasing isolation of protected areas in tropical forests over the past twenty years. *Ecological Applications* 15(10):19–26.

Dudley, N., J. Morrison, J. Aronson, and S. Mansourian. 2005. Why do we need to consider restoration in a landscape context? In: *Forest Restoration in Landscapes: Beyond Planting Trees*, ed. S. Mansourian, D. Vallauri and N. Dudley. Springer, New York: 51–58.

Ewers, R.M. and R.M. Didham. 2006. Confounding factors in the detection of species responses to habitat fragmentation. *Biological Reviews* 81:117–142.

Fahrig, L. 2001. How much habitat is enough? *Biological Conservation* 100:65–74.

Fischer, J. and D.B. Lindenmayer. 2002. Small patches can be valuable for biodiversity conservation: two case studies on birds in southeastern Australia. *Biological Conservation* 106:129–136.

Fischer, J., D.B. Lindenmayer, and A. Cowling. 2004. The challenge of managing multiple species at multiple scales: reptiles in an Australian grazing landscape. *Journal of Applied Ecology* 41:32–44.

Fischer, J., D.B. Lindenmayer, and A.D. Manning. 2005. Biodiversity, ecosystem function and resilience: ten guiding principles for commodity production landscapes. *Frontiers in Ecology and the Environment* 4(2):80–86.

Forman R.T.T. 1995. *Landscape Mosaics: The Ecology of Landscape and Regions*. Cambridge University Press, Cambridge.

Forman R.T.T. and J. Baudry. 1984. Hedgerows and hedgerow networks in landscape ecology. *Environmental Management* 8(6):495–510.

Freemark, K. 2005. Farmlands for farming and nature. In: *Issues and Perspectives in Landscape Ecology*, ed. J. Wiens and M. Moss. Cambridge University Press, Cambridge: 193–200.

Freemark, K., D. Bert, and M.A. Villard. 2002. Patch-, landscape- and regional-scale effects on biota. In: *Applying Landscape Ecology in Biological Conservation*, ed. K. Gutzwiller. Springer-Verlag, New York: 58–83.

Fuller, R.J., S.A. Hinsley, and R.D. Swetnam. 2004. The relevance of non-farmland habitat, uncropped areas and habitat diversity to the conservation of farmland birds. *Ibis* 146(Suppl 2):22–31.

Gascon, C., T.E. Lovejoy, R.O. Bierregaard Jr., J.R. Malcolm, P.C. Stouffer, H.L. Vasconcelos, W.F. Laurance, B. Zimmerman, M. Tocher, and S. Borges. 1999. Matrix habitat and species richness in tropical forest remnants. *Biological Conservation* 91:223–229.

Grashof-Bokdam, C.J., and F. van Langevelde. 2004. Green veining: landscape determinants of biodiversity in European agricultural landscapes. *Landscape Ecology* 20:417–439.

Grice, P., A. Evans, J. Osmond, and R. Brand-Hardy. 2004. Science into policy: the role of research in the development of a recovery plan for farmland birds in England. *Ibis* 146(Suppl 2):239–249.

Harris, L.D. 1988. Edge effects and the conservation of biotic diversity. *Conservation Biology* 2(4):330–332.

Harvey, C.A., F. Alpizar, M. Chacón and R. Madrigal. 2005. *Assessing Linkages between Agriculture and Biodiversity in Central America: Historical Overview and Future Perspectives*. Mesoamerican and Caribbean Region, Conservation Science Program. The Nature Conservancy (TNC), San José, Costa Rica.

Harvey, C.A., A. Medina, D. Merlo Sánchez, S. Vílchez, B. Hernández, J.C. Sáenz, J.M. Maes, F. Casanoves, and F.L. Sinclair. 2006. Patterns of animal diversity associated with different forms of tree cover retained in agricultural landscapes. *Ecological Applications* 16(5):1986–1999.

Harvey, C.A., C. Villanueva, J. Villacís, M. Chacón, D. Muñoz, M. López, M. Ibrahim, R. Taylor, J.L. Martínez, A. Navas, J. Sáenz, D. Sánchez, A. Medina, S. Vílchez, B. Hernández, A. Pérez, F. Ruiz, F. López, I. Lang, S. Kunth, and F.L. Sinclair. 2005. Contribution of live fences to the ecological integrity of agricultural landscapes in Central America. *Agriculture, Ecosystems and Environment* 111:200–230.

Hilty, J.A., W.Z. Lidicker Jr., and A.M. Merenlender. 2006. *Corridor Ecology: The Science and Practice of Linking Landscapes for Biodiversity Conservation*. Island Press, Washington, DC.

Hobbs, R.J. 2002. Habitat networks and biological conservation. In: *Applying Landscape Ecology in Biological Conservation*, ed. K. Gutzwiller. Springer-Verlag, New York: 150–169.

Kennedy, C., J. Wilkinson, and J. Balch. 2003. *Conservation Thresholds for Land Use Planners*. Environmental Law Institute, Washington, DC.

Lamb, D., P.D. Erskine, and J.A. Parrotta. 2005. Restoration of degraded tropical forest landscapes. *Science* 310:1628–1632.

Lambeck, R.J. and R.J. Hobbs. 2002. Landscape and regional planning for conservation: issues and practicalities. In: *Applying Landscape Ecology in Biological Conservation*, ed. K.Gutzwiller. Springer-Verlag, New York: 360–380.

Laurence, W.F., R. Bierregaard, C. Gascon, R. Didham, A. Smith, A. Lynam, V. Viana, T. Lovejoy, K. Sieving, J. Sites, M. Anderson, M. Tocher, E. Kramer, C. Restrepo, and C. Moritz. 1997. Tropical forest fragmentation: synthesis of a diverse and dynamic discipline. In: *Tropical Forest Remnants: Ecology, Management and Conservation of Fragmented Communities*, ed. W. Laurence and R. Bierregaard. University of Chicago Press, Chicago: 502–525.

Lawton, J.H., D.E. Bignell, B. Bolton, G.F. Bloemers, P. Eggleton, P.M. Hammond, M. Hodda, R.D. Holt, T.B. Larsen, N.A. Mawdsley, N.E. Stork, D.S. Srivastava, and A.D. Watt. 1998. Biodiversity inventories, indicator taxa and effects of habitat modification in tropical forest. *Nature* 391:72–76.

Lindenmayer, D.B., J. Fischer, and R.B. Cunningham. 2005. Native vegetation cover thresholds associated with species responses. *Biological Conservation* 124:331–316.

Lindenmayer, D.B. and G. Luck. 2005. Synthesis: thresholds in conservation and management. *Biological Conservation* 124:351–354.

Margules, C.R. and R.L. Pressey. 2000. Systematic conservation planning. *Nature* 405:243–253.

Matson, P.A., W.J. Paton, A.G. Power, and M.J. Swift. Agricultural Intensification and Ecosystem Properties, July 25, 1997. *Science*. Vol. 277 no. 532S, pp. 504–509.

Mattison, E.H.A. and K. Norris. 2005. Bridging the gaps between agricultural policy, land use and biodiversity. *Trends in Ecology and Evolution* 20(11):610–616.

McIntyre, S. and R.J. Hobbs. 1999. A framework for conceptualizing human effects on landscapes and its relevance to management and research models. *Conservation Biology* 13(6):1282–1292.

McInytre, S. and R. Hobbs. 2001. Human impacts on landscapes: matrix condition and management priorities. In: *Nature Conservation 5: Nature Conservation in Production Landscapes,* ed. J. Carig, N. Mitchell, and D. Saunders. Surrey Beatty and Sons, Chipping Norton: 301–307.

McNeely, J. and S. Scherr. 2003. *Ecoagriculture: Strategies to Feed the World and Conserve Wild Biodiversity*. Island Press, Washington, DC.

Moguel, P. and V.M. Toledo. 1999. Biodiversity conservation in traditional coffee systems of Mexico. *Conservation Biology* 13(1):11–21.

Pardini, R., S. Marques de Souza, R. Braga-Neto, and J.P. Metzger. 2005. The role of forest structure, fragment size and corridors in maintaining small mammal abundance and diversity in an Atlantic forest landscape. *Biological Conservation* 124:253–266.

Parrotta, J.A. 1992. The role of plantation forests in rehabilitating degraded tropical ecosystems. *Agriculture, Ecosystems and Environment* 41:115–133.

Perfecto, I., A. Mas, T. Dietsch, and J. Vandermeer. 2003. Conservation of biodiversity in coffee agroecosystems: a tri-taxa comparison in southern Mexico. *Biodiversity and Conservation* 12:1239–1252.

Radford, J.W., A.F. Bennett, and G.J. Cheers. 2005. Landscape-level thresholds of habitat cover for woodland-dependent birds. *Biological Conservation* 124:317–337.

Rodrigues, A.S.L., S.J. Andelman, M.I. Bakarr, L. Boitani, T.M. Brooks, RM. Cowling, L.D.C. Fishpool, G.A.B. da Fonseca, K.J. Gaston, M. Hoffmann, J.S. Long, P.A. Marquet, J.D. Pilgrim, R.L. Pressey, J. Schipper, W. Seschrest, S.N. Stuart, L.G. Underhill, R.W. Waller, M.E.J. Watts, and X. Yan. 2004. Effectiveness of the global protected area network in representing species diversity. *Nature* 428:640–643.

Schroth, G., G.A.B. Fonseca, C.A. Harvey, C. Gascon, H.L. Vasconcelos, and A.M.N. Isaac. 2004. *Agroforestry and Biodiversity Conservation in Tropical Landscapes*. Island Press, Washington, DC.

Schulze, C.H., M. Waltert, P.J.A. Kessler, R. Pitopang, D. Shahabuddin, M. Veddeler, S. Mu Hlenberg, R. Gradstein, C. Leuschner, I. Steffan-Dewenter, and T. Tscharntke. 2004. Biodiversity indicator groups of tropical land-use systems: comparing plants, birds and insects. *Ecological Applications* 14(5):1321–1333.

Somarriba, E., C.A. Harvey, M. Samper, F. Anthony, J. González, C. Staver, and R. Rice 2004. Biodiversity in coffee plantations. In: *Agroforestry and Biodiversity Conservation in Tropical Landscapes*, ed. G. Schroth, G.A.B. Fonseca, C.A. Harvey, C. Gascon, H.L. Vasconcelos, and A.M.N. Izac. Island Press, Washington, DC: 198–226.

Sutherland, W.J. 2004. A blueprint for the countryside. *Ibis* 146(Suppl 2):230–238.

Taberelli, M., J.M. Cardoso da Silva, and C. Gascon. 2004. Forest fragmentation, synergisms and the impoverishment of neotropical forests. *Biodiversity and Conservation* 13:1419–1425.

Taberelli, M. and C. Gascon. 2005. Lessons from fragmentation research: improving management and policy guidelines for biodiversity conservation. *Conservation Biology* 19:734–739.

Tews, J., U. Brose, V. Grimm, K. Tielborger, M.C. Wichmann, M. Schwager, and F. Jeltsch. 2004. Animal species diversity driven by habitat heterogeneity/diversity: the importance of keystone structures. *Journal of Biogeography* 31:79–92.

Trombulak, S.C. and C.A. Frissell. 2000. Review of ecological effects of roads on terrestrial and aquatic communities. *Conservation Biology* 14(1):18–30.

Tscharntke, T., A.M. Klein, A. Kruess, I. Steffan-Dewenter, and C. Thies. 2005. Landscape perspectives on agricultural intensification and biodiversity-ecosystem service management. *Ecology Letters* 8:857–874.

Wiens, J.A. 2002. Central concepts and issues of landscape ecology. In: *Applying Landscape Ecology in Biological Conservation*, ed. K. Gutzwiller. Springer-Verlag, New York: 3–21.

Wiens, J.A. 2005. Towards a unified landscape ecology. In: *Issues and Perspectives in Landscape Ecology*, ed. J. Wiens and M. Moss. Cambridge University Press, Cambridge: 365–373.

Wilson, J.D., M.J. Whittingham, and R.D. Bradbury. 2005. The management of crop structure: a general approach to reversing the impacts of agricultural intensification on birds? *Ibis* 147:453–463.

Chapter 9

Pollinators

Barbara Gemmill-Herren, Connal Eardley, John Mburu,
Wanja Kinuthia, and Dino Martins

Introduction

Pollination is a critical ecosystem service for food production and human livelihoods, and directly links wild ecosystems with agricultural production systems. Yet, as with other ecosystem services, pollination as a factor in food production is little understood and underappreciated, in part because it has been provided by nature as a free public good. Natural habitat has disappeared in farming landscapes, and the use of agricultural chemicals has increased, harming beneficial insects such as pollinators along with the target plant pests. As a result, farmers are losing the free service, yet they are finding it difficult to recognize and adequately document the loss, much less reverse it. The domesticated honeybee (*Apis mellifera* and its several Asian relatives) has been utilized to provide managed pollination services, but for many crops, honeybees are either not effective or are suboptimal pollinators. Securing effective pollinators to "service" large agricultural fields is proving difficult, and helping nature provide pollination services is receiving renewed interest. As with other ecosystem services, however, these natural roles are receiving insufficient attention in agricultural research, economic markets, government policies, and land management practices.

Pollination is the effective transfer of pollen from male to female flower parts. Over evolutionary history, animal vectors have become increasingly important; the radiation of flowering plants (angiosperms) seems to have coincided with the presence of important insect pollinator taxa in the fossil record, such as the bee superfamily (Apoidea). Because plants are stationary, insect pollination has been a critical means of ensuring reproduction among individuals in a plant species and has enhanced the fitness of plants. Wind pollination works among the grasses, ferns, and some trees, but the majority of dicotyledonous plants depend on animal vectors, usually insects, to some degree. Given that plants rarely depend upon a single animal vector, the provisioning of a suite of pollinators visiting particular species is an ecosystem service that is truly dependent upon the healthy functioning of a biologically diverse ecosystem.

The claim that humans depend on animal pollinators for at least one-third of all food production (O'Toole 1993) may be slightly misleading because the starches and staples that provide the bulk of foodstuffs for human nutrition, such as maize and rice, are wind pollinated. A more compelling argument for the importance of animal pollinators to human nutrition is that the majority of vitamins and minerals in the human diet come from animal-pollinated fruits and vegetables.

Estimates of the annual monetary value of pollination range from $120 billion for all pollination ecosystem services (Costanza et al. 1997), to $200 billion for the role of pollination in global agriculture alone (Richards 1993). The inconsistency of these two estimates shows how inadequate the present tools are for valuing ecosystem services. Conceptually, pollination is not much different from other agricultural inputs, so it should be possible to assess the costs of reducing pesticide use and designating nesting sites and alternate floral resources for pollinators against the benefits of increased yields from higher pollination rates.

The primary threats to animal pollinators are pesticides and habitat alteration or fragmentation that leads to loss of breeding sites and food resources. These threats continue to increase in most regions, potentially leading to profound ecological impacts and losses in biodiversity. The service may be severely degraded before the loss is realized, making its restoration very difficult.

The global community has already registered an early warning of such impacts. The loss of managed and feral honeybee colonies to parasites and disease in the past decade have raised global concern about the extent to which both wild plant communities and agricultural systems depend on reliable pollination services (Allen-Wardell et al. 1998; Buchmann and Nabhan 1996). Intensive commercial agriculture depends on a handful of pollinators, the most important being the honeybee. Migratory beekeeping spreads diseases, reduces local genetic variability, and makes native honeybees more vulnerable to pests and diseases. When managed honeybees are removed from a system due to pest or disease problems (or the fact that few young people choose to learn beekeeping skills), yields of crops dependent on insect vectors for pollination decline (Kremen 2004). Close examination of the most effective pollinators of most crops usually feature nonhoneybee pollinators, with honeybees making up in their numbers and manageability what they may lack in effectiveness. Some crop plants such as papaya are moth pollinated, so bees are not relevant to them. In addition to honeybees, populations of wild pollinators appear to be declining, including native bees, flies, moths, birds, bats, and other animals. Agricultural systems generally are not being managed effectively to secure pollination services from indigenous, unmanaged pollinators, so the resistance and resilience that biological diversity provides to natural ecosystems is being lost to agricultural ecosystems.

Pollination's Contribution to Sustainable Agriculture

The contribution of pollination to plant reproduction has been long recognized; the frontispiece of Buchmann and Nabhan (1996) depicts a carving from a BC era Assyrian palace of a half-eagle, half-man hand pollinating date palms (more ancient historical references are available in Kevan and Phillips 2001). In the late 1790s, Christian Sprengel (1793) pointed out the important role of insects and the wind in the cross-pollination of plants, and described mechanisms by which flowers ensure their own cross-fertilization. Yet Sprengel's observations were largely neglected until Charles Darwin incorporated them into his theory of evolution. Darwin used the example of pollination to highlight sexual selection in life forms, but the practical implications of his message have been neglected over time. The title of Buchmann and Nabhan's book, *The Forgotten Pollinators*, seems to be extraordinarily appropriate: human society has been almost amnesiac about pollination. Perhaps because insects are so inconspicuous as they industriously visit flowers, and the system has worked well without much intervention in the past, comparatively little attention has been given to pollination as a factor in crop yields over the past 100 years of agricultural research. The two crop-by-crop compendia on pollination, which detail the diversity of pollinators for each crop (Crane and Walker 1984; Free 1995) are out of print. Distinct fields of agronomy and university departments are built on certain identified needs of crop plants, such as soil science, water and irrigation management, plant protection, and plant nutrition, but pollination is considered to be less important. Pollination research is often relegated to a small section of an entomology or honeybee congress.

Highly localized crop pollination crises have sometimes brought pollination to the fore. For example, threats to the supply of honeybees for lowbush blueberry pollination motivated the US Department of Agriculture to invest in five years of blueberry pollination research in Maine (Stubbs and Drummond 2001). Similarly, oil palm trees, native to West Africa, were taken to Southeast Asia and planted in vast plantations to satisfy the global demand for cheap, versatile palm oil. Production was disappointing until the plantation managers realized that it could be enhanced by hand pollinating the palm flowers. Yet hand pollination was laborious and inefficient, so plantation owners began to ask how the oilpalm was pollinated in its native habitat of West Africa's forests. Researchers in Cameroon found a weevil, *Elaeidobius kamerunicus*, that enables pollination while feeding on the pollen (oil palms, though technically "wind" pollinated," apparently need the weevil to stir up the pollen through its feeding activity to make it airborne). The weevil now accomplishes all the oil palm pollination needs, bringing savings that were amounting to $150 million per year by the early 1980s (Greathead 1983). Conventional wisdom has held that crops such as tomatoes and coffee are self-pollinated, and growers need not concern themselves with insect visitors. But when these crops are grown under increasingly industrialized conditions, such as in greenhouses for tomatoes or high-input sun coffee, researchers start to recognize the contribution that animal pollination can make to yield, or conversely, the losses when native pollinators can no longer reach and service the crops (Roubik 2002). This is an expensive way to identify critical pollination needs of agricultural crops, because restoration is far more difficult than conservation of existing interactions.

As mentioned earlier, statistics on the relationship between food security and pollination have been widely variable, and carefully documented data would enhance the credibility of pollinator conservation initiatives. Some available figures suggest that 71 of the 103 to 107 crops that feed most of the world are pollinated by bees (Prescott-Allen and Prescott-Allen 1990) and at least 72% of the 1330 cultivated crops surveyed by Roubik (1995) are bee pollinated. A more recent estimate of the dependence of crop production on pollination finds that pollinators such as bees, birds and bats affect 35 percent of the world's crop production, increasing outputs of 87 of the leading food crops worldwide (Klein et al. 2006).

Several of these calculations, and large compendia such as that by Free (1995), include coverage of crops that are not actually dependent on pollination for their global productivity. For example, breeders and some indigenous populations like to work with the seed from starch crops such as potatoes and manioc, but few farmers will ever produce seed from these crops and experience their pollinator dependence. Many vegetable crops need bees only for seed production. That said, it would be disastrous for world food security for any crop to lose the ability to reproduce by seed, or to lose its most effective pollinators.

In a time of decreasing commodity prices due to oversupply, yield gains may not always be the most important consideration for agricultural producers. Pollination should be a concern of farmers from the standpoint of quality, not just quantity. For pyrethrum, derived from the flowers of *Chrysanthemum cinerariifolium*, a more potent insecticide is produced when the flower heads have been visited by insects (Crane and Walker 1984). In many countries, quality is vitally important because well-shaped fruit fetches much higher prices in the selective export market. If such quality considerations can enter into market share and market prices, pollination may contribute significantly to the income per unit area for farmers conserving pollinator services.

For Africa, the authors have developed a list of crops directly dependent upon animal vectors for pollination (African Pollinator Initiative 2004). In developing this list, they considered the circumstances unique to Africa that will affect pollinator dependence; for example, seed for the many leafy *Amaranthus* greens, which form an important part of the African diet, are not produced commercially, and therefore farmers are directly dependent on adequate pollination for seed set, even if they do not consume the seed. This list is the first step in accurately estimating pollinator contributions to crop yields within the continent. More refined information is now needed on overall yields, and the percentage yield increases due to pollinator visits.

A database of the literature addressing pollination within Africa has been developed (Rodger et al. 2004). Information from this database indicates that, although coverage for a few commodity crops of major economic importance has been reasonably extensive (e.g., oil palm and cacao), the majority of other crops in Africa have not benefited from pollination studies specific to the sites where they are grown on the continent—and pollination is generally quite site-specific. Cacao and oil palm, as well as figs, whose system was discovered in Kenya, are species with quite unconventional, non–bee pollination systems, and more of these may be unearthed for those commodities whose pollination is simply not known.

Taxonomic Challenges

Bee faunas are important natural resources for sound agricultural development. As with all natural resources, inventories of their diversity and distribution are needed if they are to be managed to their best advantage (O'Toole 1993). Unfortunately, the current state of bee taxonomy imposes restraints on the realization of these goals as it does for other pollinator groups like flies, wasps, and beetles. This taxonomic impediment derives from shortfalls in investment in training, research, and collections management. It seriously limits the capability to assess and monitor pollinator decline, to conserve pollinator diversity, and to manage it sustainably. Moreover, access to sound bee taxonomy will be at a premium because research projects in pollination biology, both of native floras and of crop plants, are on the increase and require bee identifications.

The Conference of the Parties to the Convention on Biological Diversity established an International Initiative for the Conservation and Sustainable Use of Pollinators (also known as the International Pollinators Initiative) in 2000 and requested the Food and Agriculture Organization of the United Nations to coordinate the development of a plan of action, including the need for postgraduate training of bee taxonomists.

Almost all of the important taxonomic resources needed by entomologists working in the developing world are located in museums and other institutions in the developed world. Means of more effectively sharing this information need to be implemented. The International Pollinators Initiative has suggested the following measures:

- A single Web site with access to all databases of type material in the world's museums and other research institutes. An effort along these lines is underway (http://insects.oeb.harvard.edu/etypes) to enable museum collections worldwide to create and distribute digital documentation of their biological type specimen holdings.
- A mechanism to make data associated with specimens freely available to all bona fide researchers in taxonomy and pollination biology. The Global Biodiversity Information Facility is currently working on the creation of inventories of organisms and mechanisms whereby biological information on different Web sites can be accessed through one portal.
- Exchange and transfer of information, especially literature, that is not widely available outside the museums and institutions of the developed world, using photocopying and scanning facilities.
- Repatriation of data associated with type specimens to appropriate institutions in countries of origin—again, an objective of the Global Biodiversity Information Facility.

Training of support staff may provide some needed assistance. "Parataxonomists" carry out the following functions:

- Collecting and sampling, especially in monitoring or faunistic studies, often in association with the taxonomist
- Preparation and processing of specimens
- Preliminary sorting to a convenient taxonomic level (subfamily, tribe, genus)

An ideal team would be three parataxonomists per independent research taxonomist. The appropriate proportion of parataxonomists to taxonomists would facilitate and enhance the service output of every collection facility and thus contribute to eliminating the constraint of inadequate taxonomic expertise (Dias et al. 1999).

Automated and semiautomated identification systems, currently under development, offer another potential solution. Three widely used systems include Expert Taxonomic Information (ETI), the Lucid software for creating identification keys, and Discover Life (www.discoverlife.org). These use image recognition software and identify organisms from images of the organism, or parts of the organism, without the user having to make choices. Two systems are under development.

Ecological Characterization of Pollination Services

Animals do not pollinate plants deliberately. They either exploit the plants for their resources—pollen, nectar, and oils (in a few plants only)—which the plants provide to attract potential pollinators, or are tricked to visit the plants, like orchid bees and carrion flies. Further, many visitors rob the plants of their pollen and nectar without providing any service, such as large carpenter bees that bite the base of tubular aloe flowers that are too narrow for the carpenter bees to enter at the floral opening, and take the nectar from the flower without coming near the anthers and stigma. Further, many insects visit many different plant species randomly and the pollen that attaches itself to their bodies may not come into contact with a stigma of another plant of the same species. Because pollen is rich in protein, much is ingested rather than serving the plant's intended purpose. Therefore most plants produce much more pollen than is needed for fertilization. A few plant groups make pollinia (pollen bags). Here the visitors drink nectar only and the pollinium attaches itself to the appropriate spot of the visitor for it to touch the stigma of another flower. Bees are particularly good pollinators because they use the nectar for sugar (for adults and larvae) and pollen mostly for larval food (protein), and some bees also collect oils for their larval provisions. Therefore, bees repeatedly visit flowers and usually concentrate on the same species. Other visitors are sporadic, like many butterflies and many flies, and visit a variety of flower species or remain resident on the flower for some time, like hopline beetles.

Most pollination systems can be characterized as "somewhat generalized" (Waser et al. 1996). A suite of pollinators is usually critical to provide effective pollination. Pollinator populations rise and fall, as do all animal populations, in response to environmental variables, such as levels of parasitism or abundance of nesting sites. Therefore, the same plant may be pollinated by different pollinators in different places, and the same population of plants may have different pollinators in different seasons or years. Pollinators are often quite variable in relation to ambient conditions, and a species that is relatively unimportant in one year may be of greater importance in the next year (Kremen et al. 2002). Unfortunately, under anthropogenic disturbance, such as that created by industrial agriculture with larger field sizes, monocultures, and intensive use of agricultural chemicals, the largest and most efficient pollinators may be the first to be lost (Kremen 2004). Bee assemblages do not appear to show any classic density-dependent relationships, so that when large, efficient pollinators drop out of a system, they are not replaced by an upsurge in population numbers of other bees present. Highly disturbed sites do not usually harbor specialized pollinators; the weedy vegetation of disturbed sites generally attracts the very common, "weedy" pollinators. Rehabilitation of such sites will probably require the simultaneous introduction of plants and their pollinators.

Key pollinators for one plant species may depend on the presence of other plants to provide resources at other times of the season, or different resources in the same season (some bees seem to have separate nectar plants and pollen plants, and sometimes oil plants). Very little work on such possible "cascade effects" has been carried out, though Sampson (1952) noted that grazing livestock may destroy or alter riparian vegetation, which serves as a key resource to pollinators at certain times of the year. This affects the ability of those pollinators to carry out pollination services not only on the riparian vegetation but on other plants flowering at different times of the year. Pollinator declines could, through such interconnectedness, ultimately affect multiple trophic levels (Allen-Wardell et al. 1998), yet scientific understanding of those complex and diffuse relationships is still very incomplete. Memmott (1999) is studying the interactions that make up "pollination webs," similar to the complex interactions that define food webs. Early results indicate that the most critical interactions that determine reproductive success of plants are often not the most obvious ones. Conserving plants does not necessarily conserve their pollinators, so an ecosystem approach is needed.

Conserving Space and Pollinators within an Agricultural Matrix

Human livelihoods and plant reproductive success are bound together by the need for a large and diverse suite of pollinators to assure continued and reliable delivery of effective pollination services. These services cannot generally be reduced to a focus on a single "service provider." A matrix of healthy natural ecosystems, interspersed and adjacent to human settlements and agricultural fields, can best provide such services (Kremen et al. 2002).

A few case studies may best illustrate the role of a wild habitat matrix in farming landscapes to promote pollination:

History of Bee Populations in Illinois

The bee fauna in the vicinity of Carlinville, Illinois, USA, is probably the most thoroughly sampled such fauna anywhere in the world, as the result of the exhaustive, detailed collections of Charles Robertson between 1884 and 1916. He observed and collected flower-visiting insects belonging to several orders in a 16 km radius around Carlinville. This same site was revisited in 1970 and 1972, providing a picture of trends over time (Marlin and LaBerge 2001).

Carlinville's diverse native bee fauna was remarkably persistent when resampled nearly a century later, despite the dramatic changes in the landscape, as row-crop agriculture increasingly predominated as a land use with its diverse arsenal of insect control tools. Impacts did vary from specialist to generalist bees: specialist bees (oligolectic) were most likely to be excluded from the system under agricultural development, and generalist bees to be favored, but extinctions were few.

Marlin and LaBerge (2001) suggest that the bees survived because protected riparian areas dissected the farmland, and hedgerows and fencerows were common. The area provides habitat ranging from open prairie to hills with relatively dense tree cover, the creeks, and associated topographic relief providing many sites protected from intensive agricultural and residential development. Hedge- and fencerows presumably provide habitat for bees because they are unploughed, are not directly exposed to soil-incorporated maize insecticides, and harbor plants that may provide a pollen source. They offer potential nesting sites for both ground- and stem-nesting bees. Additionally, these areas are often somewhat elevated because they trap water- and wind-driven soil particles, creating a low berm of earth. Such features presumably provide protected corridors through which insects and other species can move about agricultural areas.

Although the relatively flat areas are intensively farmed and appear to provide little bee habitat, local landowners retained many other areas in woods, pasture, meadow, and other types of cover. As long as suitable amounts of host plants and such varied habitat remain, the associated bee species seem likely to persist. This long-term study suggests that pollinator decline is a more fine-grained problem than wholesale loss of species.

Eggplant Pollination in Kenya

In a forested landscape under development for horticultural crops in Kenya, the production of eggplant is entirely dependent on native bee pollinators. Not just any pollinator will do, because eggplant can only be properly pollinated by certain bees that "buzz pollinate"; that is, they bite the flower and vibrate their wing muscles at a certain frequency, such that pollen comes flying out of small pores in the flower and can be carried to another flower to produce fruit. Without this ecosystem service, no fruit will be produced. Honeybees cannot buzz pollinate, but two species of solitary bees, which occur naturally in the forest that is being cleared for farms, are very effective pollinators. The bees only get pollen from eggplant because it does not produce nectar. Thus they cannot live exclusively on agricultural land and must make use of the plots of forest that have not yet been cleared. In the dry season, they depend more heavily upon the wild ecosystem for floral resources. Farmers have recognized the importance of this "pollination service" in leaving tracts of forest standing so this pollination service can continue.

Bee Pollination in California

Historically, farmers have imported colonies of European honeybees (*Apis mellifera*) to fields and orchards in North America for pollination services. These colonies are becoming increasingly scarce, however, because of diseases, pesticides, and other impacts. Native bee communities also provide pollination services, but the amount they provide and how this varies with land management practices are largely unstudied. In a study in an agricultural valley in northern California, the individual species and aggregate community contributions of native bees to watermelon (*Citrullus lanatus*) pollination were examined on farms that varied both in their proximity to natural habitat and in their management type (organic vs. conventional). On organic farms near natural habitat, native bee communities provided full pollination services even for a crop with heavy pollination requirements such as watermelon, without the intervention of managed honeybees. All other farms, however, experienced greatly reduced diversity and abundance of native bees, resulting in insufficient pollination services from native bees alone. Diversity was essential for sustaining the service because of year to year variation in pollinator community composition. Continued degradation of the agronatural landscape destroys this "free" service. Conservation and restoration of bee habitat are potentially viable economic alternatives for reducing dependence on managed honeybees (Kremen et al. 2002).

Conclusion: Management by Species or Management by Habitat?

To date pollination management has been mostly focused on a few species, particularly the honeybee, which is the most common generalist pollinator. Other managed species include the alfalfa leafcutter bee in the United States, Osmiines for orchards in Europe and North America, weevils for oilpalm in Malaysia, and bumblebees for greenhouse vegetables. The relative abundance as well as diversity of wild pollinators is diminished in intensively cultivated areas. For effective services abundant pollinators are needed, and thus farmers that manage pollination have traditionally turned to a species-based approach. In general, management consists of a single species being cultured and released among the crops (except the oil palm weevil, which is naturalized, itself, in the plantations). This management scenario inevitably loses the resistance and resilience provided by biological diversity, and farmers become vulnerable to pests and diseases of their pollinators. On the positive side, management of these pollinators can create an industry and provide employment.

Although single-species domestication and management may be appropriate in some cases, in general, commercial and subsistence agriculture will also benefit from land management techniques that provide a diversity of pollinators. Such systems will also usually produce diverse products, enhancing security.

Natural areas are an integral part of agricultural systems because, apart from pollinators, they maintain water tables, provide clean water, reduce flood damage, and provide habitat for natural enemies of crop pests. Many pollinators live in the ground or dead wood, make nests from mud, and play multiple roles in ecosystems. Therefore, natural areas in agricultural landscapes also need to be managed for pollinator conservation. Activities like grazing by stock, impounding water, and removing firewood reduce pollinator abundance and diversity. The need to thus explicitly manage pollination services of both domesticated and wild plants through a mosaic of diverse production and natural area management will likely be a core objective in most ecoagriculture landscapes for farmers and conservationists alike.

References

African Pollinator Initiative. 2004. Crops, browse and pollinators in Africa: an initial stocktaking. http://www.fao.org/AG/aGp/agps/C-CAB/Castudies/pdf/apist.pdf.

Allen-Wardell, G., P. Bernhardt, R. Bitner, A. Burquez, S. Buchmann, J. Cane, P.A. Cox, V. Dalton, P. Feinsinger, D. Inouye, M. Ingram, C.E. Jones, K. Kennedy, G.P. Nabham, B. Pavlik, V. Tepedino, P. Torchio, and S. Walker. 1998. The potential consequences of pollinator declines on the conservation of biodiversity and stability of food crop yields. *Conservation Biology* 12:8–17.

Buchmann, S.L. and G.P. Nabhan. 1996. *The Forgotten Pollinators*. Island Press, Washington, DC; Shearwater Books, Covelo, CA.

Costanza, R., R. D'Arge, R. De Groot, S. Farber, M. Grasso, B. Hannon, K. Limburg, S. Naeem, R.V. O'Neill, J. Parvello, R.G. Raskin, P. Sutton, and M. van den Belt. 1997. The value of the world's ecosystem services and natural capital. *Nature* 387: 253–260.

Crane, E. and P. Walker. 1984. *Pollination Directory for World Crops*. International Bee Research Association, Bucks, England.

Dias, B.S.F., A. Raw, and V. Imperatriz-Fonseca. 1999. *International Pollinators*. Report on the Recommendations of the Workshop on the Conservation and Sustainable Use of Pollinators in Agriculture with Emphasis on Bee Conservation. Brazilian Ministry of the Environment, Brasilia.

Free, J.B. 1995. *Insect Pollination of Crops*. 2nd ed, Academic Press, New York.

Greathead, D.J. 1983. The Multi-Million Dollar Weevil That Pollinates Oil Palms. *Antenna* 7(3):105-107.

Kevan, P.G. and T.P. Phillips. 2001. The economic impacts of pollinator declines: an approach to assessing the consequences. *Conservation Ecology* 5(1):8.

Klein, A.M., B.E. Vaissière, J.H. Cane, I. Steffan-Dewenter, S.A. Cunningham, C. Kremen, T. Tscharntke. 2006. *Importance of Pollinators in Changing Landscapes for World Crops*. The Proceedings of the Royal Society of London, Series B, October 2006.

Kremen, C. 2004. Pollination services and community composition: does it depend on diversity, abundance, biomass, or species traits? In: *Solitary Bees: Conservation, Rearing and Management for Pollination*, ed. B.M. Freitas and J.O.P Pereira. International Workshop on Solitary Bees and Their Role in Pollination, Federal University of Ceara, Brazil.

Kremen, C., N.M. Williams, and R.W. Thorp. 2002. Crop pollination from native bees at risk from agricultural intensification. *Proceedings of the National Academy of Sciences* 99:16812–16816.

Marlin, J.C. and W.E. LaBerge. 2001. The native bee fauna of Carlinville, Illinois, revisited after 75 years: a case for persistence. *Conservation Ecology* 5(1):9.

Memmott, J. 1999. The structure of a plant–pollinator food web. *Ecology Letters* 2:276–280.

O'Toole, C. 1993. Diversity of native bees and agroecosystems. In: Hymenoptera *and Biodiversity*, ed. J. LaSalle and I.D. Gauld. CAB International, Wallingford, UK: 169–196.

Prescott-Allen, R. and C. Prescott-Allen. 1990. How many plants feed the world? *Conservation Biology* 4(4):365–374.

Richards, K.W. 1993. Non-*Apis* bees as crop pollinators. *Revue suisse de Zoologie* 100:807–822.

Rodger, J.G., K. Balkwill, and B. Gemmill. 2004. African pollination studies: where are the gaps? *International Journal of Tropical Insect Science* 24(1):5-28.

Roubik, D.W. 1995. *Pollination of Cultivated Plants in the Tropics*. FAO Agricultural Services Bulletin 118. Food and Agriculture Organization of the United Nations, Rome.

Roubik, D.W. 2002. Feral African bees augment neotropical coffee yield. In: *Pollinating Bees: The Conservation Link between Agriculture and Nature*, ed. P. Kevan and V.L. Imperatriz-Fonseca. Ministry of Environment, Brasília: 255–266.

Sampson, A.W. 1952. *Range Management, Principles and Practices*. Wiley, New York.

Sprengel, C.K. 1793. *Das entdeckte Geheimniss der Natur in Dau und in der Befruchtung der Blumen*. Viewegsen, Berlin.

Stubbs, C.S. and F.A. Drummond. 2001. *Bees and Crop Pollination: Crisis, Crossroads, Conservation*. Thomas Say Publications in Entomology Proceedings. Entomological Society of America, Lanham, MD.

Waser, N.M., L. Chittka, M. Price, N.M. Williams, and J. Ollerton. 1996. Generalization in pollinations systems and why it matters. *Ecology* 77(4):1043–1060.

Chapter 10

From Integrated Pest Management to Adaptive Ecosystem Management

Hans Herren, Johann Baumgärtner, and Gianni Gilioli

Introduction

The increasing demand for food during the second half of the 20th century led to many developments in agricultural technology and practices that greatly increased crop and livestock production. But these advances have had their drawbacks, as exemplified by inadequate pest control technologies, which were overcome through the design and implementation of integrated pest management (IPM) systems (Flint and van den Bosch 1981). The development of IPM led to its expansion into four dimensions characterized by objectives for management, space, time and institutions (Kogan et al. 1999; Baumgärtner et al. 2007). The expansion aims to enhance sustainability and apply new approaches and technologies to ecosystem study and management.

The evolution of IPM provides a good illustration of applying holistic ecological approaches to agricultural practices. Agricultural pests are defined as any organisms that harm crops or livestock, so farmers seek to limit their damage as a way of enhancing net production. This chapter focuses on how IPM evolves into ecosystem management and contributes to the development of ecoagriculture production systems. The enhancement of sustainability in ecological, economic and social dimensions is illustrated by a project undertaken in Ethiopia by the International Centre of Insect Physiology and Ecology (ICIPE).

Expansion of IPM

Yield increase of agricultural crops was traditionally sought by selecting improved varieties and further developing cultural methods. Synthetic fertilizers and pesticides became generally available only a few decades ago and have increasingly been used in crop production. The drawbacks of reliance on synthetic pesticides stimulated research and development of alternatives in the areas of pest forecasting, biological control, autocidal control, host plant resistance, and cultural and mechanical methods (Flint and van den Bosch 1981). Chemical control was complemented with pest control substances originating from plants, and with behavior-modifying chemicals. During the past few years genetically modified cultivars able to increase pest mortalities have been made and promoted for pest control. Traditionally, synthetic pesticides were applied to increase pest mortalities, whereas both old and newly developed techniques now allow targeting of additional population processes such as fecundity, development, and movements (Stark and Banks 2003). For example, neem tree (*Azadirachta indica,* syn. *Melia azadirachta* L.) produces a secondary metabolite known as azadirachtin that can affect survival, fecundity, developmental time, and behavior of pest populations. The consideration of all pest control options, beyond killing of pests and the study of population dynamics, can underscore the utility of complementing IPM with new approaches and methodologies targeting more complex systems.

Concerns about potential undesirable side-effects of synthetic chemicals also contributed to the development of IPM. IPM has been defined as a pest management system that, in the context of the associated environment and the population dynamics of the pest species, utilizes all suitable techniques and methods as compatibly as possible and maintains the pest populations at levels below those causing economic injury (Flint and van den Bosch 1981). The consideration of pest population dynamics and thresholds introduced concepts of population ecology, ecosystem science, and economics into pest management (Metcalf and Luckmann 1975; Dent 1995; Gutierrez 1996). Furthermore, study and management of populations with respect to thresholds required adopting a systems approach (Huffaker and Croft 1976; Getz and Gutierrez 1982). The consideration of these aspects and the integration of biological control as a key element in IPM schemes (van Driesche and Bellows 1996) further increases the complexity of IPM systems.

Conceptually, IPM operates at the interface of ecological and socioeconomic systems (Kogan et al. 1999). Initially, the objective of IPM was to maintain pest populations at levels below the economic threshold, i.e. a population level that produces incremental damage equal to the costs of preventing the damage (Headley 1975). Today, production systems are expected to satisfy ecological and social criteria as well (Delucchi 2001). Further IPM development led to the consideration of holistic concepts of integrated farming (ARS 2002), biointensive farming (John Jeavons, pers. comm.), permaculture (Mollison 1997), ecofarming (Kotschi et al. 1988), and organic farming (Palaniappan and Annadurai 1999). These developments also closed the gaps between IPM and management of ecosystems and made IPM entirely consistent with the principles of ecoagriculture as advanced in this book.

To deal with these different aspects, IPM was expanded into four dimensions, to reflect objectives related to management, space, time and institutions, as illustrated in figure 10.1 (Baumgärtner et al. 2007).

With respect to the objectives of management, IPM was expanded to deal with pest assemblages and population communities rather than single pest species. Early IPM specialists often focused on single pest species, although the inclusion of multiple pests has shown advantages over single-pest control. Van Lenteren (1995), for example, has shown that the success of IPM in greenhouses was possible because a complete IPM program had been developed covering all aspects of pest and disease control for the crop.

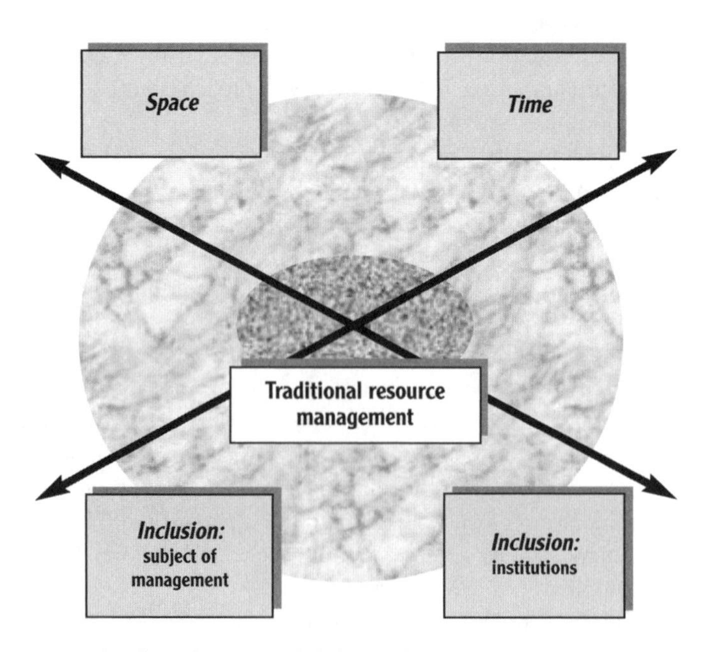

Figure 10.1. The four-dimensional framework considered in conservation ecology and used to describe the expansion of Integrated Pest Management (IPM) objectives. Expansion is into dimensions characterized by objectives for management, space, time and institutions (excerpted from Meffe and Carroll (1997), reproduced and modified with permission).

With respect to space, IPM concepts are no longer restricted to single fields but expanded into space by applying, among others, principles of landscape ecology (Turner et al. 2001). In species conservation, Comiskey et al. (1999) found a general agreement that biologically effective preservation of species, habitats, and ecological processes requires working at large spatial scales.

With respect to time, Moir and Block (2001) view adaptive management as a process of implementing land management activities in incremental steps and evaluating whether desired outcomes are being achieved at each time step. However, long-term responses should be considered when focusing on short-term management and corresponding ecosystem responses (Moir and Block 2001).

With respect to institutions, Brown (2003) advocated the establishment of institutions based on adaptive management, as well as more equitable and inclusive decision making. Baumgärtner et al. (2007) provide a summary of the structure and function of institutions involved in IPM systems.

The expansion into these four dimensions increased the complexity of IPM systems and required the consideration of new approaches and methodologies in IPM development and implementation. These follow ecosystem study and management principles, some of which are listed in the following section. Gilioli et al. (2003, 2005) have already described how biological control strategies and practices, key IPM elements (van Driesche and Bellows 1996), appear to develop into ecosystem management instruments.

Approaches and Methodologies for Ecosystem Management

Hierarchy Theory Approach to Ecosystem Management

To describe a complex system adequately, analysts often rely on hierarchical structuring (Ahl and Allen 1996). Accordingly, a level fits into the hierarchy by virtue of a set of definitions that link the level in question to those above and below (Allen 2004). IPM specialists recognize that managed ecological systems have a hierarchical structure and state that several levels need to be addressed simultaneously (Kogan et al. 1999; Baumgärtner et al. 2003a; 2007). Conway (1984) included structured decision-making institutions into the hierarchical arrangement of pest control systems and stated that, depending on the level, decision makers may be responsible for tactical, strategic, or policy measures. For example, farmers are undertaking tactical measures, the agricultural service or health board may be engaged in strategic measures, and regional, national, and international institutions may be responsible for design and implementation of policy measures. A combined effort of these institutions appears to be most promising for addressing pest problems. IPM specialists relying on hierarchy theory place a significant emphasis upon the observer in the system (Allan 2004).

The levels identified by hierarchy theory can also be seen as discontinuities arising from ecological and socioeconomic processes and conditions in the dimensions described in the previous section (Fig. 10.2). Processes resulting in discontinuities are modified or even amplified by agricultural activities (Baumgärtner et al. 2002; Baumgärtner and Schneider 2006). For example, the spatial pattern of fields arranged on a landscape may influence discontinuities resulting from basic ecological processes.

Adaptive Ecosystem Management

Good management is a dynamic process that is dependent on an array of factors, including the quality of leadership and the means to manage risk by monitoring and adjusting actions based on information acquired (Oglethorpe 2002). The outcome of management operations is characterized by uncertainties, because the dynamics of complex systems are unpredictable. Pielke and

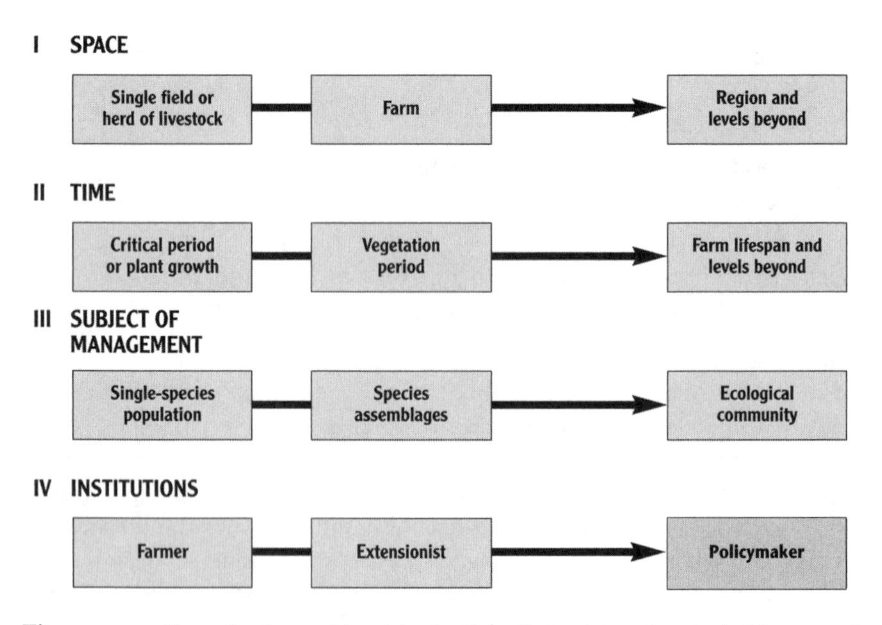

Figure 10.2. Some levels considered in the four dimensions of expanded Integrated Pest Management (IPM) systems, based on work by Conway (1984), Kogan et al. (1999) and Baumgärtner et al. (2002, 2003a, 2007) (excerpted from Koul and Cuperus (2007), reproduced with permission)

Conant (2003) remind ecologists that uncertainties about the future are more likely to be reduced by focusing on decision making rather than on prediction. This is done in the subsequent section through adaptive management, i.e. an iterative process of "optimal" decision making in the face of uncertainty that seeks to reduce insecurity through continuous acquisition and processing of monitoring data for knowledge improvement, decision-support and impact assessment (Baumgärtner et al. 2003c; Gilioli and Baumgärtner 2007).

Ecosystem Management Objectives

Management objectives are the topic of an ongoing debate in the ecological literature in which different aspects are emphasized (Gilioli and Baumgärtner 2007). The Ecological Society of America recommends the enhancement of overall sustainability as a management objective (Christensen et al. 1996). This view is supported by Berkes et al. (2002) and Goodland (1995) who provide a definition and explanation of sustainability. At its most basic level, sustainable means meeting the needs of the present without compromising the ability of future generations to meet their own needs (Brundtland Report). In the project presented in the subsequent section, we currently aim at balanced enhancement of sustainability in ecological, economic and social dimensions (Gilioli and Baumgärtner 2007).

Adaptive Ecosystem Management in Practice: Improving Human and Animal Health in Ethiopia

Among other projects, the International Centre of Insect Physiology and Ecology (ICIPE) is studying and managing Ethiopian ecosystems for human livelihood improvement. Important components of the work include human health (anopheline vectors, malaria pathogens, human hosts); animal health (tsetse vectors, trypanosomiasis pathogens, cattle hosts); plant health (arthropod pests, natural enemies, crops); and environmental health (commercial insects, biodiversity conservation). All of these elements involve pest management and contribute to ecoagriculture systems.

In the first project phase, the experience of ICIPE's Human Health, Animal Health, Plant Health, and Environmental Health research divisions was combined and their activities aimed at human health improvement and poverty alleviation (Baumgärtner et al. 2001). Technologies including tsetse control, waste management and beekeeping for income generation were successfully transferred. However, the impact of these technologies on human health improvement, poverty alleviation, and ecosystem sustainability enhancement could not be well defined.

The simultaneous consideration of human health, animal health, plant health, and environmental health created organizational problems for the institution executing the project and the communities involved. For example, the coordination of activities was difficult, at least partly because the beneficiaries were not well prepared to participate in project execution. Incomplete involvement of national institutions resulted in misunderstandings regarding project objectives and methodologies, at least partly because they considered ecosystem management as a rural development activity exclusively driven by the demand of beneficiaries, without any learning component. Limited flexibility by the project-executing institution led to delays in transfer of funds (a flexible response to project needs was not possible because of adherence to predefined work plans). Moreover, the project was oriented toward transfer of technologies, so measures of success were based on technology transfer and not on changes in the ecosystem with respect to overall objectives.

To overcome these problems, substantial changes were introduced in the second phase of project execution (Baumgärtner 2003b). First, the organizers relied on concepts developed by Becker and Ostrom (1995), who identified diversity among institutions as an important attribute of long-term sustainable agricultural systems. Accordingly, community committees were created for dealing with specific issues and the project was incorporated into Ethiopia's national development agenda (Aseffa Abraha et al. 2003). Technologies were seen as instruments to improve ecosystem qualities, and their impacts were continuously monitored. Project organizers refrained from adhering to fixed objectives and instead selected adaptive management procedures for placing the system on a trajectory toward environmental sustainability and human health improvement (Gilioli and Baumgärtner 2007).

Adaptive management was undertaken at two levels of understanding. At the first level, the team applied a procedure of immediate utility, seeking rapid improvement by precision targeting of interventions in time and space. For example, tsetse control operations are guided by information from monitoring traps (Fig. 10.3), and control traps are deployed for "hotspot" management without much insight into movements of tsetse populations (Sciarretta et al. 2005). At the second level, the accumulated experience combined with ecological knowledge permits study and management of ecosystems on the basis of improved insight into relevant processes.

Project objectives are to be achieved in two stages. In stage one, the project dealt with human and animal health systems and thereby set the stage for rural development. Specifically, the project sought gradual improvement of health of people and livestock by targeted vector control operations and by managing the vectors–pathogens–hosts system for health improvement. The participation of the community (i.e., the tsetse and mosquito control committees) and the involvement of national institutions are indispensable. The human and animal

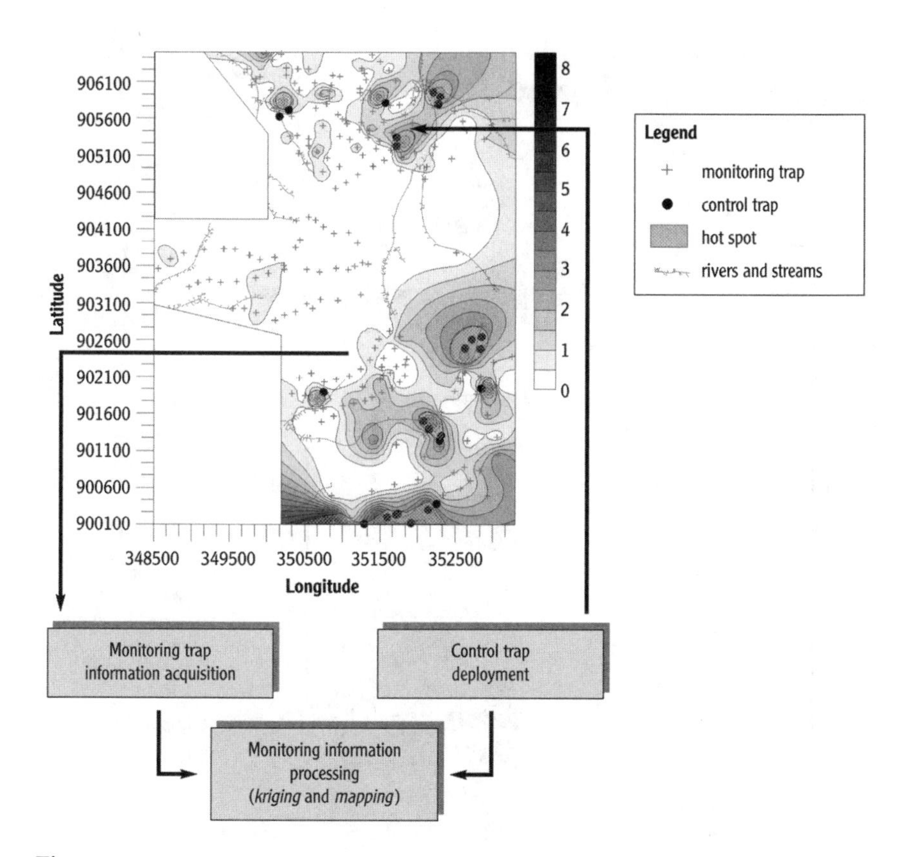

Figure 10.3. Adaptive management of tsetse populations in Luke, Ethiopia. In a biweekly rhythm, monitoring information is collected, processed and used to guide control operations directed against "hot spots" consisting of relatively high tsetse trap catches (Sciarretta et al. 2005). (Excerpted from Koul and Cuperus (2007), reproduced with permission.)

health components were selected because improved health of people and their cattle is a prerequisite for rural development. Importantly, this link was already emphasized by the villagers at the beginning of stage one.

In stage two, the ecosystem is being managed for greater sustainability (Fig. 10.4). Crop management is introduced, coordinated with livestock development and combined with attempts to maintain biodiversity on genetic, species, and ecosystem levels. The conservation of unmanaged land is an important activity to this end. A wide array of crop management technologies, including biological control and habitat management, are being integrated into the project. In this phase, the project is seeking to enhance sustainability through participating with a community whose engagement is not constrained by diseases.

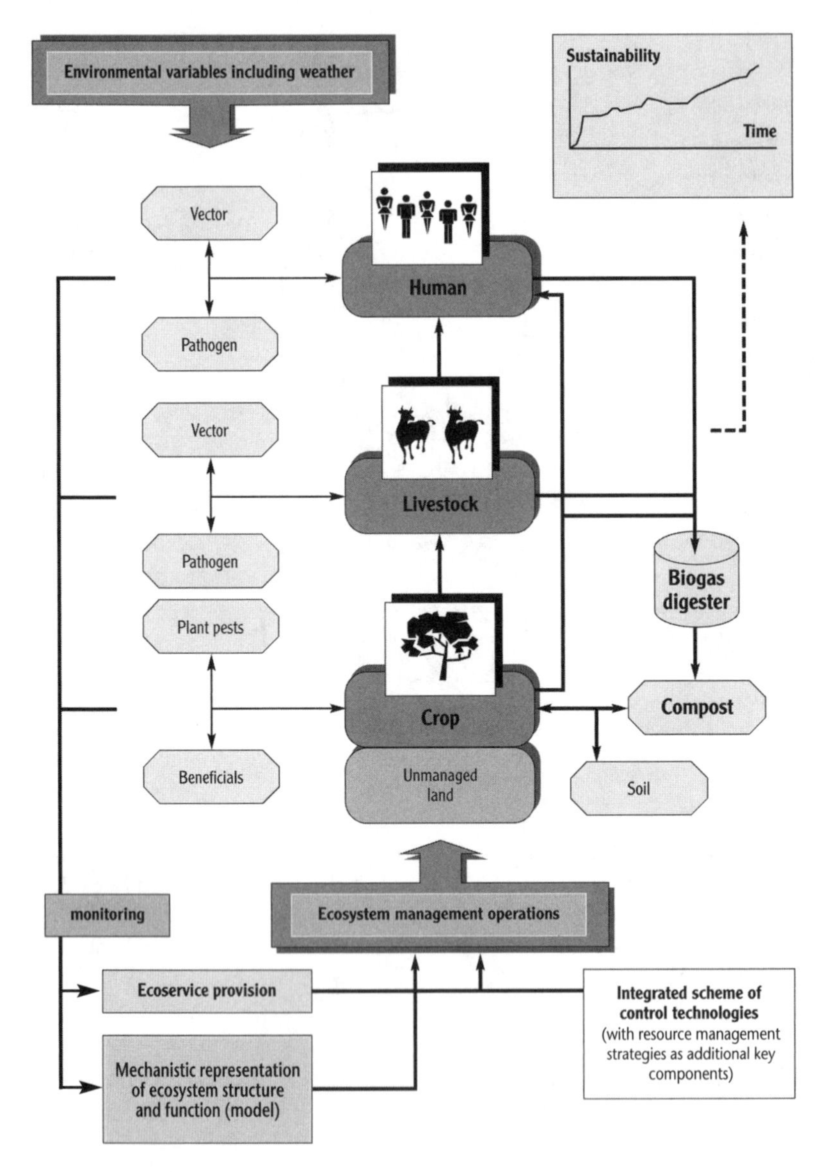

Figure 10.4. Adaptive management of an ecosystem placed on a trajectory towards enhanced sustainability. Adaptive management of subsystems constraining human and animal health is a prerequisite for further development made possible through improved ecosystem service provision (Daily and Dasgupta 2001) and use of technologies that are acceptable from ecological, economic and social standpoints. Model development and use will improve the knowledge of ecosystem structure and function, provide decision support and assist in impact assessment (Gilioli and Baumgärtner 2007). (Excerpted from Koul and Cuperus (2007), reproduced and modified with permission.)

Ecosystem management as presented here addresses all the eight Millennium Development Goals (UNDP 2003), especially Goal 1 (eradicate extreme poverty and hunger), Goal 4 (reduce child mortality), Goal 5 (improve maternal health), Goal 6 (combat HIV/AIDS, malaria, and other diseases), and Goal 7 (ensure environmental sustainability). Moreover, the incorporation of the project into activities of Praxis Ethiopia, an alliance of scientists working to assist Ethiopian institutions in their wealth-creation efforts, corresponds to Goal 8 (develop a global partnership for development). Adaptive management (learning by doing, with emphasis on participatory activities and the involvement of schools) contributes to the achievement of universal primary education Goal 2. Preliminary observations indicate that women benefit from the approach because of emphasis given to institutional diversity and participatory activities Goal 3, but further work is necessary to confirm the effectiveness of this aspect.

Conclusions

Holistic approaches to agricultural research and implementation include multilevel systems built on the presumed hierarchical organization of nature and structured institutions dealing with ecosystem management, including addressing problems of agricultural pests. These systems provide an overview of areas for management and facilitate the organization of activities as well as the assignment of responsibilities to institutions involved. They recognize the role of all elements in the ecosystems of agricultural fields, as well as the surrounding ecosystems, that contribute to the overall productivity and sustainability of the agroecosystem.

The emphasis on ecosystem management for sustainability enhancement is particularly relevant in the context of environmental degradation and depletion of natural capital. Drought, soil erosion, desertification, and other environmental problems are already driving more people from their homelands than war and political or religious persecution (Habek 2003).

The adaptive pest management approach presented in this chapter as part of a comprehensive ecosystem management approach (Fig. 10.4) is widely applicable, but appears particularly relevant to places where existing knowledge about specific ecological systems is limited and uncertainties in management are high (Holling 1978).

The proposed adaptive ecosystem management approach to addressing agricultural pests has several features of interest to research and development at international research centers such as ICIPE. First, it provides a conceptual framework for the integration of diverse technologies. For example, ICIPE's tsetse mass trapping technology as well as its biological control methods are no longer used in isolation but are inserted into an ecosystem management framework.

Because institutions are involved at multiple levels, from local to national and international, pest management is no longer executed in isolation but becomes part of a national development agenda (Aseffa Abreha et al. 2003). Thereby, the efforts of all institutions, including farmers and policymakers dealing with ecological system management, are combined and oriented toward the common goal of sustainability, a critical element of ecoagriculture.

References

Ahl, V. and T.F.H. Allen. 1996. *Hierarchy Theory: A Vision, Vocabulary and Epistemiology*. Columbia University Press, New York.

Allen, T.F.H. 2004. A summary of the principles of hierarchy theory. Unpublished document by Prof. T.F. Allen, University of Wisconsin, Madison.

ARS (Agricultural Research Service). 2002. *National Programs Integrated Farming Systems: FY 2002*. Report of The Agricultural Research Service (ARS), US Department of Agriculture, Washington, DC. www.ars.usda.gov/Research/Programs/html.

Aseffa Abreha, A., A. Getachew Tikubet, and J. Baumgärtner (eds.). 2003. *Resource Management for Poverty Reduction Approaches and Technologies*. Ethiopian Social Rehabilitation and Development Fund, Addis Ababa, Ethiopia.

Baumgärtner, J., A.O. Pala and P. Trematerra. 2007. Sociology in IPM. In: *Ecologically-based Integrated Pest Management*. Koul, O. and G. Cuperus. ed. CABI, Wallingford, UK: 154–179.

Baumgärtner, J., M. Bieri, G. Buffoni, G. Gilioli, H.N.B. Gopalan, J. Greiling, A. Getachew Tikubet, and I.M.C.J. Van Schayk. 2001. Human health management in sub-Saharan Africa through integrated management of arthropod transmitted diseases and natural resources. *Cadernos De Saúde Pública (Reports in Public Health)* 17(Suppl):17–46.

Baumgärtner J., G. Gilioli, D. Schneider, and M. Severini. 2002. The management of populations in hierarchically organized systems. *Notiziario Protezione della Piante* 15:247–263.

Baumgärtner, J. and D. Schneider. 2006. Scale and hierarchy in integrated pest management. In: *Encyclopedia of Entomology*, ed. J. Capinera. Kluwer, Dordrecht, NL: 1954–1958.

Baumgärtner, J., F. Schulthess, and Y.L. Xia. 2003a. Integrated arthropod pest management systems for human health improvement in Africa. *Insect Science and Its Application* 23:85–98.

Baumgärtner J., G. Tikubet, Gilioli, G. and M. Bieri. 2003b. Managing Ecosystems to improve human health and alleviate poverty. In: *Resource Management for Poverty Reduction Approaches & Technologies*. A. Aseffa, A. Getachew Tikubet and J. Baumgärtner (Eds.), Ethiopian Social Rehabilitation and Development Fund, Addis Ababa, Ethiopia: 179–186.

Baumgärtner, J., A. Getachew Tikubet, M. Girma, A. Sciarretta, B. Shifa, and P. Trematerra. 2003c. Cases for adaptive ecological systems management. *Redia* 86:165–172.

Becker, C.D. and E. Ostrom. 1995. Human ecology and resource sustainability: the importance of institutional diversity. *Annual Review of Ecology and Systematics* 26:113–133.

Berkes F., J. Colding and C. Folke. 2003. *Navigating Social-ecological Systems: Building Resilience for Complexity and Change*. Cambridge University Press, Cambridge, UK.

Brown, K. 2003. Integrating conservation and development: a case of an institutional misfit. *Frontiers in Ecology and the Environment* 1:479–487.

Brundland, G. 1987. *Our Common Future: The World Commission on Environment and Development.* Oxford University Press, Oxford, UK.

Christensen, N.L., A.M. Bartuska, J.H. Brown, S. Carpenter, C. D'Antonio, R. Francis, J.F. Franklin, J.A. Macmahon, R.F. Noss, D.J. Parsons, C.H. Peterson, M.G. Turner, and R.G. Woodmansee. 1996. The report of The Ecological Society of America Committee on the Scientific Basis for Ecosystem Management. *Ecological Applications* 6:665–691.

Comiskey, J.A., F. Dallmeier, and A. Alonso. 1999. Framework for assessment and monitoring of biodiversity. In: *Encyclopedia Of Biodiversity*, vol 3, ed. S. Levin. Academic Press, New York.

Conway, G. 1984. Introduction. In: *Pest and Pathogen Control. Strategic, Tactical, and Policy Models,* ed. G. Conway. International Institute for Applied Systems Analysis, Wiley, Chichester.

Daily, G., and S. Dasgupta. 2001. Ecosystem services, concept of. In: *Encyclopedia of Biodiversity*, vol. 2, ed. S. Levin. Academic Press, San Diego: 353–362.

Delucchi, V. 2001. *La Biodiversità L'Agricoltura Ecocompatibile.* Atti dell'Accademia Italiana di Entomologia. Rendiconto, Anno 49-2001.

Dent, D. 1995. *Integrated Pest Management.* Chapman and Hall, London.

Flint, M.L., and R. van Den Bosch. 1981. *Introduction to Integrated Pest Management.* Plenum Press, New York.

Getz, W., and A.P. Gutierrez. 1982. A perspective on systems analysis in crop production and insect pest management. *Annual Review of Entomology* 27:447–466.

Gilioli, G. and J. Baumgärtner. 2007. Adaptive ecosocial system sustainability enhancement in Sub-Saharan Africa. *EcoHealth* (in press).

Gilioli, G., J. Baumgärtner, and V. Vacante. 2003. Biological control as an ecosystem management tool. *Redia* 86:173–185.

Gilioli, G., J. Baumgärtner, and V. Vacante. 2005. Biological control and ecosystem services. In: *Encyclopedia of Life Support Systems (EOLSS).* EOLSS Publishers, UNESCO, Paris.

Goodland, R. 1995. The concept of environmental sustainability. *Annual Review of Ecology and Systematics* 26:1–24.

Gutierrez, A.P. 1996. *Applied Population Ecology. A Supply-Demand Approach.* Wiley, New York.

Habek, M. 2003. Hidden causalities of global warming. *Frontiers in Ecology and Environment* 1:461.

Headley, J.C. 1975. The economics of pest management. In: *Introduction to Insect Pest Management*, ed. R.L. Metcalf and W. Luckmann. Wiley, New York: 69–92.

Holling, C.S. (ed.). 1978. *Adaptive Environmental Assessment and Management.* Wiley, Chichester.

Huffaker, C.B. and B.A. Croft. 1976. Integrated pest management in the US: progress and promise. *Environmental Health Perspectives* 14:167–183.

Kogan, M., B.A. Croft, and R.F. Sutherst. 1999. Applications of ecology for integrated pest management. In: *Ecological Entomology*, ed. C.B. Huffaker and A.P. Gutierrez. Wiley, New York: 681–736.

Kotschi, J., A. Waters-Bayer, R. Adelhelm, and U. Hoesle. 1988. *Ecofarming in agricultural development*. Tropical Agroecology 2. Margraf Scientific Publishers, Weikersheim, Germany.

Koul, O. and G.W. Cuperus. 2007. *Ecologically Based Integrated Pest Management*. CABI, Wallingford, UK.

Meffe, G.K. and C.R. Carroll. 1997. *Principles of Conservation Biology*. Sinauer, Sunderland.

Metcalf, R.L. and W. Luckmann. 1975. *Introduction to Integrated Pest Management*. Wiley, New York.

Moir, W.H. and W.M. Block. 2001. Adaptive management on public lands in the United States: commitment or rhetoric? *Environmental Management* 28:141–148.

Mollison, B. 1997. *Introduction to Permaculture*. Tagari Publications, Tasmania, Australia.

Naveh, Z. and A.S. Lieberman. 1984. *Landscape Ecology: Theory and Application*. 2nd ed. Springer-Verlag, Berlin.

Oglethorpe, J. (ed.) 2002. *Adaptive Management: From Theory to Practice*. SUI Technical Series Vol. 3. IUCN, Gland, Switzerland and Cambridge, UK.

Palaniappan, S.P. and K. Annadurai. 1999. *Organic Farming: Theory and Practice*. Pawan Kumar Sharma Scientific Publishers, Jodhpur, India.

Pielke, R.A. and R.T. Conant. 2003. Best practices in prediction for decision-making: lessons from the atmospheric and earth sciences. *Ecology* 84:135–1358.

Sciarretta, A., M. Girma, L. Balayun, G. Tikubet, and J. Baumgärtner. 2005. Development of an adaptive tsetse fly population management scheme for the Luke Community, Ethiopia. *Journal of Medical Entomology* 42:1006–1019.

Stark, J.D. and J.E. Banks. 2003. Population level effects of pesticides and other toxicants of arthropods. *Annual Review of Entomology* 48:505–519.

Turner, M.G., R.H. Gardner and R.V. O'Neill. 2001. *Landscape Ecology in Theory and Practice. Pattern and Process*. Springer, New York.

UNDP (United Nations Development Programme). 2003. *Millennium Development Goals: A Pact among Nations to End Human Poverty*. Human Development Report 2003. United Nations Development Programme (UNDP), New York.

Van Driesche, R.G. and T.S. Bellows. 1996. *Biological Control*. Chapman and Hall, New York.

Van Lenteren, C. 1995. Integrated pest management in protected crops. In: *Integrated Pest Management*, ed. D. Dent. Chapman and Hall, London: 311–343.

Chapter 11

Watershed Management

Meine van Noordwijk, Fahmuddin Agus, Bruno Verbist, Kurniatun Hairiah, and Thomas P. Tomich

Introduction

Land cleared for agriculture tends to be associated with a decline in watershed quality and services. Efforts to protect forestlands from "human disturbance" can, as a side-effect, maintain clean water flows, but natural forests or systems with high biodiversity are not unique in providing this service (Bonnell and Bruijnzeel 2000). This chapter, building on earlier work (Agus et al. 2004; Susswein et al. 2001; Swallow et al. 2001, van Noordwijk et al. 2004a, 2006), will review the extent to which ecoagriculture systems that are intermediate between natural forest and intensively cropped agricultural lands can maintain most if not all of the various watershed functions and also maintain considerable biodiversity.

As explored by Grove (1995), perceptions on the relationships between deforestation, subsequent changes in rainfall, land degradation, and siltation of rivers date back more than two millennia to experiences in the Mediterranean region, with the writings of Theophrastus cited as one of the earliest sources documenting these perceptions. The European colonial expansion into the tropics, and particularly their experiences in small islands such as Mauritius, strengthened their perceptions that forests generate rainfall. Yet hard evidence of a change in documented rainfall as a consequence of deforestation still hardly exists, and the association between forests and rainfall is generally the reverse of what is perceived. A recent reanalysis of rainfall patterns for Indonesia for the 1930 to 1960 and 1960 to 1990 periods (Kaimuddin 2000; Rizaldi Boer pers. comm.), for example, indicates shifts in the isohyets (zones of equal rainfall) in Indonesia that are not obviously related to local land cover change: some areas that lost forest cover became wetter, other areas that lost forest cover became drier. For Indonesia as a whole average rainfall did not change, despite the considerable loss of forest cover, but the overall circulation pattern may have changed in ways that have affected local rainfall. For the western part of Indonesia, the Indian Ocean Anomaly influences rainfall, whereas the eastern part is dominated by El Niño Southern Oscillation (ENSO) and when long dry seasons and high fire incidence are linked to years where both phenomena peak.

Although at the local scale real changes in rainfall may well have coincided with real changes in forest cover, no convincing evidence supports hypotheses about causal relationships (Bruijnzeel 2004). The way a landscape processes the incoming rainfall, however, does directly depend on the land cover, and the total amount of water, the regularity of the flow, and the quality of the water in the streams can all be directly affected by changes in cover.

One of the reasons why debates on watershed functions are confusing is probably the extrapolation of local experience to other situations, without recognition of differences between topography, climate, forest type, and characteristics of land use after forest conversion. Both the default level of all watershed functions discussed here and their sensitivity to change vary with basic landscape properties and climate.

Based on the ratio of incoming precipitation and evapotranspiration, we can distinguish between areas that are net suppliers of water in surface or subsurface lateral flow pathways, and net sink areas where plant growth and evapotranspiration are limited by the incoming precipitation, unless surface or subsurface irrigation makes up for the difference between supply and demand. In between these source and sink areas for water, we can generally find transmission zones that can affect the quality of the water and that deliver water to oceans and seas insofar as it is not used beforehand. Because rainfall usually increases with elevation, upper watersheds are often source areas (catchments), supplying the lower parts with water via rivers or groundwater flows. Ecoagriculture practices can take place in source, transmission, or sink areas, but will obviously have a different relationship to watershed services depending on the landscape position.

Domestic and industrial use of water generally affects the quality rather than the amount of water, but it may lead to increased evapotranspiration and/or reallocation between surface and subsurface flows and thus be important for the overall hydrological cycle. A comprehensive assessment of the way land uses modify the supply and demand for water may, in this light, be expected to consider the following:

1. Land-use change in source areas and its impact on the total amount, the regularity of the flow, the allocation over surface and subsurface pathways, and the quality of the water that leaves the area
2. Land-use change as well as domestic and industrial water use in the transmission zones and their impacts on water quality, mainly via the characteristics of the rivers (channels) or groundwater flows
3. Land-use in areas that are or can be supported with additional water via irrigation

Globally the human demand for clean water and for agricultural products that are supported by irrigation is bound to increase, whereas supply is relatively constant, so the main solutions will have to come from a reduction of demand via increased efficiency of water use at the agroecosystem level. Maintaining the quality and quantity of existing source areas, however, has to be part of any holistic solution.

Forest conversion and other land-use change in source areas may affect the quality and quantity of water flows through the combined effects on cloud interception, effects on rainfall patterns as such, rainfall interception by plant canopies, infiltration of the soil surface, subsequent water use for evapotranspiration, and partitioning over surface and surface flow pathways. Of these effects, the interaction between land-use mosaics and rainfall is the least understood and most speculative one. Changes from natural forests to landscapes used for agriculture or production forestry normally involve many if not all of the elements of the water balance, with a mixture of positive and negative effects. Widely held perceptions of the overriding importance of forest cover for the maintenance of watershed functions in source areas have been questioned over the last decades in hydrological research (Bruijnzeel 2004), and rather than using a forest–nonforest dichotomy, have shown that the types of land use which follow forest conversion can make a lot of difference. Land use (including but not restricted to the protection of existing forest cover) in such source areas thus has local as well as external stakeholders and beneficiaries, and increasing demands for water in the lowlands have often led to an increased sense of conflict over what happens in the source areas. Yet upper watersheds in much of the tropics provide a living for large numbers of farmers and rural communities, who have often remained outside of the mainstream of development.

This chapter answers the following questions:

1. What are watershed functions to various stakeholders?
2. To what extent can different types of agriculturally used landscapes substitute for "natural" forest in various hydrological functions under different geophysical conditions (e.g., rainfall pattern, slope, scale)?

3. To what degree do the biodiversity conservation and watershed protection agendas overlap?

Who Is Interested in Which Watershed Function?

The main functions of watersheds from a downstream perspective are for them to provide an adequate supply of high-quality water and not to provide a medium for soil transport or flash floods. Concerns for loss of watershed functions (van Noordwijk 2005) can be a combination of various elements:

- On-site loss of land productivity as a result of erosion
- Off-site concerns about water quantity: annual water yield, peak (storm) flow (lack of buffering), dry season base flow (lack of water stocks for gradual release), and groundwater depletion or excessive recharge leading to salt movement
- Off-site concerns about water quality: sediment loads, leading to siltation of lakes and reservoirs or marine coral reefs, and transport of trees and other large debris that can destroy downstream infrastructure (bridges) and floodplain houses; organic and nutrient inputs leading to high biological oxygen demand (BOD) and its risks for fish and other aquatic life forms; agrochemical residues that affect aquatic life and restrict the safety of downstream use of water; microbes and other biota (e.g., derived from domestic water use), that restrict the safe use of water downstream (e.g., *Escherichia coli*)

Not all of these functions are equally relevant in all situations; for example, once reservoirs have been constructed or occur naturally (lakes), the downstream interests will shift from a focus on dry season river flows to concerns for total water quantity and the sediment load that affects the expected life time of the reservoir. Table 11.1 provides examples of situations where these various concerns are of particular relevance.

Many of these aspects of change may be perceived as detrimental. For example, the reduction of sediment loads due to upstream reservoirs can negatively affect aquatic productivity and the fisheries industry that depends on it in river deltas and coastal zones, whereas an increase in sediment loads of previously "clean" rivers can negatively affect coral reefs and the tourist and fisheries industries that depend on those. A range of land-use change effects on services can be traced to the various terms of the plot-level water balance.

Although forests are generally associated with positive watershed functions, forests comprise at least three separate components, each with its own relevance:

1. Trees form aboveground vegetation that intercepts rainfall and uses more water on an annual basis than other (nonirrigated) vegetation types; under

Table 11.1. Examples of situations where specific watershed functions are of relevance

Watershed function	*Importance (examples)*
Water quantity	
• Dependable (high) total water yield	• Filling up lakes and reservoirs
• High dry-season flow	• Downstream users without lakes
• Low peak flow	• Flooding risk in lowland
• Adequate groundwater recharge	• Allows sustainable groundwater use elsewhere
Soil movement	
• Low sediment load of streams	• Reservoir lifetime; marine coral reefs
• Few landslides/mudflows	• Villages and towns in valleys
Good water quality	
• Suitable as drinking water,	• Direct source of drinking water
• Adequate for fish and other biota	• Fishermen, biodiversity conservation
• Low organic pollution (low BOD = biological oxygen demand)	• *Idem,* processed drinking water
• Low nutrient load	• *Idem,* processed drinking water
• Low pesticide, heavy metals. etc.	• Use as irrigation water, processed drinking water
• No subsoil salt movement	• Salinization of valleys (e.g., dry parts of Australia)

specific conditions (cloud forest) tree canopies and their epiphytes can condense water vapor in clouds and actually increase water capture beyond the additional water they use

2. Soil that has a protective litter layer, little compaction (except for animal tracks), and many soil structure–forming soil biota, with tree root turnover as an additional source of macroporosity
3. A landscape with a considerable roughness that facilitates temporary storage of surface water and a (relative) absence of channels that can deliver surface runoff to streams

Forest conversion to forms of ecoagriculture or other land use will affect all three of these aspects, but at different time scales and with different degrees of reversibility, depending on the subsequent land-use practices. The impacts of land-cover change after forest conversion may be understood from the impacts on these three aspects of forests: creating roads, paths, and other rapid-drainage channels (Ziegler et al. 2004); compacting the soil; and changing the aboveground vegetation. These three aspects of forests have different sensitivities to disturbance, as well as likely recovery times (Table 11.2).

Table 11.2. Effects of human disturbance on forests and water balance

	Terms of water balance affected	*Effects of disturbance*	*Recovery time*
Trees	Interception, transpiration	Logging and fire reduce tree cover, slash-and-burn land clearing reduces it to zero	Water use can recover in 1–3 years, Leaf Area Index (LAI) and interception in 4–10 years; tree biomass will take decades and species composition a century or more
Forest soil	Rate of surface infiltration, percolation, and subsurface lateral flow, surface evaporation	Compaction, decline due to loss of macropore formation	Surface permeability can be restored in < 1 year, soil macroporosity may take decades depending on the degree of compaction
Forest landscape	Time available for surface infiltration, percolation, and subsurface lateral flow	Paths, tracks, and roads lead to quickflow, leveling and swamp drainage reduce surface buffer storage capacity	Drains, channels, and gullies can be closed, and surface roughness restored rapidly through specific actions

The various functions thus depend on the landscape position and perspective of the downstream stakeholder, with a single intervention often ramifying to both positive and negative effects, with different time frames.

How Do Various Watershed Functions Depend on the Land-Use Mosaic?

Causal Hierarchy of Landscape Topography and Vegetation in Watershed Functions

Unlike water movement in unsaturated soil, which is mainly vertical, surface runoff and groundwater move mainly laterally. Thus any change in land use at a plot scale that affects infiltration or recharge is likely to have effects at the landscape scale, beyond the plot, via runoff and groundwater movement. Standard representations of the water balance at plot scale include connections to lateral flows over the surface, through the upper layers of the soil profile, and as groundwater. These three types of lateral flow hydrologically connect any plot to its landscape context. Movement of water leads to lateral movement of soil, nutrients, and other solutes such as salts, which can cause a range of generally negative environmental effects downhill or downstream (although under some circumstances inflows of soil and nutrients are perceived as positive). The three lateral flows mentioned in fact represent a continuum of flow pathways, with very different residence times. Surface flows of water runoff and run-on are directly visible, can lead to substantial redistribution of soil and light-fraction organic residues, and are generally considered under the headings *erosion* and *sedimentation*. These surface flows respond quickly (in seconds to minutes) to current rainfall intensity, and the water's pathway can be easily modified through surface roughness and management of surface litter (Coughlan and Rose 1997).

By contrast, groundwater movement is measured in days, months, years, or decades and responds to the cumulative balance of rainfall and evapotranspiration, rather than to extreme events. The pathways of groundwater movement can be much less influenced than those of surface flows, and there is generally a considerable time lag between any management intervention and its effects. This means that problems are not directly apparent, but once they appear, little can be done about the problems in the short run. These characteristics of groundwater problems at the landscape scale have consequences for the degree to which and the way in which natural resources can be managed (Lovell et al. 2002). Situated between the extremes of surface and deep subsoil movement of water, issues of subsurface flows of water and solutes have received relatively little attention. The spatial and time scales at which these flows operate, however, make them more amenable to management interventions than groundwater flows, yet less obvious than surface movements.

In a review of the literature, Ranieri et al. (2004) focused on the biophysical aspects of lateral water movement, its consequences for the movement of solutes and soil, and how different types and arrangements of land use can affect these types of lateral flow. The authors distinguished four levels of intervention in the causal pathway between plot-level land cover and external environmental effects:

1. Influences via the interrelated terms of the water balance equation that determine the total amount of water leaving a plot (rainfall + lateral inflows – evapotranspiration – changes in storage), and its partitioning over surface, subsurface, and deep pathways
2. Partial decoupling of the flow of water and that of soil, nutrients, or salt through forms of bypass flow
3. Filters or interception of the lateral flows of soil, nutrients, or salt through changes in the rate of flow of the carrier (water flow) or concentration by processes such as sedimentation, uptake, sorption, and precipitation;
4. Interventions that mitigate the environmental effects of the lateral flows at their point of reemergence at the surface.

At level 1, land cover influences the pathway for the excess of rainfall over evapotranspiration. Infiltration depends on characteristics of the soil surface and topsoil and hence on the balance between soil structure formation due to root turnover and soil biological activity fed by litter inputs, as well as on water use by plants that increases the amount of water that can infiltrate to fill up to field capacity.

At level 2 the dynamic aspects of soil structure also influence the degree of bypass flow that decouples nutrient transport from the mass flow of water. For surface flows such decoupling may happen if water is directed through channels with a firm bed. Bypass flow for groundwater may occur once all salt in preferential flow pathways is washed out, and will last as long as the amount of groundwater flow remains unchanged.

Level 3 involves filters of various types. The term *filter* is used here in a generic sense of anything that can intercept a vertical or lateral resource flow (van Noordwijk et al. 2004a).

Erosion and Loss of Land Productivity

The potential role of agroforestry interventions in open-field agricultural systems as a way to reduce the rate of soil loss has been well documented for many situations (Garrity 1996; Penning de Vries et al. 1998; Young 1997). Extrapolation of such experimental results to other soils, topographies, and rainfall regimes, however, is still difficult because much of the variation in sediment loss between rainfall events remains unaccounted for (Rose and Yu 1998; van Roode 2000). Because most of the soil loss tends to occur during extreme events on an above-annual (e.g., decadal) time scale, the assessment has to be one of risk factors, rather than predicting actual rates of soil loss.

Quantity and Timing of Water

For watershed functions on quantity and timing of water delivery, we have to predict how annual water yield, peak (storm) flow, dry season base flow, and groundwater recharge or depletion change with land use.

For total water yield, a simple water balance that includes (spatially averaged) rainfall and (spatially averaged) water use by vegetation (plus evaporation from bare soil) will provide realistic estimates when sufficiently long periods of time are considered. For forests we may expect an additional annual water use compared to grassland or cropped fields of the order of 300 mm $year^{-1}$ (Calder 2002). A first estimate of total water yield can be based on a proportional average over the land cover types, with little impacts of the spatial pattern. Lateral flows of hot air (advection) can, however, increase the transpirational demand per tree in semiopen landscapes, whereas lateral flows of water can provide the water to meet this demand. Total water use may therefore be fairly independent of tree cover over a substantial range.

The partitioning between quick flow and base flow essentially depends on the rate of infiltration. Three types of control can determine the pathway of water flows, and can be modified by land use:

1. If rainfall intensity exceeds current infiltration rate, water will accumulate on the surface, and as soon as the storage capacity determined by local roughness and slope is exceeded, it will start to run off over the surface ("infiltration-limited runoff").
2. The soil profile with its macropores allows water to reach deeper layers, recharge the soil to field capacity, and percolate to the subsoil; in many tropical soils clay content increases with depth, and saturated conductivity decreases. This situation leads to the possibility of lateral subsurface flows on slopes, which can contribute to the quick flow of streams.
3. Saturation overland flow occurs if the local storage and percolation capacity of the soil is exceeded. Under certain conditions a "perched water table" may be formed, leading to saturation overland flow before the whole soil profile is rewetted, leading to base flow of streams and/or recharge of deep aquifers.

The frequency and intensity of variations in river flow can have a plot-level explanation, based on a slow transmission of water to streams, or a landscape-level explanation, based on low spatial correlation of rainfall events within the catchment. At scales where the second type of explanation dominates, the potential influence of land-use change on river flow will on average be reduced.

Sediment Load and Water Quality

Sediment loads can lead to siltation of reservoirs and transport of trees and other large debris that can destroy downstream infrastructure (bridges) and flood-plain houses. The potential for agroforestry or other land uses to reduce sediment loads to streams by providing vegetative filter zones may exceed its potential to reduce on-site erosion. Filter effects for suspended sediments in overland flow can be based on infiltration of water or on a reduction in speed. It thus depends on the impacts of land use on soil macroporosity and potential infiltration rates (as discussed earlier), on the presence of a litter layer with good contact cover (fraction of the soil surface directly covered) to slow down the rate of overland flow, and on the absence of channels. Footpaths and motorbike trails can cut through any vegetative filter and provide pathways for direct sediment transfers to streams in any land use, indicating that some controls are outside of the typical agricultural domain.

Modeling Watershed Functions from Plot to Landscape Scale

Rather than aiming at general conclusions about watershed functions, landscape managers need tools to help them assess whether specific land uses in a specific climate and topography can maintain valued watershed functions. Models that include slope, differential soil properties, measured or simulated rainfall sequences, and the above- and below-ground changes that can be attributed to agroforestry systems, compared to natural vegetation or intensively cropped lands, can be used for such an assessment. Many modeling approaches exist, some with a strong emphasis on spatial patterns (like most geographic information system software), some strong on dynamic processes (including soil physical models that operate on a time step of seconds to redistribute water after rainfall events), and some with a more balanced attention for patterns and process. On the basis of current understanding, Table 11.3 specifies the type of inputs that are probably needed for prediction of the various watershed functions discussed so far. No single model can deliver all of these outputs for the relevant range of scales, so we have pieced it together from different models with their respective strengths and weaknesses. The following section discusses some agroforestry results for three models that operate from plot to landscape scale.

Predictions of Watershed Functions of Agroforestry Systems at the Plot Scale

The WaNuLCAS model, a model of water, nutrient and light capture in agroforestry systems (Fig. 11.1; van Noordwijk and Lusiana 1999; van Noordwijk et al. 2004b; http://www.icraf.cgiar.org/sea/AgroModels/WaNulCAS/index.htm), can be used to explore agroforestry options for long-term landscape filter functions because it includes runoff–run-on and the relevant subsurface flows of water, retention on absorption sites, uptakes by trees and other vegetation, and an organic matter balance. The model can be used to explore impacts of the width and spacing of the filter strips on capture and residence time and to describe the nutrient balance on a medium time scale (5–20 years), depending on tree management practices.

The dynamics of soil structure under the influence of land use may have to be taken into account in evaluating models of watershed functions in

Table 11.3. Relating watershed functions to the main "drivers"

Watershed function	*Mean (monthly) rainfall*	*Intensity frequencies of rainfall*	*Spatial correlation of rainfall*	*Land cover fractions => water use*	*Spatial organization of landscape => filters*	*Infiltration, subsurface flows*	*Streambed characteristics*
Total water yield	★★★			★★★			
Dry-season flow	★★	★		★★	★	★★	
Low peak flow	★★	★★	★★	★	★	★	★★
Groundwater recharge	★★★			★★		★★	
Sediment load of streams	★★	★★	★	★	★★	★★	★★
Risk of landslides, mudflows	★★	★★	★★	★★	★★	★★	★★
Water quality	★★	★★	★	★★	★★	★★	★★
Primary determinant	Climate, topography	Climate, topography	Climate, topography	Land use	Land use	Topography and land-use history	Topography and land use

★★★ = column heading is a primary driver of row function; ★★ = column heading is a secondary driver of row function; ★ = column heading is a minor driver of row function; blank = column heading is not relevant to row function

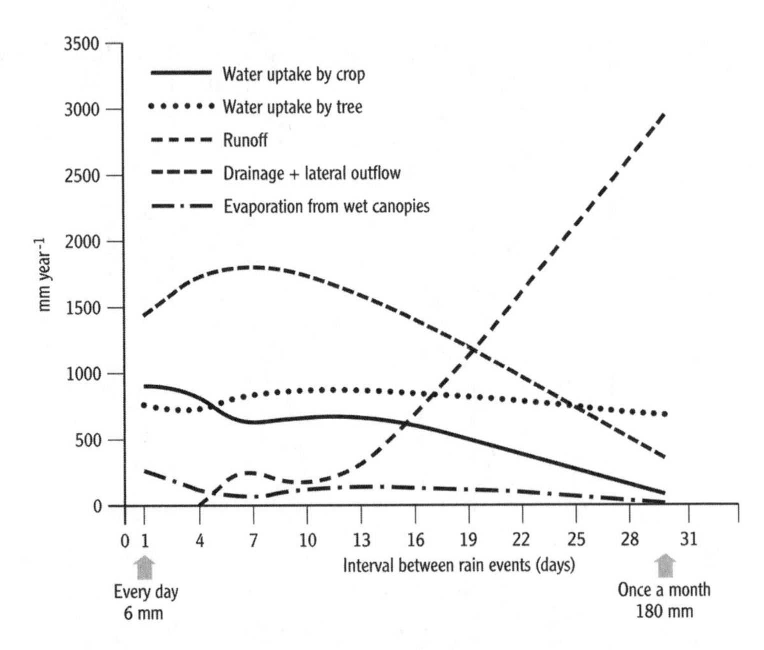

Figure 11.1. WaNuLCAS model predictions of water balance in response to simulated rainfall patterns

ecoagriculture landscapes. Factors reducing water infiltration include collapse of soil aggregates, internal slaking fills in channels in the soil, and compaction.

Water Erosion Prediction Project Model of Watershed Functions of Agroforestry Systems at the Hillslope Scale

The Water Erosion Prediction Project (WEPP) model (Flanagan and Nearing, 1995) is a physically based erosion model that can predict on-site and off-site effects of erosion. WEPP simulates plant growth and climate during the year, based on user specifications. Some of the most sensitive parameters for WEPP are the soil cover by dead or live biomass. WEPP has been calibrated and validated for "standard" crop fields and cannot yet deal with the complexities of multiple-component agroforestry systems. However, it is relatively easy to simulate spatial sequences of different land uses on a single hillslope for given soil and climate properties (Ranieri et al. 2004).

Besides the quantity of trees, their arrangement on the hillslopes also affects sediment yield, soil loss, and runoff control (van Noordwijk et al. 1998). Using this model, scenarios were created with hillslopes 500 m long and a 40% slope. Seven percentages of forest coverage on the hillslopes were tested (0, 6, 12, 25,

50, 75 and 100% coverage). Zero percent coverage represents a hillslope that has only clean-weeded coffee plantations, and 100% represents a hillslope with only natural forest. The intermediate values of forest coverage each had four types of tree arrangements. The first one divided the forest in contouring strips in the middle of the coffee trees. The second located the forest on the top and on the bottom slope, with coffee on the middle. The other arrangements were forest only on the bottom slope (riparian buffer) or only on the top slope. The effect of quantities and arrangements of forests were analyzed for on-site (soil loss) and off-site (sediment yield) effects of erosion, besides runoff control (Figs. 11.2a and 11.2b).

The best arrangement to control sediment yield was found for forest concentrated on the bottom slope (riparian forests), probably due to its capacity to trap sediments (filter effect). Arrangements with forest inserted between coffee plantations had almost the same efficiency as the best arrangement to control sediment yield. On the other hand, forests only on the top slope had the worst result in terms of sediment yield control.

Under the best arrangement, just 25% of forest coverage on the bottom slope could trap 93% of the sediments produced on the hillslope with clean-

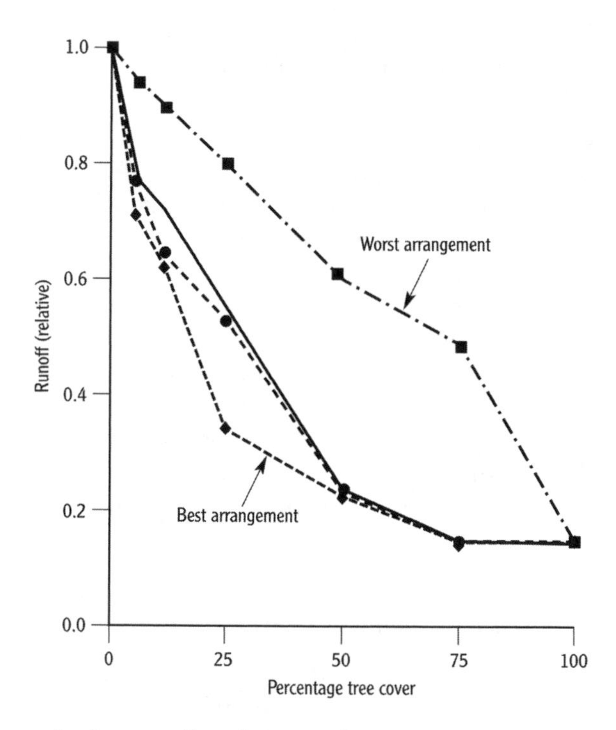

Figure 11.2a. Surface runoff as a function of percentage tree cover

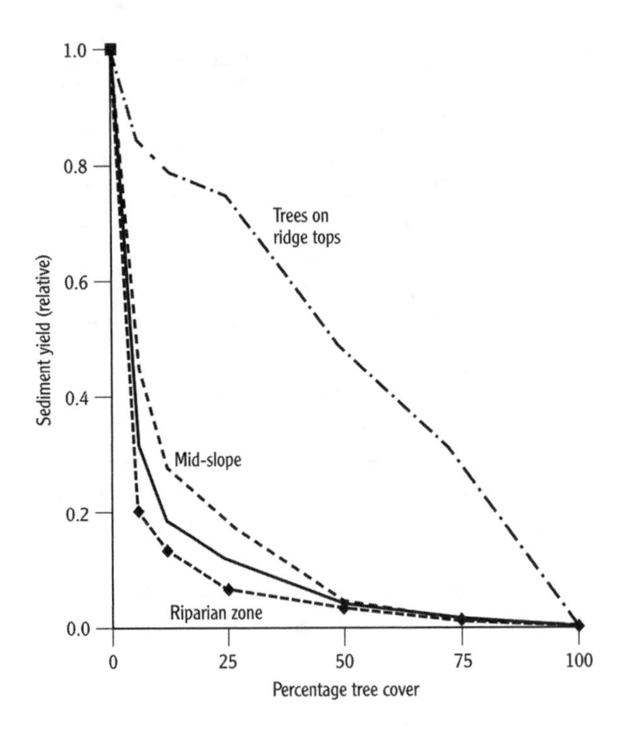

Figure 11.2b. Sediment yield as a function of percentage tree cover (modified from Rameri et al. 2004)

weeded coffee. Efficiency of tree strips on erosion control (on-site effects) had different behavior than for sediment yield control (off-site effects). For 25% forest coverage, the arrangement "forest after coffee" was more effective to control sediment yield, but the arrangement "coffee after forest" had best results in terms of soil loss (internal soil movement).

Outcomes for runoff reduction, like for sediment yield control, were best with trees on bottom slopes, except for two cases. At 50% coverage, the best arrangement was trees on the middle slope (three strips), and at 6% coverage, the best one was trees on the middle slope (five strips). Also, for runoff reduction, almost all the arrangements with 75% tree coverage were as efficient as those with 100% coverage, except for forest on the top slope.

FALLOW Predictions of Watershed Functions of Agroforestry Systems in Landscape Mosaics

The FALLOW (forests, agriculture, low-value lands or waste) model (van Noordwijk 2002; available on http://www.icraf.cgiar.org/sea/AgroModels/FALLOW/Fallow.htm) represents a landscape mosaic with shifting cultivation

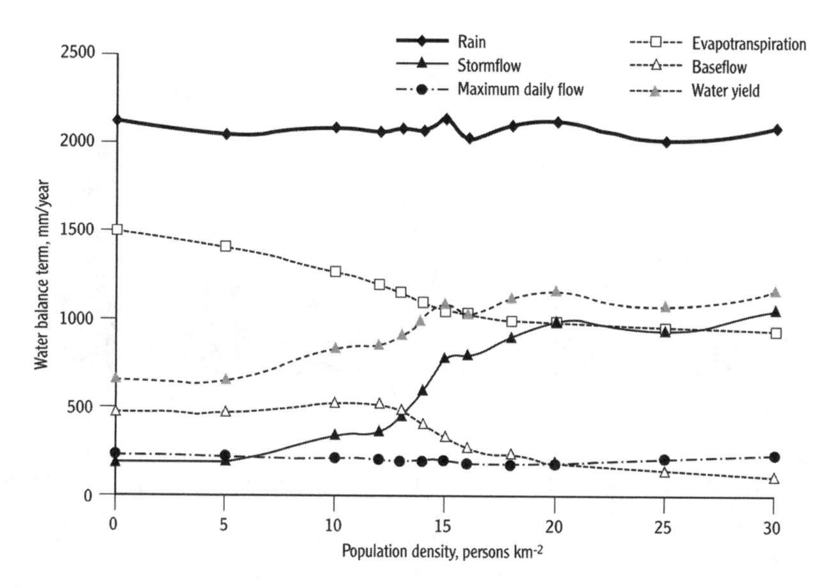

Figure 11.3. FALLOW model predictions on watershed functions in relation to population density

and crop–fallow rotations, that predicts food self-sufficiency, soil fertility, carbon stocks, plant species richness, and watershed functions, on the basis of several biophysical and management parameters, for a 100 grid-cell landscape. The example here explores the watershed functions, at a range of population densities, and under the influence of the "physical degradation" parameter for cropped fields (Fig. 11.4).

An example of the model output (Fig. 11.3) across a range of population densities shows that predicted total water yield of a subwatershed will increase if more people live there, but that this increase is based on a slight initial increase in baseflow due to a decrease in water use by the vegetation while the infiltration capacity in the landscape is still intact, and a more drastic increase due to stormflow, with a decline in baseflow, at higher population densities.

Net sediment loss increases along with stormflow as the filter functions decrease with increasing cropping intensity. The maximum daily peakflow, however, is virtually independent of land cover. The switch from baseflow to stormflow depends on the physical degradation of the soil during the cropping period. The shift from baseflow to stormflow in the model output is accompanied by an increase in predicted net sediment loss.

The model is a first approximation only, but it demonstrates that the basic concepts can be operationalized, and it points at sensitivity for specific parameters.. A hypothetical landscape was constructed in such a way that four strategies could be compared for a 25% forest reserve:

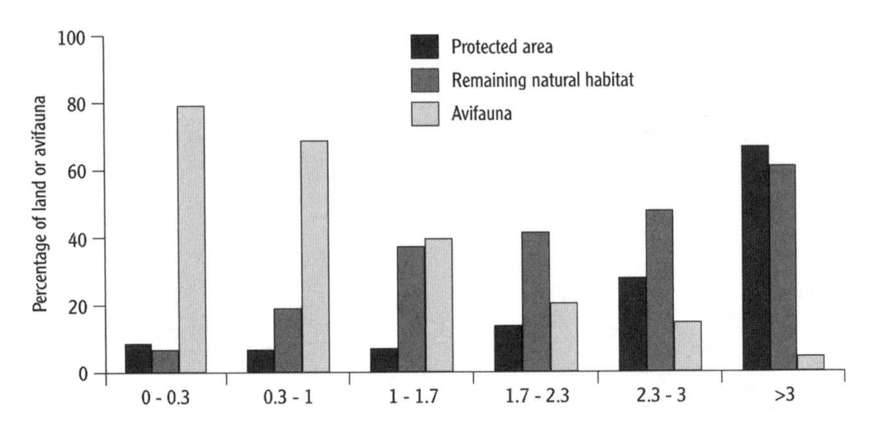

Figure 11.4. Mismatch in the Philippines between protected areas and remaining natural habitat (as percentage of land area in that elevation class) and bird species richness (percentage of Philippine fauna by elevation class) (based on MacKinnon, 2001)

1. Protection of the riparian zones
2. Protection of the steepest slopes
3. A forest reserve on the poorest soils of the landscape
4. A random allocation of 25% cells to the protection class

Simulations were made at two population densities: 12 and 15 persons km^{-2}, because these are values just within and just exceeding the carrying capacity at the technological and yield level of the default parameter settings of the model. Results recorded include the gross erosion per plot, filtered amounts, and net sediment loss.

The results were surprising at first sight, in that the lowest net sediment loss was predicted for scenarios where the protection forest was allocated on the poorest soils of the landscape, rather than on the riparian zone (maximizing filter effects) or the steepest slopes (minimizing gross erosion). Random allocation of forest reserves (only three replicates are shown) predicts a considerable variation in outcome. Only allocation of forest reserves to the poorest soils appears to be consistently better than a random allocation at 12 persons km^{-2}while at 15 persons km^{-2}all three strategies tested led to less net sediment loss than a random choice of forest reserves.

On further analysis, however, it appears that the average cropping ratio achieved has a dominant influence on the gross erosion rates. The relatively small effect of riparian forest may derive from the measures used, which assign similar potential filter capacities to most noncropped land-use classes; thus a bushy early fallow phase may be as effective a filter as an old-growth forest in the current default settings.

Allocation of forest reserves to the poorest soils allows the farming households to raise the food they need on the smallest fraction of the land and therefore minimizes net sediment loss, according to the model. The model, which includes a reasonable set of responses of farm households to the environmental conditions that may develop during the simulation, thus predicts that the immediate impacts on reducing erosion or maximizing filter functions can be easily overruled. Of course, the way in which erosion and filter effects are described in the model, both as regards model structure and the parameter settings used for the default version of the model, may need further scrutiny before these simulation results are accepted. Qualitatively, however, results point to an important feedback loop by which landscapes with farmers can differ from the expected results of land-use planning.

Partial Overlap between Watershed Functions and the Biodiversity Agenda

These examples of models of land-use systems at plot, hillslope, and landscape mosaic level have indicated that the potential role of agroforestry in maintaining watershed functions varies with the specific function of interest, but that much of it hinges on the changes of soil physical properties over time under various types of land use. Because soil biota play a key role in re-creating soil macroporosity, to counteract the continuous decline by physical collapse, we may expect some relationship between below-ground biodiversity and the maintenance of watershed functions in agroforestry. However, it is more logical to seek relationships with presence or absence of specific groups of soil biota, rather than with aggregate properties such as biodiversity.

The manifestation of biodiversity and watershed concerns across scales (from plot to global) shows important differences (Table 11.4).

Biodiversity conservation is primarily a global problem, in that it can be achieved only if all ecoregions that contribute unique elements to the global diversity have a sufficiently effective level of local conservation. For watershed functions, however, global relations, outside of the lateral flow domains of river systems, are relatively weak. The rate and timing at which rainfall is recycled to the atmosphere via evapotranspiration may have some feedback on rainfall outside of the catchment area of the same river, but such effects are as yet poorly understood. Modeling studies that assume very radical land-cover change over vast areas (such as conversion of the whole Amazon Basin to treeless grasslands) still produce relatively small effects on rainfall that would be difficult to detect in the real world given the inherent variability of rainfall patterns.

At the other side of the scale spectrum, we see that plot-level changes of infiltration rates and evapotranspiration are the dominant features controlling

Table 11.4. Biodiversity conservation and watershed functions

	From a watershed perspective	*From a biodiversity perspective*
Global/continental	Feedbacks to global climate of changes in watershed areas are weak and as yet uncertain	Global conservation requires conservation in every ecoregion regardless of watershed functions (Olson et al. 2001)
National/river system	Criteria for "critical watersheds" depend largely on downstream interests; impacts of sedimentation on freshwater biota and coral reefs may be the main direct linkage between watershed and biodiversity	Underrepresentation of lowlands in protected area networks (MacKinnon 2001, Fig. 14)
Landscape	Spatial organization of the landscape is more important than relative land cover fractions	Riparian forests and ridge protection can provide corridors
Plot	The key role for the surface litter layer and the soil biota this supports and that generates soil porosity and infiltration capacity	Differential tolerance by organisms to human disturbance determines the biodiversity value of a plot

many watershed functions, whereas plot-level species richness does not tell us much about the relevance of that plot for biodiversity conservation, nor of its stability or level of ecosystem functioning. Interestingly, the landscape scale is the one where the biodiversity and watershed agendas have the most in common: the spatial organization of a landscape has a major modifying effect on a number of watershed aspects (especially where overland flow of water and sediment is concerned, and to a smaller extent where subsurface lateral flows can be intercepted by rows of deep-rooted trees with high evaporative demand), while it also provides the corridors and dispersal/migration pathways for many biota that are important for aspects of biodiversity. Between landscape and global levels of scale there is some convergence between a watershed and a biodiversity agenda, in that increased sediment loads from rivers can be a major threat to freshwater biota downstream and to biodiversity-rich coral reef systems in the sea near its estuary. However, we also see a de facto separation in that in many if not most countries the current emphasis for the protected area system is biased toward mountains and other areas of low potential agricultural productivity and high presumed importance for watershed functions, but not necessarily to the places that would be richest in biodiversity values.

Figure 11.4 illustrates this point for the Philippines, where the fraction of lands conserved by elevational class is inversely related to the relative bird richness of these classes. Lowland tropical rain forest is, or used to be, the ecosystem with highest species richness (at average plot, but also at ecosystem scale), but is poorly represented in the protected area network, at least in Southeast Asia and Africa.

Discussion

Watershed functions are easier to maintain in agricultural landscapes than the more sensitive parts of local flora and fauna, which rate highest in the potential contribution of an area to global biodiversity conservation. The watershed function and biodiversity conservation agenda thus require separate attention, and we cannot expect ecoagriculture or agroforestry to automatically contribute to both types of environmental service.

Over the past few decades integrated approaches to the social, economic, institutional, and biophysical aspects of watershed protection in source areas have been developed and tested, with various degrees of participation by the respective stakeholders. Although substantial progress has been made in a number of cases, the overall effectiveness of interventions has stayed below expectations. Factors contributing to this include:

- Overoptimistic views on land-use planning in situations with weak governance and without an incentive structure
- The need for communal action in protecting the overall water resource in situations where local institutions are weak
- Complex social relations between lowland and upland beneficiaries, with unequal access to national institutions and government, and cultural perceptions that complicate situations
- Incomplete biophysical understanding of the real impacts on watershed functions of land-use change, because most research and public debate have focused on a forest–nonforest dichotomy and have failed to recognize the diversity of (agro)ecosystems

Biophysical aspects that have received inadequate attention of research and models so far include:

- Scaling rules for watershed functions in landscape mosaics, including forests, agroforestry, and agriculture, where properties such as infiltration rates and

transport capacity for subsurface flows as well as evapotranspiration vary according to different land uses and their watershed functions
- The time dimension of changes, both for the gradual degradation that may be the result of land-use change but that is less noticeable than immediate effects, and for the process of rehabilitation that is likely to be much slower for the below-ground aspects than it is for the above-ground impacts of vegetation
- The important role of extreme events in the biophysical functioning of watershed areas as well as the real and perceived risks for downstream (floodplain) stakeholders, whereas research tends to focus on processes that occur within its funding horizon
- The elusive interactions between land use and rainfall patterns as part of local climate change, the scope for and likely hydrological consequences of increasing tree densities in landscapes, as suggested as part of a mitigation strategy for reducing the increase in atmospheric CO_2concentrations

Watershed management within an "ecoagriculture" context should be able to meet all reasonable downstream expectations, but it does involve a number of tradeoffs that may require landscape level reward schemes. Indicators for adaptive management need to be combined with appreciation for the "slow variables," mostly based on soil changes.

References

Agus, F., Farida, and M. van Noordwijk (eds.). 2004. *Hydrological Impacts of Forest, Agroforestry and Upland Cropping as a Basis for Rewarding Environmental Service Providers in Indonesia.* Proceedings of a workshop in Padang/Singkarak, West Sumatra, Indonesia. World Agroforestry Centre (ICRAF), Bogor, Indonesia http://www.worldagroforestry.org/sea/Publications/Files/proceeding/PR0024-04/PR0024-04-1.ZIP.

Bonnell, M. and L.A. Bruijnzeel. 2000. Forests–water–people in the humid tropics: past, present and future. Hydrological research for integrated land and water management. Presented at the International Symposium and Workshop, Kuala Lumpur, Malaysia, July 30 to August 4 2000, http://www.nwl.ac.uk/ih/help/kl/index.html.

Bruijnzeel, L.A. 2004. Hydrological functions of tropical forests: not seeing the soil for the trees? *Agriculture, Ecosystems and Environment* 104:185–228.

Calder, I.R. 2002. Forests and hydrological services: reconciling public and science perceptions. *Land Use and Water Resources Research* 2:2.1–2.12. www.luwrr.com.

Coughlan, K.J. and C.W. Rose. 1997. *A New Soil Conservation Methodology and Application to Cropping Systems in Tropical Steeplands: A Comparative Synthesis of Results Obtained in ACIAR Project PN 9201I.* ACIAR Technical Reports No 40. ACIAR

Flanagan, D.C. and M.A. Nearing. 1995. *USDA-Water Erosion Prediction Project: Hillslope Profile and Watershed Model Documentation.* 1st ed. USDA-ARS-MWA-SWCS, West Lafayette, IN.

Garrity, D.P. 1996. Tree–soil–crop interactions on slopes. In: *Tree–Crop Interactions: A Physiological Approach*, ed. C.K. Ong and P.A. Huxley. CAB International, Wallingford, UK: 299–318.

Grove, R.H. 1995. *Green Imperialism: Colonial Expansion, Tropical Island Edens and the Origins of Environmentalism, 1600–1860.* Cambridge University Press, Cambridge.

Kaimuddin, 2000. *Dampak perubahan iklim dan tataguna lahan terhadap keseimbangan air wilayah Sulawesi Selatan.* PhD thesis, Program Pascasarjana, Institut Pertanian Bogor, Bogor, Indonesia.

Lovell, C., A. Mandondo, and P. Moriarty. 2002. The question of scale in integrated natural resource management. *Conservation Ecology* 5(2):25. http://www.consecol.org/vol5/iss2/art25.

Olson, D.M., E. Dinerstein, E.D. Wikramanayake, N.D. Burgess, G.V.N. Powell, E.C. Underwood, J.A. D'amico, I. Itoua, H.E. Strand, J.C. Morrison, C.J. Loukes, T.F. Allnutt, T.H. Ricketts, Y. Kura, J.F. Lamoreux, W.W. Wettengel, P. Hedao, and K.R. Kassem. Terrestrial Ecoregions of the World: a new map of life on earth. *BioScience.* November 2001. Vol. 51 no. 11.

Penning de Vries, F.W.T., F. Agus, J. Kerr (eds.). 1998. *Soil Erosion at Multiple Scales: Principles and Methods for Assessing Causes and Impacts.* CABI and IBSRAM, Wallingford, UK.

Ranieri, S.B.L., R. Stirzaker, D. Suprayogo, E. Purwanto, P. de Willigen, and M. van Noordwijk. 2004. Managing movements of water, solutes and soil: from plot to landscape scale. In: *Belowground Interactions in Tropical Agroecosystems*, ed. M. van Noordwijk, G. Cadisch and C.K. Ong. CAB International, Wallingford, UK: 329–347.

Rose, C.W. and B. Yu. 1998. Dynamic process modelling of hydrology and soil erosion. In: *Soil Erosion at Multiple Scales: Principles and Methods for Assessing Causes and Impacts*, ed. F.W.T. Penning de Vries, F. Agus and J. Kerr. CABI, Wallingford, UK: 299–318.

Susswein, P.M., M. van Noordwijk, and B. Verbist. 2001. Forest watershed functions and tropical land use change ASB_LN 7. In: *Towards Integrated Natural Resource Management in Forest Margins of the Humid Tropics: Local Action and Global Concerns*, ed. M. van Noordwijk, S.E. Williams and B.J. Verbist. ASB-Lecture Notes 1–12. International Centre for Research in Agroforestry (ICRAF), Bogor, Indonesia. http://www.icraf.cgiar.org/sea/Training/Materials/ASB-TM/ASB-ICRAFSEA-LN.htm.

Swallow, B.M., D.P. Garrity, and M. van Noordwijk. 2001. The effects of scale, flows and filters on property rights and collective action in watershed management. *Water Policy* 3(6):457–474.

Van Noordwijk, M. 2002. Scaling trade-offs between crop productivity, carbon stocks and biodiversity in shifting cultivation landscape mosaics: the FALLOW model. *Ecological Modelling* 149:113–126.

Van Noordwijk, M. 2005. *RUPES Typology of Environmental Service Worthy of Reward.* RUPES Working Paper. World Agroforestry Centre (ICRAF), Bogor, Indonesia. http://www.worldagroforestry.org/sea/Networks/RUPES/download/Working%20Paper/RUPES_Typology.pdf.

Van Noordwijk, M., Farida, P. Saipothong, F. Agus, K. Hairiah, D. Suprayogo, and B. Verbist. 2006. Watershed functions in productive agricultural landscapes with trees. In: *World Agroforestry into the Future*, ed. D.P. Garrity, A. Okono, M. Grayson and

S. Parrott. World Agroforestry Centre–ICRAF, Nairobi, Kenya: 103–112. http://www.worldagroforestry.org/sea/Publications/searchpub.asp?publishid=1482.

Van Noordwijk, M., B. Lusiana, N. Khasanah. 2004b. *WaNuLCAS 3.01: Background on a Model of Water Nutrient and Light Capture in Agroforestry Systems*. International Centre for Research in Agroforestry (ICRAF), Bogor, Indonesia.

Van Noordwijk, M. and B. Lusiana. 1999. WaNuLCAS, a model of water, nutrient and light capture in agroforestry systems. *Agroforestry Systems* 43:217–242.

Van Noordwijk, M., J. Poulsen, and P. Ericksen. 2004a. Filters, flows and fallacies: quantifying off-site effects of land use change. *Agriculture, Ecosystems and Environment* 104:19–34

Van Noordwijk, M., T.P. Tomich, B. Verbist. 2002. Negotiation support models for integrated natural resource management in tropical forest margins. *Conservation Ecology* 5:21.

Van Noordwijk, M., M. Van Roode, E.L. McCallie, and B. Lusiana. 1998. Erosion and sedimentation as multiscale, fractal processes: implications for models, experiments and the real world. In: *Soil Erosion at Multiple Scales, Principles and Methods for Assessing Causes and Impacts*, ed. F. Penning de Vries, F.W.T., F. Agus and J. Kerr. CAB International, Wallingford, UK: 223–253.

Van Roode, M. 2000. *The Effects of Vegetative Barrier Strips on Surface Runoff and Soil Erosion in Machakos, Kenya: A Statistical versus a Spatial Modelling Approach*. Koninklijk Nederlands Aardrijkskundig Genootschap/Faculteit Ruimtelijke Wetenschappen Universiteit Utrecht, Utrecht.

Young, A. 1997. *Agroforestry for Soil Conservation*. 2nd ed. CAB International, Wallingford, UK.

Ziegler, A.D., T.W. Giambelluca, R.A. Sutherland, M.A. Nullet, S. Yarnasarn, J. Pinthong, P. Preechapanya, S. Jaiaree. 2004. Toward understanding the cumulative impacts of roads in upland agricultural watersheds of northern Thailand. *Agriculture, Ecosystems and Environment* 104:145–158.

Chapter 12

Exploiting Ecological Disturbance

Jan Sendzimir and Zsuzsanna Flachner

Introduction

Diversity is commonly regarded in agriculture as a fundamental strategic hedge against sources of environmental variability ranging from pest outbreaks to extreme events and trends in weather. However, diversity may also be a resource that emerges *from* ecological disturbance. Ecoagricultural strategists may have opportunities to move beyond defensive reconciliation of production systems with landscape-scale ecosystem services by reconsidering fundamental attributes of production systems and their relation with dynamic ecological processes.

This chapter discusses the challenges of managing and adapting to diversity and disturbance in ecoagriculture landscapes, especially those prone to flooding. The chapter begins with a discussion of key issues, followed by a case study from agriculture in the Tisza River floodplains of Hungary. The next section examines the production and livelihood impacts of systems that respond to disturbance by riding it rather than by controlling it. In closing the chapter draws conclusions for the further development of ecoagriculture landscape management strategies.

Agricultural Diversity as a Response to Ecological Disturbance

Diversity in agricultural systems can play a number of roles and take different forms over space and time. Of course, farmers often simply seek to provide diverse products for consumption and income. But diversity also helps farmers and agricultural communities to reduce short-term risks associated with variability in weather conditions, pests, and markets and long-term risks to ecosystem services important for production and livelihoods. Much less widely appreciated is agricultural diversity as a strategic response to ecological disturbances, such as flooding, pest outbreaks, intense storms, or fire, that create diverse niches in the landscape.

Diversity to Manage Risks

Spatial diversity, or "crop heterogeneity," has been employed for millennia in almost every agricultural society. Prominent examples include hillside farmers in Peru who may plant as many as 150 potato varieties in small (30 ha) farms in the Andes (Brush et al. 1995; Roach 2002), and farmers in Bangladesh who developed thousands of rice varieties to deal with the vagaries of flooding and dispersion of the Bhramaputra and Ganges rivers in their delta (Nishat Ainun, IUCN, pers. comm.). Crop heterogeneity (of as little as two varieties interspersed in a plot) has been shown to vastly reduce pest damage at the field or plot level (Finckh and Mundt 1992, 1998; Finckh et al. 2000; Garrett and Mundt 1999; Mundt 2002; Wolfe 1985, 2000; Zhu et al. 2000).

Some strategies incorporate multiple varieties at the plot level and integrate them within designs that establish heterogeneity at the landscape level. Agroforestry does this by increasing integration of trees and/or agroforestry practices over time into land-use systems as a parallel to natural succession (Leakey 1996). This pushes young ecosystems toward more mature stages with "higher ecological integrity" and enhances biological diversity and ecological stability by creating a complex land-use mosaic (Leakey 1999). For example, one can mix wild trees into orchards to increase local access to fuel and construction materials or cash income through timber production (Leakey and Tchoundjeu 2001). Wild tree "domestication" within an orchard or garden mosaic increases the quality, number, and diversity of products and increases the range of environmental processes serving the farmer (see Leakey, chapter 5, this volume).

Diversity of crop varieties and practices can be achieved in time as well as in space. Agroforestry strategies can shift annual agricultural plots within forest systems. In tropical climates, subsistence crops (cassava, maize, rice, sorghum, sweet potato, bananas, taro) and agroforestry provide a diversity of fruits, herbs, pharmaceuticals, and spices (Ewel 1999). Crops can be moved to benefit from brief and intermittent episodes of resource availability. For example, in desert landscapes the Papago Indians plant fields where summer thunderstorm-driven surface flows concentrate detritus as a wild green manure (Nabhan 1982). Agriculture in areas of low human density can be shifted to avoid the arrival of stressors (grass and grasshoppers or soil impoverishment) or to coincide with infrequent rain events (Ewel 1999). In arid regions like the Sahel, nomadic herders can move their livestock herds such that their grazing or browsing patterns match the variable and patchy patterns of rainfall and plant growth on thin and fragile soils. Natural variability in resource availability is integrated at two scales here: in fluctuations of grazer biomass and in movement from the patch to landscape to regional levels.

Diversity as a Strategic Response to Natural Disturbance

Although many people see diversity as a means to minimize impacts from disturbance, some ecosystems exist where disturbance processes actually maintain appreciable levels of both natural and agrobiodiversity. Obviously extreme disturbances (massive volcanic, meteorite, earthquake, flooding, or drought events) are most commonly associated with biodiversity loss and even extinction. However, ecosystems do exist where flora and fauna have evolved within a regime of disturbance processes at more moderate levels.

Connel (1978) proposed the intermediate disturbance hypothesis (IDH) to explain how biodiversity peaks at disturbance levels between extreme highs (too disruptive) and lows (insufficient stimulation). Although the IDH has not proven to be predictive of biodiversity–disturbance relationships in most ecosystems (Mackey and Currie 2000, 2001), it serves as a metaphor to explain flooding as a disturbance associated with high levels of biodiversity (in both natural and agroecosystems).

This chapter illustrates the phenomenon with the example of the Tisza River Basin (TRB) floodplain. In the TRB, preindustrial society employed a variety of production approaches to exploit, rather than resist, the structures and ecological functions generated by the natural flooding regime. Preagrobiodiversity was already high, but it appears that preindustrial agriculture in the TRB could skillfully steer flooding energies to sustain high agrobiodiversity as well. Lessons learned from this study may have relevance for managing ecoagriculture landscapes in regions subject to flooding and other types of intermediate ecological disturbances.

Responses to Natural Disturbance: Evolution of Floodplain Agriculture in the Tisza River Basin

The Hungarian reach of the TRB is one of the larger tributaries of the Danube River, with a total catchment area of 157,200 km^2. Starting in the Ukrainian Carpathian Mountains, the Tisza River cuts west along the border of Romania and then down to the southwest across the great Hungarian plain (*Alföld*), eventually issuing into the Danube River in the Serbian Republic (Fig. 12.1).

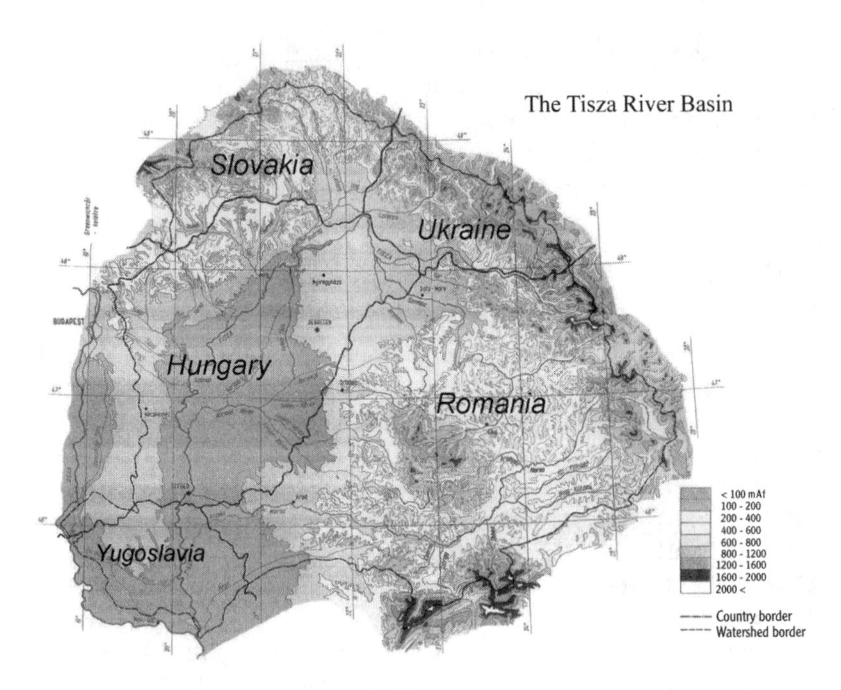

Figure 12.1. Topographic scheme of the Tisza River Basin, Hungary (adapted from Jolankal and Pataki 2005)

The extensive mountain catchment swiftly concentrates rain events over a vast area, shooting the runoff from a relatively short and steep outfall onto a broad, flat floodplain, and driving some of the most sudden (24–36 hours) and extreme (up to 12 m) water level fluctuations in Europe (Halcrow Group 1999; Koncsos and Balogh 2005; Koncsos and Szab 2004; Kovács 2003). Such extreme floods occur on average every 10 to 12 years in the TRB (Wu 2000; Jolankai and Pataki 2005), but the last century has seen rising trends in all facets of flooding: flood crest or peak height, flood volume, and flooding frequency.

This chapter contrasts agricultural systems in the TRB throughout the modern and premodern periods.

Modern Floodplain Agriculture: Evolution, Achievements, and Losses

Starting in the 18th century, in response to an exploding market for bread during the Industrial Revolution, the Tisza was deepened to hasten water flow, shortened by 400 km to facilitate export from the northern part of the TRB, and bracketed with dikes to prevent flooding of wheat fields and habitations (Fig. 12.2). The dikes also provided footpaths for draft animals to pull barges, but

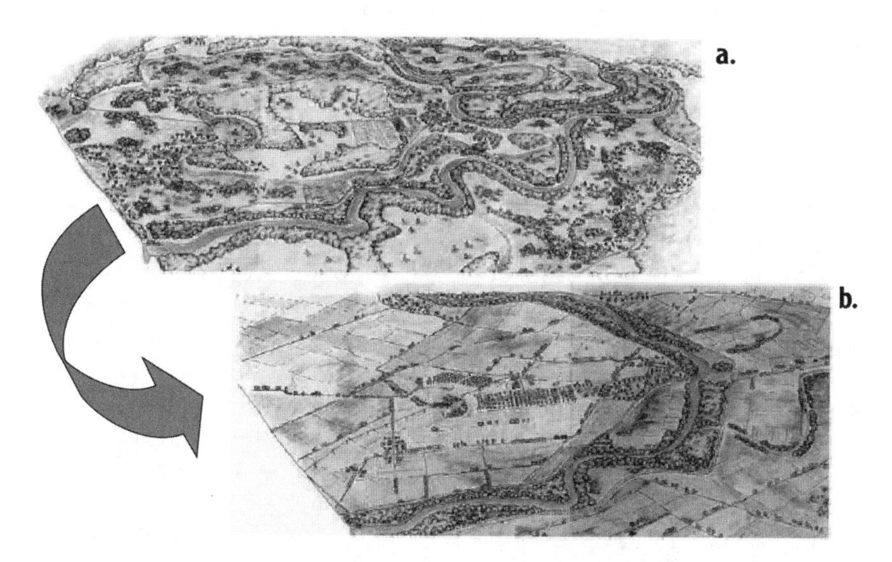

Figure 12.2. Artist's interpretation of different landscape mosaic patterns in the Tisza River floodplain (a) pre- and (b) post-execution of the original Vásárhelyi River engineering plan in 1871 (illustration courtesy of WWF-Hungary)

these embankments also cut off the *foks* or notches that used to connect the floodplain with the river channels. As a result, sediments previously flushed by floods back into the river began to accumulate as mud layers on the floodplain. Frequent dike collapses also hampered operation of the new system, which was further complicated by political developments. Complaints by local farmers eventually provoked an agreement between the empress, Maria-Theresa, and the local aristocracy to solve the problem by expropriating the entire floodplain, which had previously been common land, for grain production. The drive to acquire new sources of grain to feed horses and soldiers was further stimulated by frequent military actions in the 19th century.

Improvements in engineering technology and science were repeatedly called for to address complications resulting from earlier river basin restructuring. The Vásárhelyi plan in 1870 was designed to address the problems of collapsing dikes and floodplain sedimentation, and to revitalize the basin with side channels. Unfortunately, the latter goal was abandoned once the initial reinforcing of the dike system was accomplished. The local farmers' complaints of lost biodiversity and productivity were silenced politically when they lost access to the floodplain; the revitalization they petitioned for was never accomplished. Implementing this plan spread the hydroengineering experiment from the northern section to the entire Hungarian TRB. Essentially, during the last 130 years, there was massive investment in the TRB to maintain a static river channel by using steam- and human-powered shovels. Over the period this process added 4200 km of dikes and reduced the floodplain area by 90%.

Practically in step with mounting flood statistics, regional development has also climbed since the mid-19th century, and the clash between these two rising trends has created ever larger losses. The infrastructure of towns and row crop farms burgeoned and spread into the flood danger zone, the TRB floodplain, residents having been reassured by the apparent security of a dike and canal flood defense system that reduced the area of the active floodplain. With less lateral space in the floodplain to absorb them, flood volumes had only one direction left: up and over the dikes. Hydroengineering promised security, which might hold for a decade or two, but ultimately would be breached by ever-larger floods that devastated homes, roads, and crop fields. Damage to built capital and commerce from one major flood event could reach as high as approximately 25% of the gross domestic product (GDP) of the riverine basin or 7 to 9% of the national GDP (Halcrow Group 1999).

These sudden, catastrophic losses stand out against a backdrop of steady, regional decline in all forms of capital that contribute to biocomplexity: natural capital (biodiversity and aesthetics lost, rising flood statistics), economic (previous industry gone, region no longer prosperous but impoverished, apathy about farming), and sociopolitical (cities, schools, businesses disappearing, political apathy as power concentrates in Budapest) (Sendzimir et al. 2004). These trends of declining capital have increased interest in the preindustrial patterns of production and habitation (Molnár 2003) that sustained local populations and even exported products regionally. The added challenge of facing novel forms of uncertainty arising from climate change and globalization has made the search for sustainable livelihoods even more urgent.

Preindustrial Flooding—Dynamic Source of Useful Structure and Water Buffers

The initial success of the industrial strategy to "jacket" the river channel with dikes and minimize floods has given way to a rising wave of negative outcomes that have increased curiosity about preindustrial floodplain management. Prior to the 19th century the river flood regime was the dominant force that defined the basin and its rhythms of growth and rest. The preindustrial river landscape was anything but static, shaped and reshaped by periodic flood surges. Sometimes the flood wave's power to shape the terrain or remove vegetation was amplified when winter floods ripped deeper gouges by ramming trees, ice floes, and jams through the channel and the floodplain. Pulses of rapid, high-volume flows over weeks alternated with years of low flows, carving the floodplain into a shifting mosaic of ponds interspersed by meadows and terraces of various elevations. Like most natural shapes, the floodplain's topography, specifically its cross-sectional profile, is fractal, and the length one measures depends on the dimension of the measuring device. However, the diversity and structural complexity of its elevation pattern can be reasonably clustered into four categories (Bokartisz 2002):

DEEP FLOOD PLAIN

Deep-lying areas of the flood plain, around the low water level of the river, adequate as permanent reservoirs (existing, permanent, or old stream beds, ponds between and behind the dikes), which were critical water buffers against droughts and nurseries that supported the regional fishery.

LOW FLOOD PLAIN

Low areas of the flood plain that are in most years below the high water level of the river. In these areas water coverage might exceed 1 m. These areas were traditionally used for grazing and special orchards (jungle orchards) that could stand flooding for periods of up to six weeks, so long as the crown of the tree was not covered. The orchards acted as gene banks of ancient local varieties of apple, plum, pear, and nut, which were hardy to disease, and protected by many natural enemies of pests harbored in the natural undergrowth.

HIGH FLOOD PLAIN

Parts of the flood plain, which are below the high water level of the river in approximately 6 years out of 10, where the elevation of water cover remains below 0.7 m. The high ambient humidity and moist (but not too wet) soil supported good vegetable and crop production.

DRY ELEVATION

The highest zones in the floodplain almost always remained dry, a fact suggested by their unique soil content, which has significantly less clay than that found in surrounding floodplain habitats. Traditional settlements were located here along with areas for cultivation of corn and winter wheat and forest groves that, as suggested by Molnar (2003), may have pumped water up during high floods and then slowly released it during late summer drought. Oak groves also supported swine grazing and provided timber and fuel.

Loosely hemmed in by the bowl shape of the *Alföld*, the flood pulses rushing out from the Carpathians spread as sharp crests of flood waves over vast areas of floodplain, suspended for weeks by thick, impermeable sediments over hundreds of square kilometers. With the passing of each flood wave, water would begin returning to the river channel, first as sheet flow over the entire plain and then through the natural minor channels that breached embankments through notches linked to lateral channels (*fok* in Hungarian) connecting the floodplain

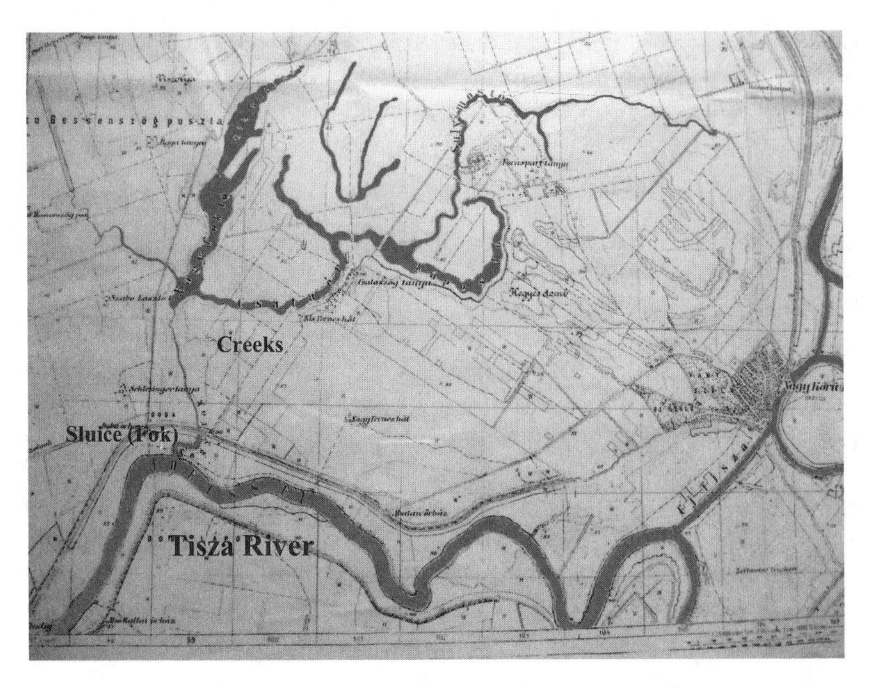

Figure 12.3. A version of a historical, military map (Austro-Hungarian Army ca. 1880) showing the Tisza River main channel in the vicinity of the village of Nagykörü, Hungary with tributary creeks connecting the river channel through sluices to ponds and low marsh areas in the floodplain. After thirty years of dike construction some areas of the Hungarian/fok/system still remained intact.

with the main channels of the Tisza River (see Fig. 12.3). Considerable amounts of water were stored in the deeper swales and ponds and released slowly over many months back to the river, though some deeper areas might retain surface water for years.

Molnar (2003) posits that tree copses and thickets on hummocks and higher sections of the floodplain could sustain evapotranspiration that would draw and accumulate water underneath as a more ephemeral water storage, which would slowly release water to the surrounding environment during times of drought. Topography and photosynthesis could mound water in myriad points across the floodplain, providing a distributed network of buffers against the main chronic disturbance of the region: drought. An alternative hypothesis might be that tree patches and ecotones provide buffering against dehydration at two scales in the floodplain. At the scale of hundreds of square meters, temporary (overnight) buffering might result through hydraulic lift and release of tree transpiration to neighboring plants (Burgess et al. 1998; Caldwell et al. 1998). Over the longer term (weeks), buffering against desiccation might result when tree patches act as shelter belts that sustain a moister microclimate for the agricultural areas they border or surround (Ryszkowski 1988; Hillbricht-Ilkowska1 et al. 1990). This water network also offered sanctuary and nursery functions for invertebrates and fish as well as their avian and mammal predators, and then provided new cohorts that annually emerged from backwater nurseries with passage through the *fok* network of notches and sluices, replenishing the fish stocks of the main Tisza channels. For the preindustrial Tisza basin, flood was not a disturbance but a process that provided and renewed structure, water storage, and production.

Traditional Livelihoods on the Tisza Floodplain

Opinion varies as to whether the preindustrial society of the TRB was wealthy (Molnar 2003), but it certainly was self-sustaining in agricultural produce (fruits, nuts, and grains) and protein (fish, cattle, and swine). Over the past millennium these agrarian societies managed to develop cultures that could utilize and thrive on massive, periodic flooding in order to build fisheries and fruit enterprises that sometimes produced surpluses they could export. The forest landscape they originally faced was a mix of climatic–zonal associations. The warmer forest association, found in areas with a southern aspect, was the Pannonic–Balkanic turkey oak (*Quercetum petraeae-cerris*). The extrazonal association, found on the cooler, northern side and in the valleys, is the Pannonic oak-hornbeam (*Querco petraeae-Carpinetum*). There used to be alder gallery forest (*Alnetum glutinosae-incanae*) on the upper side of the brook valleys, and willow gallery forest (*Salici albae-fragilis, Salicetum-triandrae*) along the downstream slower watercourses. Although they did not have the concentrated energy sources (fossil fuels) nor the technologies that emboldened later societies to try to dominate the flooding regime, they did use their renewable resources to alter the landscape, opening the forest with meadows that turned eventually to plots of grain, vegetables, or vineyards, depending on the plot's elevation.

As a result of the appearance and development of settlements, the territory of the floodplain forest shrank and was slowly transformed for agricultural use (crop production, husbandry, meadow, and hay grazing), which fragmented the original mosaic of forest types. At the same, time new habitats and land-use types appeared: slope steppe, hillside meadow, and hay meadow along the brooks and the vineyards and orchards as a part of the cultural landscape. With time, a land-use mosaic emerged that balanced the distribution between forests (30%), meadows (30%), arable land (30%), and extensive (low-input agriculture as opposed to intensive high-input agriculture) orchards (10%). The land use of the community contributed to this balance: it was common to have orchards and vineyards around the houses. These orchards formed an ecological buffer zone between the village and the meadows. Meadows were located along the draws and gullies, which drained into the creeks. The forests were located in the upper parts of the hills, which, similarly to the meadows, were not conducive to ploughing. These land-use patterns were adjusted to the soil types and water-flow regimes of the floodplain.

The balance between land uses is suggested not only by old maps but also by the habitat requirements of keystone bird species like the great bustard (*Otis tarda*) that have long been common to the region. Present efforts to conserve this endangered species are based on evidence that they survive where their habitat is diverse, and all the desired types of habitat (grassland, grains, and plants from the Papilionaceae—beans, peas, and chick peas) are equally divided (Flachner and Kovác 2000).

Changes in land use also reflected the increasing influence of individuals and families. By the 15th and 16th centuries, local communities in the Hortobagy region of the TRB regulated floodplains primarily for common use. Two people, elected by the community, acted as officers to guarantee equal access to all. However, the community ceded special rights to anyone who invested special efforts in cultivating the land. Practices such as ploughing, which required special skills and facilities (cows, horses, ploughs, and related equipment), were rewarded by granting the farmer exclusive choice of how a plot would be used. However, the community did require that the farmer share some of his produce (grain) with people in the community. The common ground of the floodplain began to fragment into individual plots as farmers were increasingly drawn by the power to decide and the profitability of grain production. The fragmentation process accelerated with the introduction of vineyards, where the higher-quality fermented product could be profitably exchanged for salt or money. The community sustained its right of control by removing access and user rights to anyone who did not maintain their vineyard. In addition, to maintain some equity and slow the concentration of power, the community rotated the land's access by lottery every seven years to reallocate the ownership. However, power did increasingly concentrate as certain families managed to control ever-larger areas, eventually being incorporated in the nobility. The community lands remained common in the floodplain much longer than in other areas. This delay suggests that the floodplains remained common so long as the land's maintenance needs (given the technology of that era) required cooperation at the community level.

Implications for Managing Floodplain Agriculture

River engineering has drastically changed or eliminated water flows on floodplains worldwide. On all continents rivers are constrained by hundreds of thousands of kilometers of dikes, and their "flows" are so punctuated by dams and locks that water moves more as a cascade of higher pools spilling into lower pools. In western Europe only a handful of "natural" rivers remain. Within this engineered context, the question of whether productivity emerges as crops, electricity, animal products, biodiversity, fisheries, timber, or aesthetics is salient on almost any floodplain in the developed world and many in the developing world. Floodplain conversion to dry monocultures bordering but isolated from river channels has drastically reduced biodiversity (Ward et al. 1999), particularly for molluscs (Lydeard et al. 2004; Pringle et al. 2000), fish (Dudgeon 2000), and migratory waterfowl (Brown et al. 1995).

The modern challenge is to build on the potential of renaturalization of floodplains to increase flood storage capacity by extending it to increase ecosystem functions and, hence, the diversity of species and agricultural products. This challenge begs the question as to what kind of agriculture could thrive in a floodplain that floods more frequently but at shallower depths. Traditional practices of several centuries ago have received the most scrutiny. The picture of preindustrial TRB agriculture can only be assembled from maps, literature, and folklore. But research in the TRB over the past 10 years suggests that benefits may arise from using these traditional land patterns and practices as templates to develop new agroecosystems, including restoring hydrological connectivity and increasing biodiversity, in ways that support economic diversification and stabilization.

Hydrological Connectivity

No part of the TRB has a floodplain that is hydraulically linked with the river channel, so direct observations of a fully functional (in terms of hydrology, ecology, or ecoagriculture) floodplain are currently impossible. This lack motivates current plans to hydrologically reconnect specified sections of the river floodplain in the Bodrogköz region and study the performance of hydrology, ecology, and even agriculture in a renaturalized ecosystem. High-intensity industrial agriculture evolved to become a macroscale disturbance that so completely reconfigured the basin morphometry that it eliminated the possibility to steer the effects of flood waves for plant propagation or for water storage in drought. That leaves modern floodplain farmers in greater peril of floods and droughts, forcing them to rely on the national government to maintain the dikes and on international companies to generate and sustain the vigor and robustness of their crop varieties. The question is whether such peril is clearly and broadly understood enough to overcome a number of logistic, institutional, and psychological barriers (Sendzimir et al. in review) to floodplain naturalization. Centuries behind the dikes have deeply ingrained the paradigm of "water as a threat to be shed from the landscape" in the minds of many local farmers and government officials at all levels. Such attitudes could defeat outright any pilot experiments or create the potential for "policy resistance" (Sterman 2000) where initial success is reversed by long-term efforts to sabotage policy initiatives.

Increasing Biodiversity

These pilot projects are also motivated by evidence of considerable biodiversity and productivity generated by traditional floodplain agriculture. Our modern sense of diversity, tuned to the three or four varieties of any domestic fruit a supermarket might offer, would be amazed to find the hundreds of varieties of fruits and nuts traditionally grown in the TRB. Local varieties of apple (402) and pear (470) have been officially registered, but the unregistered varieties run well past a thousand for each (Dr. Dezső Surány pers. comm.). In addition, the fishery of the modern Tisza, isolated by dikes from a dehydrated floodplain, appears impoverished compared to preindustrial production that could often export over the entire region. For example, currently the Tisza can support one commercial fisherman (as a part-time job) for every 10 km of river reach in the area around Nagykörü (Peter Balogh, pers. comm.). In direct contrast to the intensification of resource and machinery use in floodplain monoculture, traditional uses were extensive—extending through a rich web of diverse agricultural practices and land uses (Fig. 12.4).

Traditional Extensive Agricultural Practices in the Tisza River Floodplain

Without direct observation we can only surmise what processes and structures contributed to preindustrial biodiversity and productivity. High biodiversity is documented in unaltered river systems in many ecosystems in Europe (Amoros and Bornette 2002; Bornette et al.1998; Poux and Copp 1996; Schmutz and Jungwirth 1999).The capacity of floodplain water storage to sustain productivity by buffering water dynamics during drought is not controversial, though this capacity is not fully measured for a renaturalized floodplain in the TRB. The source of such significant agrobiodiversity (thousands of fruit varieties) also needs investigation, but we propose that it was driven by flood dynamics. Flooding promoted biodiversity in several ways. First, it created a diversity of test bed contexts by shaping and reshaping a very heterogeneous landscape mosaic into a variety of patches at different elevations. Soil moisture levels were dynamic across the landscape because the rise and fall of floodwaters wetted and dried these patches at different rates. Second, flood waves provided the energy to increase the variety of genetic combinations on these different test beds. It did so by picking up different fallen fruit and propagules and mixing them in a myriad of ways as it spread them at varying distances and deposited them in

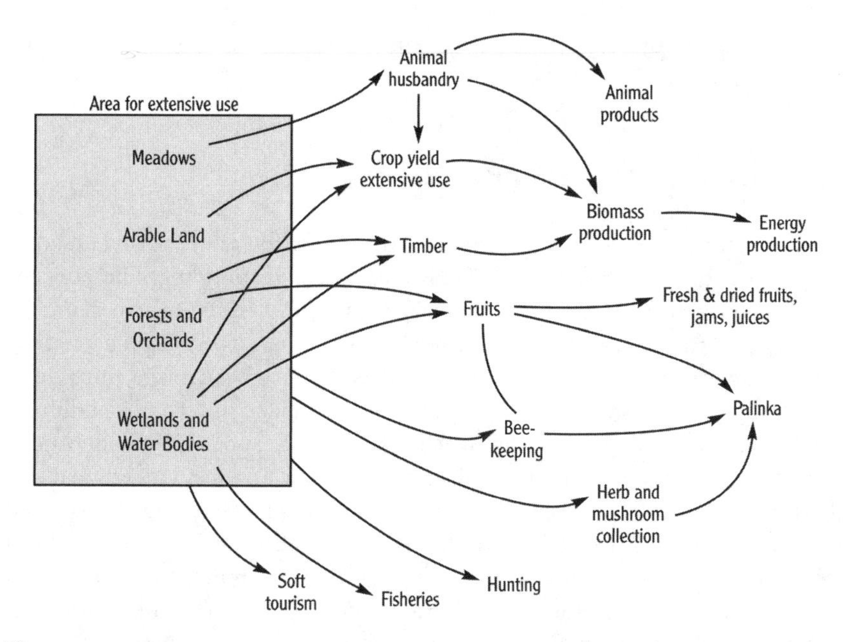

Figure 12.4. Practices and products associated with extensive land uses in the Tisza river floodplain, Hungary

novel combinations across the landscape. Eventually the trees growing from these propagules would bloom and be cross-fertilized by the novel combinations of other varieties in the vicinity. The new hybrid varieties that survived under different flooding conditions would be recognized by local farmers who might then select the fittest for propagation elsewhere. Thus humans could exploit the way floods generated diversity at two scales (landscape patches and genetics) by spreading and cultivating the successful survivors to other parts of the basin with similar elevation and hydroperiod characteristics. The increased biodiversity at landscape scales was further amplified when subsequent floods again spread and mixed fallen propagules over a range of distances.

In this manner the preindustrial TRB self-organized into a highly diverse landscape mosaic through the interaction of human selection and cultivation practices, landscape structure, and flooding. Low-intensity agriculture constituted a very minor perturbation that acted more as a steering mechanism for the chronic disturbance of major flood pulses that redistributed the fruits of natural propagation experiments. Although preindustrial floods often constituted intermediate level disturbances that stimulated biodiversity, species levels have since declined under the massive restructuring of basin morphometry required for crop agriculture and export. Dikes became entry vectors for exotic invasive species, and plantation (silviculture and row crop) agriculture all but eliminated the diversity of ancient varieties.

Conclusions and Recommendations

This case study suggests new ways of looking at how to balance form and function in ecoagriculture systems in landscapes prone to flooding. The conventional impulse to redesign river systems grows from a legacy of two centuries where fossil fuel and industrial technology made massive reshaping of landscapes achievable. So long as the accumulated feedbacks (damage from flood and drought) were small, it was often highly profitable. Under the industrial paradigm, river basin form followed the functions of export agriculture (transport and intensified productivity). Should our reliance on fossil fuel dwindle in the face of energy and climate uncertainty, then river basins might be shaped more as they were prior to industrialization—by flood waves. Then river form might follow the functions of the hydrological cycle (pulses of flooding and drought). Although agroforestry and other kinds of ecoagriculture provide insightful and clever design alternatives, the history of the Tisza suggests that ecoagriculture need not actively engineer alternative landform designs. Engineering can be left mostly to the hydrological cycle if agricultural strategies are designed to thrive on the productivity that emerges as natural functions periodically reshape the basin. Disturbances can be steered and ridden rather than resisted. Nepalese farmers can preemptively set off landslides and control their impacts (Ives and Messerli 1989; Johnson et al. 1982), just as foresters can prescribe burns to reduce biomass and, ultimately, fire impacts (Johnson and Miyanishi 1995; Moritz and Odion 2004).

Such large-scale ecophysical questions feed into the broader consideration of the sustainability of social–ecological systems in river basins. Will river basin communities be resilient to global stresses and shocks from climatic, economic, and political turbulence? The economic goals that reshaped the TRB now seem too narrow as mounting flood damages make the whole venture seem ultimately unprofitable. Similar questions arise for the millions of people along major rivers, especially in Asia, where dams and dikes boost economic capital at the expense of natural capital. For example, along the Mekong River, hydropower production continues to expand as local fisheries die off (Bakker 1999; Hill and Hill 1994).

How can the sustainability and resilience of the TRB be assessed in ways that inform river managers and offer alternative data and perspectives to farmers and communities? Experimenting with alternative configurations of the river valley and new practices is expensive and requires broad support and understanding in society. Present pilot studies of floodplain naturalization are narrowly focused on assessing how much floodplain area must be naturalized to provide sufficient storage capacity for the larger and more intense floods anticipated under certain climate change scenarios. The challenge posed by multiple, global sources of uncertainty stresses the need to examine how the community's resilience is affected not simply by one sector, such as technology, but by interactions within and between all sectors (technical, ecological, economic, and social). Such experiments must expand to look at such interactions so as to identify what institutions and policies can help integrate practices of agricultural and water management in the TRB. To foster broader understanding to support such experiments we are currently modeling such complex interactions (Sendzimir et al. in review) as a basis for games, exercises, and interactive computer tools that allow a diversity of stakeholders to explore policy options in managing agroecosystems in river basins.

Summary

It is naive to suggest that any society could perceive major floods not to be a disturbance. The power of floods to violently reshape the landscape and remove vegetation was perhaps clearer to preindustrial populations than to current society, though combinations of intense rain events, forest cover loss in the Carpathians, and narrowing of the floodplains may have increased the speed and size of current flood waves over the past 50 years. However, preindustrial agriculture in the TRB illustrates the development of skillful practices to utilize flooding as an engine of productivity and biodiversity. Their fisheries were regionally famous simply from their prodigious export, and the diversity of their orchards beggars our modern imagination. These may well be the skills modern society must rediscover to live profitably in river basins increasingly beset by the uncertainty of climate change.

References

Amoros, C. and G. Bornette. 2002. Connectivity and biocomplexity in waterbodies of riverine floodplains. *Freshwater Biology* 47(4):761–776.

Amoros, C. and A.L. Roux. 1998. Interaction between water bodies within the floodplains of large rivers: function and development of connectivity. In: *Connectivity in Landscape Ecology*, ed. Schneider. Münstersche Geographische Arbeiten.

Bakker, K. 1999. The politics of hydropower: developing the Mekong. *Political Geography* 18(2):209–232.

Bokartisz K.H.T. 2002. Rationale of floodplain landscape management: philosophy and concrete steps at the Bodrogköz area, Phare CBC project closing report. Karcsa, Hungary.

Bornette, G., C. Amoros, and N. Lamouroux. 1998. Aquatic plant diversity in riverine wetlands: the role of connectivity. *Freshwater Biology* 39(2):267–283.

Brown, C., C. Baxter, D.N. Pashley, 1995. The ecological basis for the conservation of migratory birds in the Mississippi Alluvial Valley. In: *Strategies for Bird Conservation: The Partners in Flight Planning Process.* The Nature Conservancy: Baton Rouge, Louisiana. http://www.birds.cornell.edu/pifcapemay/brown.htm.

Brush, S., R. Kesseli, R. Ortega, P. Cisneros, K. Zimmerer, and C. Quiros. 1995. Potato diversity in the Andean center of crop domestication. *Conservation Biology* 9(5): 1189–1198.

Burgess, S.O., M.A. Adams, N.C. Turner, D.A. White, C.K. Ong. 1998. The redistribution of soil water by tree root systems. *Oecologia* 115: 306–311.

Caldwell, M.M., T.E. Dawson, J.H. Richards. 1998 Hydraulic lift: consequences of water efflux from the roots of plants. *Oecologia* 113:151–161.

Connell, J.H. 1978. Diversity in tropical rainforests and coral reefs. *Science* 199:1302–1310.

Ewel, J. 1999. Natural systems as models for the design of sustainable systems of land use. In: *Agriculture as a Mimic of Natural Ecosystems*, ed. E.C. Leroy, R.J. Hobbs, M.H. O'Connor, J.S. Pate. Kluwer Academic, Dordrecht: 57–109.

Finckh, M.R., E.S. Gacek, H. Goyeau, C. Lannou, U. Merz, C.C. Mundt, L. Munk, J. Nadziak, A.C. Newton, C. Vallavieille-Pope, and M.S. Wolfe. 2000. Cereal variety and species mixtures in practice, with emphasis on disease resistance. *Agronomie* 20:813–837.

Finckh, M.R. and C.C. Mundt. 1992. Plant competition and disease in genetically diverse wheat populations. *Oecologia* 91:82–92.

Finckh, M.R. and M.S. Wolfe. 1998. Diversification strategies. In: *The Epidemiology of Plant Disease*, ed. D.G. Jones. Kluwer Academic, Dordrecht: 231–259.

Flachner, Z. and E. Kovác. 2000. Cooperation between the Kiskunság National Park Directorate and the local farmers: a case study for the Organization for Economic Cooperation and Development (OECD) Working Group on Economic Aspects of Biodiversity. OECD Electronic Working Papers Series. Budapest.

Garrett, K.A. and C.C. Mundt. 1999. Epidemiology in mixed host populations. *Phytopathology* 89:984–990.

Halcrow Group, Ltd. 1999. Feasibility study of flood control development in Hungary. Vituki Consult Plc, Budapest.

Hill, M. and S. Hill. 1994. *Fisheries Ecology and Hydropower in the Mekong River: An Evaluation of Run-of-the-River Projects.* Don Chapman Consultants.

Hillbricht-Ilkowska, A., L. Ryskowski, A.N. Sharpley. 1995. Phosphorus transfers and landscape structure: riparian sites and diversified land use patterns. *SCOPE.* Vol. 54:200–228.

Ives, J. and B. Messerli. 1989. *The Himalayan Dilemma: Reconciling Development and Conservation.* Routledge, London.

Johnson, E. and K. Miyanishi. 1995. The need for consideration of fire behavior and effects in prescribed burning. *Restoration Ecology* 3(4):271–278.

Johnson, K., E.A. Olson, S. Manandar. 1982. Environmental knowledge and response to natural hazards in mountainous Nepal. *Mountain Research and Development* 2(2):175–188.

Jolankai, G. and B. Pataki. 2005. *Description of the Tisza River Project and Its Main Results: A Report on The Tisza River Project: Real-Life Scale Integrated Catchment Models for Supporting Water- and Environmental Management Decisions*. Contract No: EVK1-CT-2001-00099, http://www.tiszariver.com/index.php?s=results.

Koncsos, L. 2007, in review. A Tiszai árvíizi szbályozás a Kárpát medencében. MTVSZ, Budapest.

Koncsos, L. and E. Balogh. 2005. Optimum operation of flood control reservoirs. *Environmental Modelling and Software*.

Koncsos, L. and G.C. Szabó. 2004. *Entwicklung eines physikalischen, numerischen Hochwasserabflussmodells*. Wasserbau und Wasserwirtschaft, No. 101. 192–2003.

Kovács, Z.C. 2003. Sustainability along the Tisza River, Budapest. Master's thesis, Oxford Brookes University, Oxford, UK.

Leakey, R.R.B. 1996. Definition of agroforestry revisited. *Agroforestry Today* 8(1):5–7.

Leakey, R.R.B. 1999. Agroforestry for biodiversity in farming systems. In: *Biodiversity in Agroecosystems*, ed. W.W. Collins and C.O. Qualset. CRC Press, New York: 127–145.

Leakey, R.R.B. and Z. Tchoundjeu. 2001. Diversification of tree crops: domestication of companion crops for poverty reduction and environmental services. *Experimental Agriculture* 37:279–296.

Lydeard, C., R.H. Cowie, W.F. Ponder, A.E. Bogan, P. Bouchet, S.A. Clark, K.S. Cummings, T.J. Frest, O. Gargominy, D. Hebert. 2004. The global decline of nonmarine mollusks. *BioScience* 54(4):321–330.

Mackey, R. L. and D. J. Currie. 2000. A re-examination of the expected effects of disturbance on diversity. *Oikos* 88:483–493.

Mackey, R. L and D. J. Currie. 2001. The diversity–disturbance relationship: is it generally strong and peaked? *Ecology* 82(12):3479-3492.

Molnár, G. 2003. A Tiszánál. Ekvilibrium, Budapest.

Moritz, M. and D. Odion. 2004. Prescribed fire and natural disturbance. *Science* 306(5702):1680.

Mundt, C.C. 2002. Use of multiline cultivars and cultivar mixtures for disease management. *Annual Review of Phytopathology* 40:381–410.

Nabhan, G.P. 1982. *The Desert Smells Like Rain: A Naturalist in Papago Indian Country*. North Point Press.

Poux, A.L. and G.H. Copp. 1996. Fish populations in rivers. In: *Fluvial Hydrosystems*, ed. G.E. Petts. Kluwer Academic, Dordrecht: 64-88.

Pringle, C., M.C. Freeman, B.J. Freeman. 2000. Regional effects of hydrologic alterations on riverine macrobiota in the New World: tropical-temperate comparisons. *BioScience* 50(9):807-823.

Roach, J. 2002. Saving the potato in its Andean birthplace. *National Geographic News*, 28 April 2003. http://news.nationalgeographic.com/news/2002/06/0610_020610_potato.html.

Ryszkowski, L. 1988. Landscape management for nature-friendly farming: shelter belts in Poland. *Naturopa* 86.

Schmutz, S. and M. Jungwirth. 1999. Fish as Indicators of large river connectivity: the Danube and its tributaries. *Large Rivers Arch. hydrobiol.* (Suppl b)11(3):329–348.

Scurrah, M., R. Canto, N. Zuniga. 2003. Self-sufficiency and sustainability of the native potato cultivation in the community of Quilcas, in the Mantaro Valley, in the central Andes of Peru. Grupo Yanapai. http://www.unu.edu/env/plec/cbd/abstracts/Scurrah.doc.

Sendzimir, J., P. Balogh, A. Vari, and T. Lantos. 2004. The Tisza River Basin: slow change leads to sudden crisis. In: *The Role of Biodiversity Conservation in the Transition to Rural Sustainability*, ed. S. Light. ISO Press, Amsterdam: 261–290

Sendzimir, J., P. Magnuszewksi, Z. Flachner, P. Balogh, G. Molnar, A. Sarvari, and Z. Nagy. In review. Assessing transformability from the shadows: informal learning in the Tisza river basin. *Ecology and Society*.

Siposs, V. and F. Kis. 2002. *Living with the River: LIFE-Nature Project in the Tisza Floodplain*. World Wide Fund for Nature (WWF), Hungary, Budapest.

Sterman, J. 2000. *Dynamics: Systems thinking for the modern world*. McGraw Hill, New York.

Ward, J., K. Tockner, and F. Schiemer 1999. Biodiversity of floodplain river ecosystems: ecotones and connectivity. *Regulated Rivers: Research & Management* 15(1):125–139.

Wolfe, M.S. 1985. The current status and prospects of multiline cultivars and variety mixtures for disease resistance. *Annual Review of Phytopathology* 23:251–273.

Wolfe, M.S. 2000. Crop strength through diversity. *Nature* 406:681–682.

Wu, F. 2000. *The Tisza River Crises: Integrating Stakeholder Views for Policy Decisions*. IIASA Internal Report, Laxenburg, Austria.

Zhu, Y., H. Chen, J. Fan, Y. Wang, Y. Li, J. Chen, J. Fan, S. Yang, L. Hu, H. Leung, T.W. Mew, P.S. Teng, Z. Wang, and C.C. Mundt. 2000. Genetic diversity and disease control in rice. *Nature* 406:718–722.

Chapter 13

Irrigation

David Molden, Rebecca Tharme, Iskandar Abdullaev, and Ranjitha Puskur

Introduction

Developing and managing water for agriculture has been and will remain instrumental in providing food and livelihood security, but at a high environmental price. The costs range from progressive degradation, fragmentation, and drying up of aquatic and terrestrial ecosystems through to their irreversible loss, affecting valuable ecosystem services on which people also depend for their livelihoods. Given this context, how can water be managed in agriculture to meet the goals of food production, livelihoods, and environmental security?

This chapter focuses on managing water in agriculture and the impacts of this management on the environment, with special attention to biodiversity, and consequently on livelihoods. The chapter will show key trends and conditions in the use of water for food production; discuss how water-for-food actions affect biodiversity; indicate several ways in which biodiversity can be maintained or enhanced in agricultural systems; and identify several issues that remain difficult to resolve.

Within the scope of water and agriculture, large-scale irrigation—large hydraulic infrastructure to divert and store water, and then deliver it to high-intensity agricultural producers—is often at the center of the discussion because of the high levels of water withdrawals and high production levels. However, a range of water and food activities, for instance groundwater pumping, water harvesting, and flood recession agriculture, operating at a much smaller scale, support livelihoods and food security. To understand the present situation and to identify ways forward it is essential to consider the full range of options for managing water for agriculture, starting with rainfall as a primary input.

Water for Food Production: Key Trends and Driving Forces

Food consumers are key drivers of agricultural practices. To produce one kilogram of grain, plants must transform between 400 and 5000 liters of water—based on the crop type, climate, and management practices—into water vapor through evapotranspiration. Depending on diet and where food is grown, each person is responsible for the conversion of 2000 to 5000 liters of water each day into water vapor (Renault and Wallender 2000). The 2 to 5 liters for drinking and 50 to 200 liters per day for household use seem insignificant when compared to the amount of water required for producing food.

As living standards improve, people consume more food of greater diversity, usually shifting to more meat and vegetable consumption. For example, in China meat contribution to calorie intake has quadrupled over the last 20 years. Improving diets for those who were living in poverty can be seen as positive, but this also means more pressure on water resources. The Millennium Development Goals target the 840 million malnourished people, for whom a better diet would require more water consumption. It is also true that many people in developed countries could be much more conscious about their water-consuming diet habits. In summary, adding 2 to 3 billion people with more diversified diets will require greater water consumption by agriculture.

Provision of water for agriculture is central for the livelihood security of many rural poor in developing countries. Approximately 70% of the world's malnourished live in rural areas with limited livelihood opportunities outside of agriculture (FAO 2004). Providing these people with reliable water, whether it is from small-scale water harvesting or large-scale irrigation, makes it possible for them to move beyond subsistence farming. It prevents yield losses due to short-term drought, which in sub-Saharan Africa may claim one out of every five harvests. It gives farmers the "water security" they need to risk investing in other productivity-enhancing inputs, such as fertilizers and high-yielding varieties. It also enables them to grow high-value crops, such as vegetables, which are more sensitive to water stress and which have higher input costs. Although water resource development has been significant in reducing poverty (Rijsberman 2003), investments in this area have declined, partly because people do not know how to invest or what to invest in.

Irrigation has provided employment, income, and lowered food prices, all of which benefit both urban and rural poor (Lipton et al. 2003). In India, Shah and Singh (2003) found that more irrigation means fewer people below the poverty line. Hussain and Hanjra (2003), in a review of 120 studies, found higher wage rates in irrigated than in rain-fed areas. From a broader national perspective, Bhattarai et al. (2002) estimated a multiplier of 3.15 for irrigation in India, meaning that for every $1 generated by irrigated crop production that directly benefits farmers, another $2.15 indirectly benefits economic development. Poor communities, women in particular, also benefit from irrigation as a source of water for domestic uses, small-scale industry, and fishing (Bakker et al. 1999, Meinzen-Dick and van der Hoek 2001).

The environmental costs of irrigation development and other forms of agriculture are also well established. Over half of the world's more than 45,000 large dams, some two-thirds of which are located in the developing world, have been built for irrigation and water supply purposes. Their environmental impacts are legion (World Commission on Dams 2000). Lemly et al. (2000), in a global study of wetlands, sum it up by stating, "The conflict between irrigated agriculture and wildlife conservation has reached a critical point at a global scale." Moreover, Acreman et al. (2000) and others note that in some cases the values generated by irrigation are less than the values generated by the ecosystems they replaced. According to a recent global study of environmental water requirements conducted as a part of the Comprehensive Assessment of Water Management in Agriculture (www.iwmi.cgiar.org/Assessment/), over 1.4 billion people already live in river basins where current high levels of water use

Box 13.1. Positive impacts of agriculture on wetlands: small tank systems of Sri Lanka

Sri Lanka has one of the oldest traditions of irrigated agriculture in the world, dating back to 500 BC. A hydraulic civilization has evolved since that time of localized irrigation schemes, predominantly for rice cultivation, centered on cascades of small tanks (water storage reservoirs). Some 600 small tanks in the Kala Oya Basin (the northwestern dry zone of Sri Lanka) occur in conjunction with a large-scale irrigation scheme, with the latter the recipient of 65% of basin water. Some 400,000 rural people are engaged in farming as their principal livelihood source, half of whom are poor (monthly income below US$15). A socioeconomic assessment of 429 tanks revealed that they are used for cultivation of rice and other crops, domestic purposes such as household water supply and bathing, fishing, and provision of other edible plants and lotus flowers for ceremonies, each with significant value. Other services that are difficult to evaluate include mitigation of downstream flooding, replenishment of groundwater reserves, nutrient retention, and purification. Benefits were greatest for the poorer families for whom alternative sources of income and domestic water supply were limited (Vidanage and Kallesøe 2004).

Figure 13.1. Water stress in major basins taking into account environmental water requirements

threaten freshwater ecosystems (Smakhtin et al. 2004). This view of environmental water scarcity shows that many countries are already making serious environmental trade-offs to grow food, often without adequate recognition of this fact, and that many more will be facing the same dilemma in the next 25 years (Fig. 13.1). It seems likely, therefore, that the main competition for water over the next century will be between agriculture and the environment (Rijsberman and Molden 2001).

The dilemma is clear. Water is essential for food and livelihood security. Increasing population and incomes and diversifying diets exert pressure to increase amounts of water for food production. Growing cities and industries compete for additional water. Water is also essential for ecosystems, yet taking water out of nature affects ecosystem character and the delivery of ecosystem services to humans and, as a result, the long-term viability of food-producing systems is threatened. Biodiversity is increasingly under pressure as people reallocate water away from supporting terrestrial and aquatic ecosystems to agriculture. New and different ways of developing and managing water in agriculture are needed if this dilemma is to be resolved. Indeed, the authors of this chapter hold that it is possible to design, implement, and manage irrigation systems to maintain socially acceptable and, in some instances, even enhanced, biodiversity.

Water-for-Food Drivers of Ecosystem Change

It is necessary to consider biodiversity broadly, in terms of its primary attributes of composition, structure, and function, which operate across genetic to population-species (the most widely known aspect of biodiversity), community-ecosystem, and finally, regional-landscape level. Conservation of biodiversity in agricultural systems, as elsewhere, would thus encompass all human actions ranging from the total preservation of any individual component of biodiversity, in its own right, to the use of biological resources within sustainable limits, so that no biodiversity is lost (World Conservation Monitoring Centre 1992).

Replacing Natural Systems with Agricultural Lands

Agricultural expansion has had profound impacts on rivers and other wetland systems (MA 2005). Notably, agriculture likely affects over 50% of the world's 1669 wetlands of international importance (e.g., Wiseman et al. 2003). Drainage and conversion for agricultural use is the principal and most direct cause of wetland loss globally, with around half of the world's wetlands already destroyed. By 1985, an estimated 56 to 65% of available wetland had been drained for agriculture in Europe and North America, 50% in Australia, 27% in Asia, 6% in South America, and 2% in Africa.

Removing Water from Aquatic Systems and Changing Hydrologic (Flow) Regimes

When, how much, how often, and in what distributional pattern water flows in rivers or moves through other wetlands affects ecosystem character and aquatic biodiversity (Poff et al. 1997). Major hydraulic structures to store and divert water completely change the character of parts of rivers, in some cases changing a river system to a reservoir cascade system. In other cases, rivers dry out immediately downstream of dams, or in their lower reaches, simply because too much water has been abstracted for agriculture.

A further 160 to 320 new large dams are being constructed annually (WCD 2000). An estimated 60% of the world's major rivers are now fragmented to some extent by dams and other forms of hydraulic infrastructure and associated flow regulation such as run-of-river abstraction (Nilsson et al. 2005). Aquatic ecosystem integrity is underpinned by the natural hydrological regime, so it is not surprising that such alteration of water regimes for irrigation is the most serious continuing threat to wetland biodiversity.

Hydrological alteration has led to a variety of impacts ranging from genetic isolation through habitat fragmentation, to channel erosion and declines in biodiversity and floodplain fisheries (Bunn and Arthington 2002). It has contributed to extinction or increased risk thereof in a significant proportion of freshwater species globally, including at least 20% of the world's 10,000 freshwater fish species. Species extinction rates for US fresh-water fauna are projected to increase from 0.5% to 3.7% per decade in the future, five times higher than that for terrestrial species and comparable to losses observed in tropical rainforests (Postel and Richter 2003). River flow regulation associated with dam operations has been identified as one of three leading causes of the imperilment of aquatic fauna (along with invasive species and non–point source pollution; Richter et al. 1997). Lower-input rain-fed systems undoubtedly also contribute significantly to such biodiversity losses through extensive conversion of natural terrestrial habitats and direct agricultural activities in wetlands.

In addition to reducing surface flows, agriculture has triggered substantial groundwater exploitation in regions with many groundwater-dependent ecosystems. In the United States, about one-fifth of the total area under irrigation is extracting groundwater at rates exceeding recharge. Water tables are declining by up to a meter per year in places on the North China Plain (Kendy et al. 2003) and by 25 to 30 m per decade in parts of India, with largely unassessed ecological impacts.

Alteration of Water Quality: Diffuse Agricultural Pollution and Eutrophication

Water quality has deteriorated in almost all agricultural regions worldwide, with increased concentrations of dissolved salts (salinization), suspended solids, pesticides, fertilizers, and other agrochemicals, as well as livestock wastes, all of which may degrade wetland and other ecosystems. These can be toxic to plants and animals through processes of bioaccumulation and also have negative impacts on human well-being. Reduced capacity of inland waters to filter and assimilate waste has led to declines in biodiversity. Eutrophication of inland waters, the associated distortion of natural nutrient cycles, and groundwater contamination have often been associated with nutrient enrichment due to fertilizer runoff (most significantly nitrogen and phosphorus); for example, human-generated nitrogen currently totals about 210 MT/yr, of which agriculture accounts for some 86% (WRI 1998).

On-Site Impacts

Development and management practices within irrigated agricultural systems affect biodiversity by various means:

- Replacement of natural systems with semi- or nonnatural ones (trees to crops), with concomitant changes in biodiversity and ecosystem character, and altered biodiversity levels either through promotion of monocropping or alternatively through the promotion of floral production, especially in dry areas ("making the desert green"). In the dry zone of southern Sri Lanka, for example, increased water to the area through irrigation allowed wet-zone flora and fauna to flourish (Renault et al. 2001). Water harvesting structures in watersheds create humanmade wetlands.

- Habitat fragmentation and reduction in the area of habitats suitable to plants and animals, principally through clearing and physical alteration of habitat types. Problems include a loss of natural corridors or vegetation belts for the movement of animals or dispersal of plants, and increased edge effects, for instance where small remnant patches of natural habitat are surrounded by agroecosystems.
- Disturbance to fauna during the actual construction of irrigation schemes, for example, leading to the displacement and outward migration of species (and even localized extinction within an irrigation area), or their redistribution within the system.
- Changing the microclimate of the area, groundwater conditions, and air circulation, leading to changes in flora and fauna, including microbiota within the irrigated areas.
- Changing soil quality, through replacing or enhancing natural fertility with artificial fertilizers, resulting in the transformation of the soil environment and consequent changes in biodiversity.

Off-Site Impacts

Flowing water links irrigated areas with agricultural, ecological, and other systems both upstream and downstream. The impacts of irrigation are felt over a far broader area because irrigation activities change flow paths of water in terms of quality, quantity, and timing. Changes in upstream use can quickly influence downstream water uses. Additionally, there may be long lag periods from the time of effect to response when considering natural systems. Such impacts therefore need to be considered in relation to issues of scale.

Scale Issues

One of the biggest difficulties in understanding water resources is the prevalence of cross-scale effects that often have counterintuitive impacts. For example, drip and sprinkle irrigation are widely touted as water conservation practices because less application to crops is needed and drainage flows are reduced from surface practices. The story is not complete, however. Farmers often use this saved water to irrigate more area, and thus overall evaporation increases and downstream flows decrease. Over time, downstream farmers and ecosystems may become dependent on return flows from agriculture. A reduction in drainage flows, without compensating flows, may cause serious downstream impacts. The lesson is that actions taken at one scale, say farm, field, or irrigation scheme, typically have broader basinwide impacts that must be clearly understood before actions are taken. The Aral Basin is an excellent example of how upstream irrigation development caused unanticipated and substantial downstream adverse impacts to the Aral Sea.

In another example, in the semiarid tropics of India, which have predominantly supported dryland cropping systems, small-scale water conservation and harvesting have led to irrigated cropping in pockets. Watershed development geared toward improving agricultural productivity on rain-fed lands has tended to rely on a range of soil and water conservation techniques, including structural and vegetative barriers, field bunds, check dams, as well as erosion control through gully plugs. Widespread adoption of watershed development programs has altered the water balance generally in favor of the uplands, but also in the form of increased groundwater recharge that is most evident in water table increases in the valley portions of the watersheds. Several studies have shown how the intensification of land use resulting from such interventions has increased the availability and utilization of runoff water and soil moisture in the upstream reaches, which in the semiarid agroecosystems has tended to result in diminished runoff flows to downstream users (including irrigation reservoirs and wetlands) (Puskur et al. 2004). Whether this is positive or negative can only be evaluated in a broader basin context.

Similarly, in water for food production, time is an important but complex dimension. Agricultural water management builds on slow-response variables—changes in groundwater levels, salinity, and pollution buildup, all of which influence long-term sustainability of these systems. Sustainability of major food production systems—the Aral Sea Basin, the North China plains, northwest India—is in question. Natural ecosystems similarly show lagged responses to agriculture-driven change over many years, such as long-term changes in the geomorphological character of rivers with flow regulation, or progressive declines in inland fisheries with combined pollution, overharvesting and water abstraction effects, leading to their eventual, sometimes sudden, collapse.

Implications of On-Site and Off-Site Impacts for Livelihoods

Intensified use of natural resources by some members of the community and/or decreased reliance on such resources by others tends to occur with a shift toward the use of land for agriculture. In many instances, the rural poor are directly dependent on these components of wild biodiversity for their nutrition and broader livelihood strategies. This sector of the population is potentially the most vulnerable to change; the rural poor are often landless and not adequately compensated within agricultural systems, and are affected most by even relatively small declines in biodiversity. Use of the scarce water resources for one reduces the availability for the other. With the reduction in the heterogeneity and thereby the resilience of the agricultural systems, the vulnerability of the farming households increases.

Irrigation introduces changes in farming systems that may lead to reduced resilience. Increased incidence of pests and diseases may increase in monocropping systems, leading to higher costs of production for crop protection. As a result of the increasing intensity and costs (e.g., due to unduly high groundwater abstraction) of irrigation, households shift to high-value crops, which have minimal or no fodder value, negatively affecting livestock production. The livestock population (per farm) declines, creating dependence on import of nutrients, which drives up the costs of production. In addition, farmers have to contend with the market risk of cash crops. The farming households may perceive an increased risk and difficulty in farming as a result, even as their incomes are increasing.

Responses to Impacts

Given the inevitable increasing use of water for agriculture, is it possible to irrigate in a way that better supports biodiversity and healthy ecosystems? There are indeed many ways to do this, often not costly or difficult to implement. Major challenges are to mainstream these solutions into design and management processes, and to devise incentives for their implementation. Here we list some possible options to mitigate damage or promote more sustainable ecosystems.

Improving Water Productivity

In terms of freeing up water for the environment and other users, improving basin water productivity seems the best solution. In its broadest sense, improving water productivity means obtaining more value from each drop of water—whether it is used for crop-based agriculture, fisheries, livestock, industry, domestic purposes, creation of energy, or the environment. By improving productivity, it is possible to reduce the need for investments in new water withdrawals—investments that many countries cannot afford in terms of financial and ecosystem costs. Moreover, with a pro-poor focus, it can contribute to the incomes and food security of some of the world's poorest people. Ideally, improving water productivity should allow farmers to grow the same amount of food with less water. The water that is "freed up" can go to ecosystems and cities, or it can be used to grow more food. Of course, water productivity gains can have an environmental cost. They will in many cases require increased agrochemical inputs, with negative consequences for the environment. However, such negative outcomes can be minimized through better water and land management practices, so that the overall benefits for the environment outweigh the costs.

A range of practices and technologies on rain-fed and irrigated land can lead to improved water productivity (Kijne et al. 2003). At farm and field scales, improved crop varieties and improved soil fertility boost yields and water productivity. More precise irrigation application using sprinkler or drip technologies or improved surface systems, such as laser leveling, can also enhance yield and require fewer diversions of water. In dry areas, deficit irrigation—applying a limited amount of water, but at a critical time—can boost productivity of scarce irrigation water. Giving farmers better access to water for irrigation through groundwater development and small-scale technologies can increase productivity and reduce poverty in both irrigated and rain-fed areas. Within basins, allocating supplies to various uses to enhance values improves water productivity. Modifying the landscape, for example, through livestock grazing practices or changing land use, influences water flows (Falkenmark 2003) and thus water productivity.

Along with technical solutions, strong supporting policies are needed. For example, agricultural subsidies in industrialized countries may discourage farmers in Africa from investing in productivity-enhancing inputs because crop prices are too low for them to earn a return on their investment. Firm land and water rights are needed so people will invest in long-term improvements.

Trade in Virtual Water

At the global level, trade can reduce water consumption in agriculture if exporters are able to achieve higher water productivity than importers. More reliance on trade could mean that food is grown in environments that minimize damage, such as rice-growing in wet areas instead of in deserts. In most cases, the major exporters (United States, Canada, and the European Union) have highly productive rain-fed agriculture, whereas most importers have relied on irrigation. Currently, cereal trade reduces annual global crop water depletion by 6% and irrigation depletion by 11%. Estimates that take into account trends in virtual water trade forecast 19% less irrigation use in 2025 than those that do not include trade (de Fraiture et al. 2004).

However, the economic and political interests associated with agricultural trade cannot be ignored. Is it realistic to assume that countries will change trade policies because of emerging global water scarcity issues? Current data show no relationship between countries' available water resources and their volumes of trade (Ramirez-Vallejo and Rogers 2004). A rough estimate indicates that only 20 to 25% of all cereal trade takes place from water-abundant to water-short countries (de Fraiture et al. 2004).

Extremely water-short countries have no choice but to import food. Others with a choice between increasing pressure on water resources and increasing dependence on imports are opting to face the environmental consequences of the former, rather than the political and economic consequences of the latter. For countries such as China and India, with large, growing populations and increasing water problems, a certain degree of food self-sufficiency is still a national priority. Moreover, the question remains whether the countries that will be hardest hit by water scarcity will be able to afford "virtual water." Even if they can afford it, what about the impact it will have on livelihoods and food security for rural poor? Without money or market infrastructure to distribute food, a heavy reliance on trade would simply not work for many rural communities. For many countries, improving agricultural productivity and rural incomes will remain a priority and may even be a prerequisite for virtual water flows.

On-Site Alternatives for Mitigation of Impacts

There are several actions to be taken within irrigated areas.

Maintain Habitat Integrity and Connectivity

This can be done by incorporating ecoagriculture strategies in the system, such as hedgerows and corridors of natural vegetation interconnecting parcels of irrigated land (McNeely and Scherr 2003 provide numerous case examples). In large irrigated areas canals and roads, often lined with trees, are dominant features of landscapes and can serve as important passages for biotic movement and as habitats. They need to be viewed and managed as such, meaning that at critical times, canals should have water, gates or sluices should not necessarily be closed, and water-consuming trees should not necessarily be cut down.

Promote Diversity in Landscapes

One reason that traditional tank cascades (a series of small dams and reservoirs connected by water flows) in Sri Lanka support biodiversity is that they provide mixed, heterogeneous landscapes—small tanks, irrigated paddy fields, forests, and villages in small areas. In many larger irrigated landscapes, it is important to break up large monocropped areas by identifying, protecting, and linking natural habitat patches that provide for elevated biodiversity.

Choose the Right Infrastructure and Operation

Most infrastructure considerations are based on functions of delivering water to crops—flexibility in delivery and storage so that water can be used when crops need it. More attention needs to be paid to infrastructure that supports the multiple uses of water. For example, canal designs that allow for movement of fish may not require so many weirs and gates. Unlined canals may better support flora and fauna. Dual canal systems can allow delivery of different quantities of water over variable time intervals to two types of crops. This will be the case in southern Sri Lanka, where the Walawe Left Bank irrigation canal system currently under construction is intended to provide for different water delivery regimes for paddy and banana crops. Modification of larger-scale hydraulic infrastructure, such as through the provision of dam multiple-release outlets rather than single near-bottom releases, as well through changes in the operation of dam releases generating downstream flow regimes, can result in enormous reductions in ecological impacts on-site, as well as off-site.

Social Mobilization

Engaging local communities (including farmers) and other stakeholders in the process of biodiversity conservation is a critical element of managing water sustainably for meeting food and environmental needs. Raising awareness among them of the implications of alternative regimes of water use and their tradeoffs will greatly enhance their contributions to conserving biodiversity.

Institutional Development

Creation and development of institutions play a strong role in ensuring effective, sustainable and equitable management of water for agriculture. Relatively more effort has been placed in building institutions to manage irrigation delivery and maintenance than in overall natural resource management. But effective institutions could certainly be engaged in maintaining healthy ecosystems within the irrigated area.

Off-Site Alternatives for Mitigation of Impacts

Because agricultural water management impacts other uses outside of the immediate irrigated areas, further measures are required to mitigate negative impacts.

Irrigation alters hydrologic flow paths to deliver water to crops and drain it away, thus affecting the quality and quantity of river and wetland flows, all of which affect the ecological character of systems. Managing irrigation systems so as to maintain environmental water requirements for wetlands can serve to greatly minimize the negative impacts of irrigation.

The hydrological regime and its natural variability are recognized as central to the structure, functioning, and biodiversity of wetlands (Poff et al. 1997). For wetlands to be able to retain their character or some level of degradation acceptable by society, and thereby provide services to humans, both their water quantity and quality requirements must be defined and met. The science of environmental flow assessment is designed to meet this need (Dyson et al. 2003; Tharme 2003). Because ecosystems and their species are known to be dependent on complex combinations of higher flows/flood pulses as well as low flows, it is increasingly recognized as insufficient to deliver a single bulk quantity of water to rivers, distributed in an unvarying manner over time. Rather, water needs to be managed to make sure that natural flow patterns are mimicked to the extent possible or agreed upon in terms of the desired future condition of the ecosystem (Richter et al. 1997a). Considerable work has been done on approaches for determining such ecosystem water requirements (environmental flows) to support biodiversity and healthy ecosystems, in both of their roles, as users of water and as the base of the freshwater resource. Tharme (2003) describes a range of methodologies in use globally and highlights their relative merits and limitations.

Even if the environmental flow required to maintain a certain level of ecosystem functioning is known, getting that water to the right place at the right time is another matter. Two additional procedures are required: first, a formal, negotiated allocation of water to the ecosystem; and second, an actual distribution of supplies. Basin water allocation rules are required to ensure that various parties receive a portion of developed water supplies. Within river basins, allocations to cities, industries, and agriculture have commonly been in place, but recognition of the need to allocate some water to the environment is growing. A first step in allocating water equitably and in a sustainable way that addresses the concerns of all users, including ecosystems and users of their services, is for all users to get a seat at the table and negotiate how much allocation is required. Once the allocation has been made, the next step is to deliver the water to the user. Very few examples of operational implementation or of follow-up exist (Postel and Richter 2003; Tharme 2003).

Integrated Water Resources Management

As impacts occur at a larger scale, institutional arrangements need to incorporate means to deal with both on-site and off-site effects. Nested institutional structures can help deal with the problem. Irrigation organizations that have the responsibility of delivering water service to farmers may not by themselves be responsive to downstream problems they cause, unless another authority is dealing with the problem. Institutional arrangements that cover the river basin scale should be able to manage these larger-scale upstream–downstream issues while not having to manage the details of irrigation service delivery. Real implementation of institutional arrangements using principles of integrated water resources management (Jonah-Clausen 2004) is a priority action.

Activities like livestock production and fisheries are often integral components of irrigated agriculture that seem to be neglected in the development and management of irrigation. In this sense, fishers, herders, use of water for drinking and small industrial activities, and people dependent on other ecosystem services are "off-site," external to their management regime. An action then is to internalize these noncrop irrigation uses of water into irrigation management and include these stakeholders in water allocation and overall management decisions.

Issues Emerging from Impact-Response Options

Several questions arise from on- and off-site responses that are difficult to resolve.

How Much Water for Agriculture and How Much Water for Wetlands?

As water development for food, cities, energy, and industries intensifies, many basins will simply run out of water to allocate to various uses. Competition increases between cities, industries, and agriculture, with water often moving to the more highly valued or more politically powerful uses. Agricultural and environmental uses, often perceived to have the lowest value, are forced to vie for the last drops. The issue is how much water should agriculture get, and how much water should remain in environmental uses—an issue that generates a variety of opinions. Some people put priority on water for agriculture to promote economic development and are willing to give up some water-supported environments. They argue that only rich countries can afford environmental flows. Others argue for the need to preserve environmental flows and point out that the services they provide have significant value for sustainable development (Nichter et al. 2003). One future challenge is to resolve this issue, probably on a basin by basin basis. Required for adequate resolution of this issue is an institutional base for negotiation, and tools to help people better understand the complex trade-offs involved.

Is Replacing Natural with Artificially Created Irrigation Environments Acceptable?

Intensive irrigation dramatically alters river flows and landscapes and also leads to a change in the state of ecosystems. Irrigated agriculture can result in the creation of wetland habitat for biodiversity or enhancement of important wetland ecological resources. For instance, irrigating in desert environments has the effect of creating oases. Irrigation converts natural wetlands to artificial ones. This is well established, for example, for the 1.3 million km^2 of rice fields that exist globally, which provide staple food for over half of the world's population and harbor a rich biological diversity. Is the changed environment unacceptable? Artificial oases may create habitat and enhance biodiversity. In some cases, however, large tracts of irrigation may irreparably damage natural landscapes.

Although many of these changes may be acceptable for livelihoods and may support biodiversity, significant trade-offs may become evident at larger scales. For example, irrigation may have acceptable and positive changes within irrigated areas in Central Asia, but the water that enabled these changes did not reach the Aral Sea—an end user of the resource. One conclusion is that environments created artificially by irrigation may have positive environmental consequences, though this point is the subject of ongoing debate. But possible trade-offs at larger scales, especially downstream, need to be considered.

Is Small-Scale Irrigation Infrastructure Environmentally Friendly?

The environmental consequences of large dams and canals discussed earlier are clearly evident. Given this, it is important to remember that the reason for large-scale infrastructure development is to target many peoples and stimulate local economies in a cost-effective manner.

Small irrigation infrastructure has distinctly different features, advantages, and some possible pitfalls. Technologies such as pumps and small water harvesting structures are highly divisible as opposed to large dams. They do not necessarily require government subsidies, and individuals or small communities can invest and quickly reap returns on investments. They are excellent for targeting smallholders and poverty and for providing an important means of improving livelihoods.

Because of their small scale, the corresponding ecological impacts of individual systems are also small. Small-scale irrigation infrastructure tends to support a more diverse landscape. Yet the cumulative impact of many small-scale works can significantly alter environments in different ways. Groundwater levels are falling in the North China plains because of the cumulative effect of millions of pumps. Upstream water harvesting is not usually targeted to tap main river stems but does change small tributaries and groundwater, which may ultimately lead to decreased river discharge and desiccated wetlands in dry environments. Communities and individuals can manage their own small scale irrigation but may not be easily set up to influence the cumulative impact of several small-scale structures. For example, it is probably much easier for a single organization to release environmental flows from a large dam than it is from hundreds of small water harvesting structures. The bottom line is that large and small infrastructure have a place, and small scale is not automatically better than large scale. What is important in either situation is to manage what exists to match livelihood and environmental needs.

Conclusions

It is possible to design, construct, and manage irrigation systems to maintain and, in some instances, even enhance biodiversity. This conclusion is important because of the significant negative environmental, social/livelihood, and other impacts from irrigation development offsetting significant economic and poverty reduction gains. Irrigation systems that are designed without enough consideration of the ecological consequences can seriously alter environmental balance. As a consequence, biodiversity, livelihoods, and even long-term productivity are threatened. Dealing with the negative effects of irrigation development and management is a priority issue.

A high degree of biodiversity can and does exist in irrigated landscapes. Roles of biodiversity need to be better understood in terms of agroecosystems and the maintenance of ecological character of natural systems so as to provide ecosystem services (CA 2005). There is a need for continued assessment of natural resource use and its economic and broader livelihoods value, based on the increasing evidence of the importance of access to and use of natural resources by rural poor. There are a number of ways to maintain and protect biodiversity, and there are bright spots from which lessons can be learned on how to enhance biodiversity in irrigated landscapes. Traditional and locally managed irrigation practices have much to offer in best-practice management. Although a few examples are listed in this chapter it would be worthwhile to compile further examples of these practices to disseminate lessons learned.

In many instances it is likely that the indirect positive benefits and negative costs of irrigation are of the same order of magnitude as production and income benefits from growing irrigated crops. This will require different views to be considered in the design of irrigation from different stakeholders representing affected communities, and a multidisciplinary team, including ecologists.

Unfortunately, to date, scientific assessments of ecological disruption resulting from agricultural development projects are generally post hoc, with the predevelopment status unknown or only superficially examined in the course of environmental impact assessments. An equally important aspect of sustainable development is assessment of the socioeconomic status of the communities living or resettled in the project area. This should include the degree to which the livelihoods of the rural poor, in particular, are dependent upon the products and services provided by the various ecosystems they inhabit, given the clearing of natural habitats and loss of biodiversity with the landscape shift to a mosaic of natural and agroecosystems.

The challenge in the very near future is to bring about some fundamental changes in how we think about, design, and manage irrigation (CA 2007). Methods of irrigation can be greatly improved, but for this to happen we need to better understand the positive and negative impacts and why they occur, to develop practical solutions, to spread the knowledge, and to get the incentives and institutions in place that will support ecoagriculture in irrigated landscapes.

References

Acreman, M.C., F.A.K. Farquharson, M.P. McCartney, C. Sullivan, K. Campbell, N. Hodgson, J. Morton, D. Smith, M. Birley, D. Knott, J. Lazenby, R. Wingfield, and E.B. Barbier. 2000. Managed flood releases from reservoirs: issues and guidance. Report to DFID and the World Commission on Dams. Centre for Ecology and Hydrology, Wallingford, UK.

Bakker, M., R. Barker, R. Meinzen-Dick, and F. Konradsen (eds.). 1999. Multiple uses of water in irrigated areas: a case study from Sri Lanka. SWIM Paper 8. International Water Management Institute, Colombo, Sri Lanka.

Bhattarai, M., R. Sakthivadivel, and I. Hussain. 2002. Irrigation impacts on income inequality and poverty alleviation: policy issues and options for improved management of irrigation systems. IWMI Working Paper 39. International Water Management Institute. Colombo, Sri Lanka.

Bunn, S.E. and A.H. Arthington. 2002. Basic principles and ecological consequences of altered flow regimes for aquatic biodiversity. *Environmental Management* 30(4):492–507.

CA (Comprehensive Assessment of Water Management in Agriculture). 2007. Water for food, water for life: A comprehensive assessment of water management in agriculture. Earthscan, London and International Water Management Institute, Colombo, Sri Lanka.

De Fraiture, C., X. Cai, U. Amarasinghe, M. Rosegrant, and D. Molden. 2004. *Does Cereal Trade Save Water? The Impact of Virtual Water Trade on Global Water Use.* Comprehensive Assessment Research Report no. 4. International Water Management Institute, Colombo, Sri Lanka.

Dyson, M., G. Bergkamp, and J. Scanlon (eds.). 2003. *Flow: The Essentials of Environmental Flows.* World Conservation Union (IUCN), Gland, Switzerland, and Cambridge, UK.

Falkenmark, M. 2003. *Water Management and Ecosystems: Living with Change.* GWP TEC Background Report no. 9. Global Water Partnership, Stockholm.

FAO (Food and Agriculture Organization). 2004. *The State of Food Insecurity in the World 2004.* FAO, Rome.

Hussain, I. and M.A. Hanjra. 2003. Does irrigation water matter for rural poverty alleviation? Evidence from South and Southeast Asia. *Water Policy* 5(5–6):429–442.

Jonch-Clausen, T. 2004. *Integrated Water Resources Management (IWRM) and Water Efficiency Plans by 2005: Why, What and How?* TEC background papers no.10. GWP, Stockholm.

Kendy, E., D. Molden, S.T. Steenhuis, and L. Changming. 2003. *Policies Drain the North China Plain: Agricultural Policy and Groundwater Depletion in Luancheng County, 1949–2000.* Research Report 71. International Water Management Institute, Colombo, Sri Lanka.

Kijne J.W., R. Barker, and D. Molden (eds.). 2003. *Water Productivity in Agriculture: Limits and Opportunities for Improvement.* Comprehensive Assessment of Water Management in Agriculture Series. CABI Publishing, UK, in association with the International Water Management Institute, Sri Lanka.

Lemly, A.D., R.T. Kingsford, and J.R. Thompson. 2000. Irrigated agriculture and wildlife conservation: conflict on a global scale. *Environmental Management* 25:485–512.

Lipton, M., J. Litchfield, and J.M. Faurès. 2003. The effects of irrigation on poverty: a framework for analysis. *Water Policy* 5(5–6):413–427.

MA (Millennium Ecosystem Assessment. 2005. *Ecosystem Services and Human Well Being: Wetlands and Water Synthesis.* World Resources Institute, Washington, DC: 68.

McNeely, J.A. and S.J. Scherr. 2003. *Ecoagriculture: Strategies to Feed the World and Save Wild Biodiversity.* Future Harvest and IUCN (World Conservation Union). Island Press, Washington, DC.

Meinzen-Dick, R.S. and W. van der Hoek. 2001. Multiple uses of water in irrigated areas. *Irrigation and Drainage Systems* 15(2):93–98.

Nilsson, C., C.A. Reidy, M. Dynesius, and C. Revenga. 2005. Fragmentation and flows regulation of the world's largest river systems. *Science* 308:405–408.

Poff, N.L., J.D. Allen, M.B. Bain, J.R. Karr, K.L. Prestegaard, B.D. Richter, R.E. Sparks, and J.C. Stromberg. 1997. The natural flow regime. A paradigm for river conservation and restoration. *BioScience* 47:769–784.

Postel, S.L. and B.D. Richter. 2003. *Rivers for Life: Managing Water for People and Nature.* Island Press, Washington, DC: 253 pp.

Puskur R., J. Bouma, and C. Scott. 2004. Sustainable livestock production in semi-arid watersheds. *Economic and Political Weekly* 39(31):3477–3483.

Ramirez-Vallejo J. and P. Rogers. 2004. Virtual water flows and trade liberalization. *Water Science and Technology* 49(7):25–32.

Renault, D., M. Hemakumara, and D. Molden. 2001. Importance of water consumption by perennial vegetation in irrigated areas of the humid tropics: evidence from Sri Lanka. *Agricultural Water Management* 46(3):215–230.

Renault, D. and W.W. Wallender. 2000. Nutritional water productivity and diets. *Agricultural Water Management* 45:275–296.

Richter, B.D., J.V. Baumgartner, R. Wigington, and D.P. Braun. 1997a. How much water does a river need? *Freshwater Biology* 37:231–249.

Richter B.D., D.P. Braun, M.A. Mendelson, and L.L. Master. 1997b. Threats to imperiled freshwater fauna. *Conservation Biology* 11:1081–1093.

Rijsberman, F. 2003. Can development of water resources reduce poverty? *Water Policy* 5(5– 6):399–412.

Rijsberman, F.R. and D.J. Molden. 2001. *Balancing Water Uses: Water for Food and Water for Nature.* Thematic Background Papers. International Conference on Freshwater, Bonn, 3–7 December: 43–56.

Rosenburg, D.M., P. McCully, and C.M. Pringle. 2000. Global-scale environmental effects of hydrological alterations: introduction. *BioScience* 50(9): 746–751.

Shah, T. and O.P. Singh. 2003. *Can Irrigation Eradicate Rural Poverty in Gujarat?* Water Policy Research Highlight 10. IWMI TATA Water Policy Program, Gujarat, India.

Smakthin, V., C. Revenga, P. Döll. 2004. *Taking into Account Environmental Water Requirements in Gobal-Scale Water Resources Assessments.* Comprehensive Assessment Research Report no. 2. International Water Management Institute, Colombo, Sri Lanka.

Tharme, R.E. 2003. A global perspective on environmental flow assessment: emerging trends in the development and application of environmental flow methodologies for rivers. *River Research and Applications* 19:397–441.

Vidanage S.P. and M. Kallesøe. 2004. *Kala Oya River Basin, Sri Lanka: Integrating Wetland Economic Values into River Basin Management.* Environmental Economics Programme, IUCN, Sri Lanka Country Office, Colombo, Sri Lanka.

WCD (World Commission on Dams). 2000. *Dams and Development: A New Framework for Decision-Making: The Report of the World Commission on Dams.* Earthscan Publications Ltd., London, and Sterling, VA.

Wiseman R., D. Taylor, and H. Zingstra (eds.). 2003. Proceedings of the workshop on agriculture, wetlands and water resources: 17th Global Biodiversity Forum, Valencia, Spain, November 2002. National Institute of Ecology and International Scientific Publications, New Delhi, India.

World Conservation Monitoring Centre (WCMC). 1992. *Global Biodiversity: Status of the Earth's Living Resources.* Chapman and Hall, London.

WRI (World Resources Institute). 1998. *World Resources 1998–99: A Guide to the Global Environment.* World Resources Institute, Washington, DC.

Chapter 14

Remote Sensing

Aaron Dushku, Sandra Brown, Tim Pearson, David Shoch, and Bill Howley

Introduction

With remote sensing and geographic information systems (GIS), new assessment and measurement techniques can be applied to a wide range of ecoagriculture landscapes. McNeely and Scherr (2003) suggest six key strategies to manage the coexistence of biodiversity and agriculture. Innovative techniques in analysis of remote sensing products and application of GIS tools could play an especially valuable role in planning and tracking progress for three of these situations:

- Enhancing wildlife habitat on farms and establishing farmland corridors that link uncultivated spaces
- Establishing protected areas near farming areas, ranch lands, and fisheries
- Mimicking natural habitats by integrating productive perennial plants

Some strategies for soil, water, and vegetation management, and for reducing agricultural pollution can also be tracked remotely.

Remote sensing tools can (1) allow participating human communities to play a part in a transparent and interactive planning process, (2) allow participants to visualize anticipated project results at a landscape scale, and (3) enable measurement of project success according to the principles of ecoagriculture. Quantitative performance measures for ecoagricultural integrity can be derived through remote sensing of land cover, biomass, or carbon stocks, or be supported by direct measurement in the field through proxy indicators in the case of biodiversity or habitat quality (e.g., through the use of spatially explicit bird inventory data or GIS-based habitat fragmentation models).

This chapter describes applications of GIS-based decision support in ecoagriculture landscape planning, landscape performance monitoring, and assessment of spatial integrity of habitats. It then discusses the limitations of remote sensing, and future directions.

GIS-Based Decision Support in Ecoagriculture Landscape Planning

McNeely and Scherr (2003) suggest that the existing network of global protected areas is insufficient to adequately conserve biodiversity, especially in areas of high human population pressure. They suggest that farmers, conservationists, and policymakers should manage landscapes in a manner that considers the objectives of all, resulting in the creation of a mosaic landscape of agricultural and conservation land uses. Integration of sophisticated GIS decision-making tools greatly aids in participatory rural land use and resource planning, such as protected area establishment near farming areas, ranch lands, and fisheries.

Multicriteria Evaluations

Multicriteria evaluation (MCE) uses a variety of data inputs as factors to evaluate a phenomenon, such as a potential land use in a specified area (e.g., a specific agriculture type). It has been recognized by the United Nations as a valuable tool for involving communities in their land-use planning. In 1990, the Idrisi Project signed a memorandum of understanding with the United Nations Institute for Training and Research (UNITAR) to provide scientific assistance to its training programs through the development of curricula and training materials related to specific application areas like decision support (Eastman et al. 1993). These methodologies have been applied to land-use allocation studies, for example, in Malawi by Orr et al. (1998) and in Italy by Villa et al. (2002). They have also been used in more specific analyses such as the United Nations Food and Agriculture Organization (FAO)-sponsored aquaculture suitability analyses in Latin America and Africa by Aguilar-Manjarrez and Nath (1998) and Kapetsky and Nath (1997) or in reservoir siting in Malaysia and in South Africa by Serwan and Wan-Yusof (2003) and Woods (1997).

The spatial modeling techniques used to model the suitability and the susceptibility of lands for deforestation in many of the world's largest conservation-type carbon emissions offset projects rely heavily on the logic of MCE (Brown 2003; Dushku and Brown 2003; Hall et al. 1995; Pontius et al. 2001). Crediting schemes that are designed to allow projects to sell certified emissions reduction units to their funders must prove that forests would have been deforested if the funding had not arrived to create conservation lands. To model the likelihood of the occurrence of deforestation, the suitability of an area for nonforest land uses is analyzed using a kind of empirical MCE. Such techniques have also been used as a planning tool for biodiversity conservation projects (Menon et al. 2000).

To make suggestions for potential future projects, the best possible understanding of the available lands and the pressures exerted on them is essential. Using a combination of available digitized maps, a separate analytical map can be produced to identify the areas where a certain activity is most likely to occur or would occur with the best success. This process can be repeated for opposing objectives, such as human and wildlife resource needs, and the analytical data can be used to generate optimal solutions for both groups. By mapping all the related factors and assessing the current and potential needs through a stakeholder-involved process, MCE allows better-educated steps to be made toward common goals of efficient land use and natural resource management.

These analyses provide further options for future natural resource planning activities and conflict resolution as envisioned by McNeely and Scherr (2003) in ecoagriculture landscapes. For example, an MCE can be conducted to analyze watershed health and to map potential productivity for a given agricultural land use. Such suitability modeling is done by combining empirical knowledge of historically good areas for certain activities, input from stakeholders at the local level, and information and input from the academic or technical sector. These can be accounted for within GIS-based suitability models and given different weights through the interactive process. When objectives for using a landscape are in conflict, such as conservation of a watershed or intensive industrialized agriculture, this can also serve as a systematic decision-making tool (i.e., zoning of high-production land uses in sites that do not affect watershed health).

By using local knowledge about the actual state of a landscape and its uses, and then graphically showing to those involved the area of concern and the current and projected needs, stakeholders gain a better understanding of a situation and are able to make more educated decisions on what to do. Often, this is difficult for people who have never taken a "big-picture" point of view to conceptualize issues and thereby develop and understand longer-term strategies for change. Involving stakeholders in the MCE process enables them to get a firsthand look at the amount of land available for farming, current uses, climate change patterns, wildlife habitats, migration routes, and the like. Once they are trained to go through this process and are therefore aware of larger-scaled realities of resource pressure, they are better able to work closely with a team that is prepared to implement activities for conservation or agricultural development.

Over time, the MCE processes for identification of suitable sites for certain activities can be repeated with updates to the input data. Protocols and procedures for local team members can be taught as a means to enable the inclusion of their own unique data layers from research and conservation activities in the collective information hierarchy, and the export of information into linked knowledge repositories in web-based systems for internal and external actors to access according to necessity. This allows for monitoring and evaluation of project objectives, and it highlights areas showing the greatest successes or challenges in achieving project goals. Most important, perhaps, it can provide for transparency and replicability in other parts of the country or region.

Case Study: Identifying Priority Areas for Community Action in the Georgia Energy Security Initiative

In response to an increase in illegal wood harvesting resulting in the depletion of forest resources and biodiversity, PA Consulting and Winrock International, funded by the US Agency for International Development (USAID), recently embarked on an innovative project in the Republic of Georgia. Called the Georgia Energy Security Initiative (GESI), this project sought to implement commercially viable projects to generate income, reduce deforestation, and introduce renewable energy and energy-efficiency technologies. The first step in this effort was to design a transparent and systematic process for selecting, among all communities across Georgia, the areas and communities best suited to achieve these results (Fig. 14.1). A multiphased process was implemented, beginning with a remote MCE that drew upon a variety of data sources. From an initial data set of approximately 7500 communities, this phase produced a ranked list of the most promising communities in each region, as well as maps showing areas of particular interest. Using these maps, further exploration in the community selection phases followed (ground truthing, community surveys, and participatory assessments).

To define what would constitute the most promising communities and areas for further exploration, it was necessary to translate the primary selection criteria from the GESI work plan into variables that could be quantified, weighted in terms of importance, ranked, and mapped in GIS. For example, communities near a deforested or protected area having hydrologic characteristics suggesting small hydro potential and in an area covered by a microfinance provider would have a relatively high score and be a promising GESI partner. The rankings thus produced were not intended to be definitive or exclude any community, but rather to offer an objective standard and guidance to the GESI team in deciding which communities to visit to interview community actors and gather additional energy, natural resource, and other data.

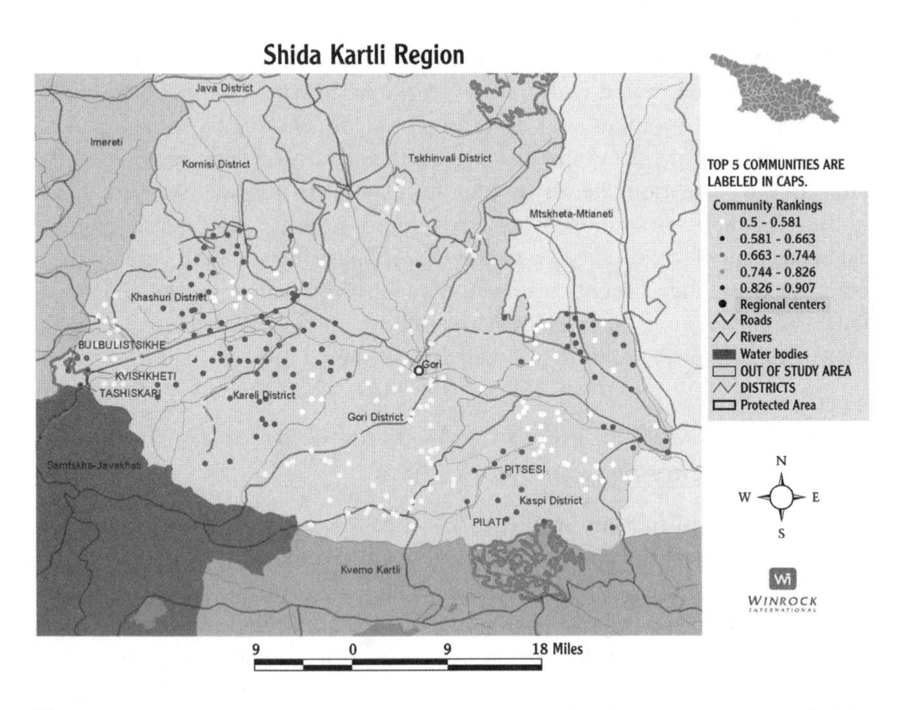

Figure 14.1. Region map with the point locations for the communities coded by their range in suitability scores

Each variable was assigned a weight, from 5 (highest) to 1 (lowest), reflecting the GESI team's consensus judgment on how critical that variable was to the selection of communities. Each variable was also assigned a "decision rule"—for example, being located close to a forested area represents deforestation danger; close to rivers represents hydro potential; an area served by a credit provider or having a strong community-based organization represents ability to implement and sustain projects. Using the decision rule, a "suitability index" for each variable was produced on a 0 to 1 scale (e.g., a community adjacent to a protected area might get a score of 0.9 for this variable). Once suitability indices had been generated for each variable (factor) and community, they were multiplied by the assigned weights and averaged across all variables to produce a weighted average score (or overall suitability score) for each community.

MCE thus offers the capability to compare the results of choosing the most strategic partners regardless of geographic location versus choosing the best partners in each region for the sake of geographic distribution. It is also possible to overlay existing data layers (rivers, roads, railroads, hydro plants, protected areas, etc.), or layers that become available in the future (pipelines, sawmills, etc.), to evaluate the effectiveness of the selected GESI partners in the implementation of their project activities and draw conclusions about the importance and weights of the factors used. As such, the GIS-based MCE is designed to serve as a flexible, ongoing selection and monitoring tool that can be applied to planning for protected area creation or for agricultural projects the same way as it was to renewable energy projects here.

Performance Monitoring in Ecoagricultural Landscapes

Land-use planning is ideally done through the engagement of stakeholders representing various interests using interactive multicriteria evaluations as described earlier. Through that process, landscapes will commonly become a mosaic of agriculture and protected areas. This section describes a number of remote-sensing tools that can be used to monitor progress toward achieving biodiversity and ecosystem service goals. Examples illustrate their application to monitoring biomass and carbon stocks as well as spatial fragmentation.

Landscape-Wide Assessments versus Sampling Regimes

For use in natural resource surveys and inventories, the acquisition and analysis of conventional remote sensing data, like the use of satellite imagery or high-resolution digital aerial imagery, are often sufficient. Whichever databases are used, it may be possible to gather data at the beginning of any project for the entire project region or, depending on the complexity of the variable to be monitored, it may only be possible to gather data from a representative sample of the region.

If the variable to be monitored is the presence or absence of forests (or other productive perennial agricultural species), remote sensing products can be acquired that will allow for a baseline snapshot of a project area to be compared to any snapshot that can be taken at the midpoint or at the end of the project. The changes in area of such land-cover types can be measured and reported. If a project that aspired to establish forested stream buffer zones in a previously denuded agricultural region wanted to demonstrate its successes after the project ended, it could compare satellite imagery from the beginning of the project to imagery at the end of the project and show exactly how much forest it had planted (depending on the duration of the project and the growth rates of the selected species). However, when using satellite imagery, the scale of the imagery is important given that sometimes narrow riparian zones may not show up in the imagery; LANDSAT or SPOT with pixel sizes of 20 to 30 m by 20 to 30 m may not be able to detect riparian zones of less than this dimension. Smaller-scale projects or narrower buffers might need to use aerial digital imagery to allow for fine-enough resolution to capture such detail. Similar kinds of analyses can be envisioned for the establishment of agroforestry systems in an agricultural landscape.

In cases where the spatial characteristics of forested areas are to be modified by an ecoagricultural project such as the "defragmentation" of a region's forests, similar snapshot analyses might be conducted. Remotely sensed data from the beginning of the project can be analyzed using the appropriate GIS software (e.g., FRAGSTATS). Indices for fragmentation can be derived for the entire landscape and compared to the same indices derived from remotely sensed data at the end of the project to communicate a change in the fragmentation. This technique is further explained following here.

Sometimes, the analysis of landscape-wide indicators, such as increase in overall biomass or certain spatial characteristics of wooded areas, is the full extent of the study. More frequently, however, whether these indicators have influenced a project's fundamental goal of achieving biodiversity conservation or other ecosystem services is the issue that needs to be illustrated. Such conclusions are often only possible through the analysis of trends in the spatial data as compared to some site-specific data gathered at sample locations. In this case, the practice of measurement and monitoring of an initiative becomes much more complicated, and statistical rigor is necessary to assure significant results. Extensive research has been conducted in the measurement and monitoring of carbon in ecosystems (Brown 2002; Roshetko et al. 2002), so this example of carbon sequestration is used to illustrate the point.

Measuring Biomass and Carbon Stocks

To estimate total biomass or the carbon stock sequestered in the landscape one could measure everything—every single tree, for example, in the project area. However, complete enumerations are almost never possible in terms of time or cost. Consequently, sampling is necessary, in which studying a subset allows generalizations to be made about the whole population or area of interest. The values attained from measuring a sample are just an estimation of the equivalent value for the entire area or population. Statistics provide some idea of how close the estimation is to reality. Once confidence is achieved in estimates of carbon stocks for a certain area of forest type, that amount can be extrapolated to the entire mapped area that was derived from remote sensing or other conventional mapping methods. Two statistical concepts must be clearly understood in this discussion: precision and accuracy. Accuracy constitutes how close sample measurements are to the actual value. Precision is how well a value is defined. In sampling, precision represents how closely grouped the results are from the various sampling points or plots.

When sampling for carbon (or any variable), measurements should be both accurate (i.e., close to the reality for the entire population) and precise (closely grouped so we can have confidence in the result). However, one must also consider the amount of resources available for generating accurate and precise estimates. Sampling on the land inevitably involves measuring a number of plots, and installation of plots by field crews can be a costly exercise. The number of plots should be predetermined to ensure accuracy and precision. The average value when all the plots are combined represents the wider population, and we can tell how representative it is by looking at the 95% confidence interval, for example. The 95% confidence interval tells us that 95 times out of 100 the true value lies within the interval. If the interval is small then the result is precise.

Projects will vary in the amount of resources they have available for implementing a measuring and monitoring plan. As mentioned earlier, poorly measured variables with unknown or very low precision will not be able to demonstrate positive impacts such a poor measurement plan has little value. Measurement methods can be tailored to fit within the resources available for assessing project impacts, yet still give credible results. The amount of resources needed to be spent on implementing a monitoring plan is highly related to the number of plots established. The number of plots needed to attain a given precision level decreases with a decreasing desired precision level.

At Noel Kempff Mercado National Park in Bolivia, for example, measurements were taken in a complex tropical forest with five different forest strata (Fig. 14.2). To attain a high precision (i.e., a 95% confidence interval of ±5% of the mean), 452 plots would be needed. However, this drops rapidly to 81 plots needed for ±10% of the mean and even further to only 14 plots for ±20% of the mean. Thus, when resources are limited, the precision level can be set lower, and fewer plots could be sampled. Even at ±20% of the mean the results would be accurate with a known precision and would likely be able to show with confidence a change in carbon stocks.

Case Example: Aerial Digital Imaging for Pine Savanna in Belize

Winrock International has developed and tested an aerial digital imagery system (multispectral three-dimensional aerial digital imagery [M3DADI] system) as a very high resolution remote measurement tool to estimate biomass and carbon sequestration in a landscape (Brown et al. 2005, Electric Power Research Institute 2000; Slaymaker et al. 1999). By flying transects over a landscape, detailed samples of the vegetation structure can be collected. Measurements of individual plant crown diameter and height can be made from sample plots established on the images. Regression equations can then be developed between tree crown diameters and heights and biomass of the trees or shrubs. Using this technique, M3DADI can cost-effectively estimate changes in carbon stocks over large, and often remote, areas.

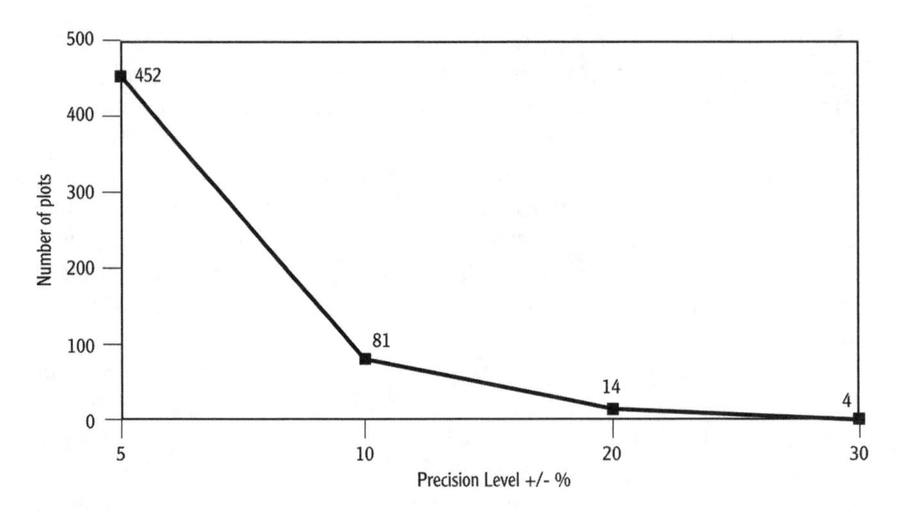

Figure 14.2. The relationship between sampling and precision

This system has been applied to a pine savanna system in Belize (Fig. 14.3) (Brown et al. 2005). The pine savanna is a system with pine trees as the dominant vegetation mixed with broadleaf trees, shrubs, palmetto, and grasses. The pine savanna is highly heterogeneous and difficult to stratify for carbon measurement and monitoring plan design. Seventy-seven plots were established on the pine savanna images, and using a series of nested plots the crown area and heights of pine and broadleaf trees, palmettos, and shrubs were digitized.

Measures of biomass or carbon can be calculated by using remotely sensed measures of tree crown and height, tree populations, and vegetative cover, together with measurements from representative field plots. Where land use is relatively homogeneous, only a small number of plots are required; where heterogeneity is high, hundreds of plots may be needed. In the case just described, the cost-effectiveness of the M3DADI approach was compared with conventional field methods; the conventional approach took about three times more person-hours than the M3DADI approach.

Evaluating the Spatial Integrity of Habitats

Forest fragmentation is detrimental to native fauna and flora survival rates and thus decreases biodiversity. Landscape ecologists have offered many approaches to quantify such phenomena (e.g., Baker and Cai 1992; McGarigal and Marks 1995; O'Neill et al. 1988; Turner 1990; Turner and Gardner 1991) and hundreds of indices have been developed. When an ecoagriculture project seeks to re-

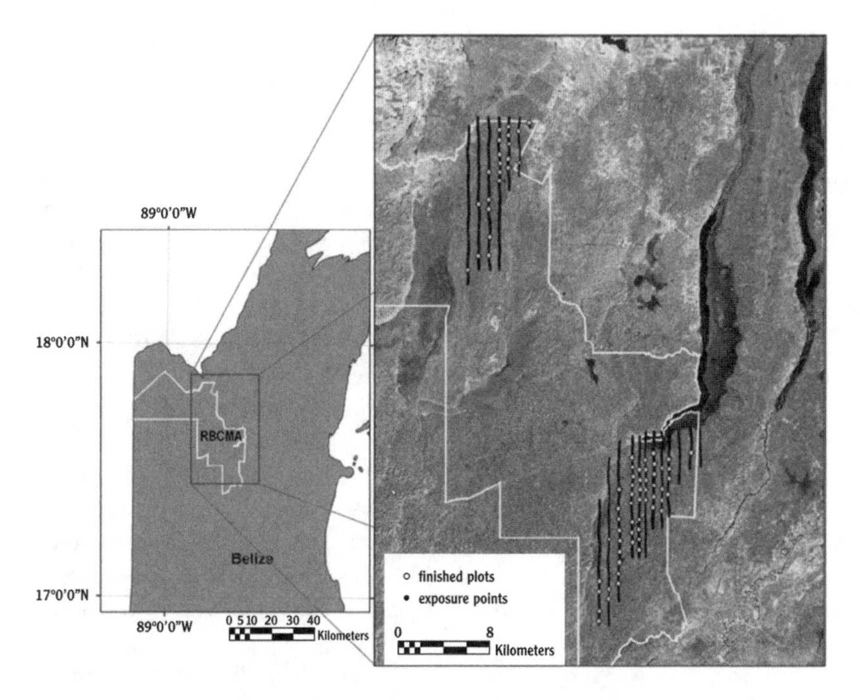

Figure 14.3. Pine savanna plots embedded in the broadleaf forest landscape

verse fragmentation by planting of trees in buffers or in corridors, remotely sensed data from the beginning of the project can be analyzed to extract measurements on preproject conditions. Quantitative indices for forest fragmentation can be derived from the remotely sensed data for the landscape and compared to the same indices at the end of the project to communicate a change in the fragmentation.

With advances in computer hardware and remote sensing for data collection, researchers at the University of Massachusetts, Amherst, have developed software called FRAGSTATS that produces many of the indices for a geographical area (McGarigal et al. 2002). FRAGSTATS can be used to plan a project and anticipate what benefits are to be expected or to assess an existing project for the benefits it has created.

Case Study: Effects of Fragmentation on Biodiversity in Appalachian Forests

To determine the effect that fragmentation of forests on a landscape can have on biodiversity, it is necessary to actually examine animal populations and to study their relationship to fragmentation indices. Forests surrounding annual bird inventories are being analyzed to examine this relationship in a pilot study in the US state of Ohio.

Site coverages and classified satellite imagery are being overlain with publicly available georeferenced US Geological Service (USGS) Breeding Bird Survey (BBS) data. The BBS was established in 1965 and consists of 145,000 annually surveyed point counts (Blondell et al. 1970) on randomly located routes in the United States and Canada. Time, season, weather, and often observer are constant from year to year, allowing for unbiased monitoring of trends in relative abundance.

Periodic counts of indicator species are being calculated from BBS data for the area, and corresponding landscape metric values describing the relative size and configuration of forest patches, as measured by FRAGSTATS, are being derived from classified satellite imagery. To date, species richness (mature hardwood forest–associated, area-sensitive birds) has been estimated for subsets of individual BBS routes in the study area. Species detection data for each point have been disaggregated from the route, with 10 consecutive points representing independent samples from a single (closed) population (Fig. 14.4). Twenty-two indicator species have been selected that are associated with mature upland hardwood forests.

Future analyses will produce landscape-level metrics for the forests in each segment and relationships will be analyzed and reported as the project progresses. At this time, preliminary correlation has been identified between forest area within the route segments and general bird species richness (Fig. 14.5).

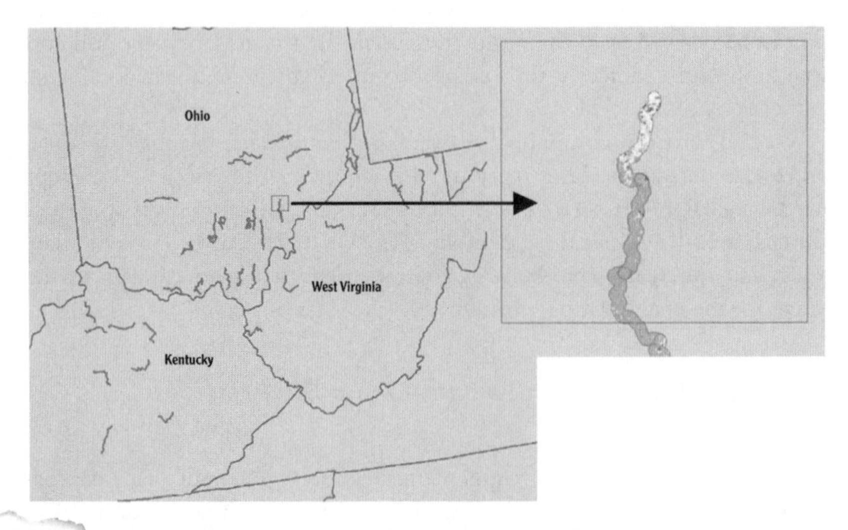

Figure 14.4. Route segmentation in the Appalachian study

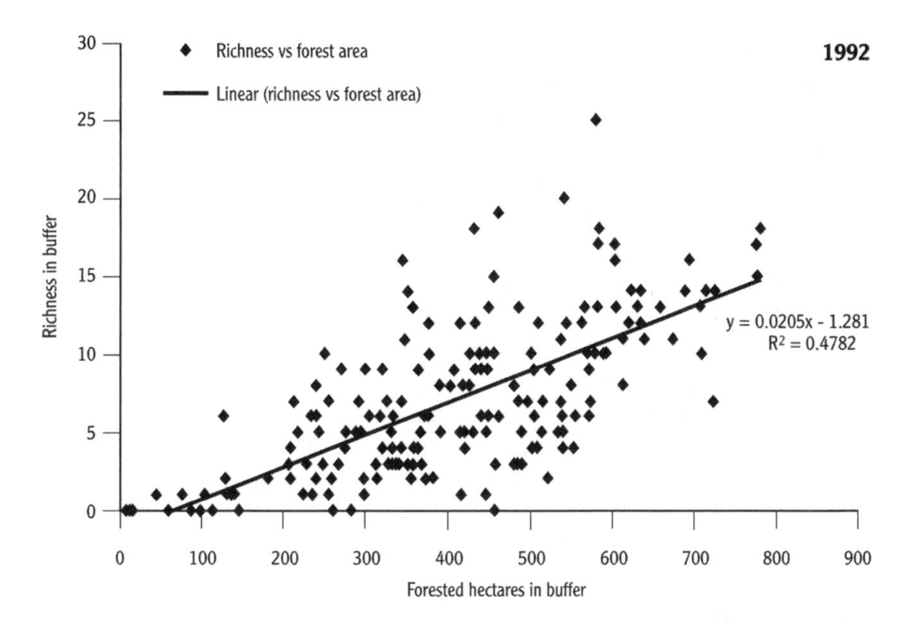

Figure 14.5. Graph of breeding bird survey species as correlated with forest area

This result alone indicates that one can accurately estimate the actual effects on certain bird populations that clearing forest cover might have. Future efforts to increase or recuperate biodiversity levels through the strategic planting of forests or woody species in the selected landscape can count on that.

Limitations of Remote Sensing Tools

Putting these monitoring tools to work involves limitations. GIS and interpretive models are only as good as the data that go into them. When available, the quality of international GIS data sets is usually coarser and more geographically inaccurate than in the United States. Therefore, the results should be interpreted carefully. For example, the BBS is a data set that is found only in the United States, Great Britain, Canada, and, recently, Mexico. Deriving actual relationships between forest fragmentation and biodiversity would be very difficult without spatially explicit field data from the desired study area.

These analyses do not necessarily capture everything that goes on in a landscape, especially if essential supporting data are not included from the outset of the analysis. In many instances, using MCE is an excellent way to convey general trends in land use and provide efficient prescriptions. However, models are approximations. Usually models should be used as the first stage in a longer project where the following stage involves firsthand assessments in the field. An incorporation of M3DADI data with any MCE analysis at a community level would add great detail to assessments, although it is not clear whether such an investment is worthwhile without some incentives for such detailed planning and monitoring, such as in the case of a carbon emissions offset project.

Using M3DADI to measure carbon in landscapes is a science in its infancy. More detailed data could undoubtedly be gathered and processed using high-tech LIDAR systems and automatic crown-delineation software, but the advantage of M3DADI is that it is much less expensive and thus accessible to many more projects. The M3DADI system can be packed into a case and checked as luggage on any conventional airline flight, and it requires very little in the way of additional software systems. Within roughly an hour of arriving on location, it can be attached to the side of any conventional aircraft, and the imagery is maneuvered and processed using Erdas Imagine's Stereo Analyst software suite.

Conclusions

The field methodology and the statistical basis behind remote sensing measurement and monitoring as described in this chapter now serve as the standards across terrestrial carbon sequestration projects worldwide. The techniques discussed using M3DADI to measure carbon in ecosystems are well developed and other applications, such as forest harvest monitoring, are being tested. By using sample data such as the USGS BBS and analyzing the characteristics of the surrounding buffer zones, quantifiable relationships of the spatial characteristics of forests to bird species richness can be extracted. Study of the spatial characteristics of riparian buffers in watershed locations may help to understand the ecological importance of trees in riparian zones. In the future, the use of M3DADI to gather high-resolution data on riparian vegetation in these kinds of situations may allow for even finer levels of analysis.

The role that M3DADI could play in community development and efficient land-use planning, perhaps in conjunction with MCE analyses, has not been fully explored, but by allowing a bird's eye view of a community's resources and land area, local leaders would benefit and make sounder decisions.

References

Aguilar-Manjarrez, J. and S.S. Nath. 1998. *A Strategic Reassessment of Fish Farming Potential in Africa.* CIFA Technical Paper. no. 32. United Nations Food and Agriculture Organization (FAO), Rome.

Baker, W.L. and Y. Cai. 1992. The role of programs for multiscale analysis of landscape structure using the GRASS geographical information system. *Landscape Ecology* 7:291–302.

Blondell, J., C. Ferry, and B. Frochot. 1970. La méthode des indices ponctuels d'abondance (IPA) ou des relèves d'avifaune par "Stations d'écoute." *Alauda* 38:55–71.

Brown, S. 2002. Measuring carbon in forests: current status and future challenges. *Environmental Pollution* 116:363–372.

Brown, S. 2003. *Finalizing Avoided-Deforestation Project Baselines*. Report prepared by Winrock International for the United States Agency for International Development. Contract No. 523-C-00-02-00032-00, Washington, DC.

Brown, S., T. Pearson, D. Slaymaker, S. Ambagis, N. Moore, D. Novelo, and W. Sabido. 2005. Creating a virtual tropical forest from three-dimensional aerial imagery: application for estimating carbon stocks. *Ecological Applications* 15:1083–1095.

Dushku, A. and S. Brown. 2003. Spatial modeling of baselines for LULUCF carbon projects: the GEOMOD modeling approach. In: *Proceedings of the 2003 International Conference on Topical Forests and Climate Change*. Manila, October 21. University of the Philippines, Los Baños: 13–27.

Eastman, J.R., P.A.K. Kyem, J. Toledano, and W. Jin. 1993. *GIS and Decision Making*. Explorations in Geographic Information Systems Technology, vol. 4. United Nations Institute for Training and Research (UNITAR), Geneva, Switzerland.

Electric Power Research Institute (EPRI). 2000. *Final Report on Assessing Dual Camera Videography and 3-D Terrain Reconstruction as Tools to Estimate Carbon Sequestering in Forests*. EPRI, Palo Alto, CA, and American Electric Power, Columbus, OH.

Hall, C.A.S., H. Tian, Y. Qi, R.G. Pontius, J. Cornell, and J. Uhlig. 1995. Modeling spatial and temporal patterns of tropical land-use change. *Journal of Biogeography* 22:753–757.

Kapetsky, J.M. and S.S. Nath, 1997. *A Strategic Assessment of the Potential for Freshwater Fish Farming in Latin America*. COPESCAL Technical Paper no. 10. United Nations Food and Agriculture Organization (FAO), Rome.

McGarigal, K., S.A. Cushman, M.C. Neel, and E. Ene. 2002. FRAGSTATS: spatial pattern analysis program for categorical maps. Computer software program produced by the authors at the University of Massachusetts, Amherst. www.umass.edu/landeco/research/fragstats/fragstats.html.

McGarigal, K. and B.J. Marks. 1995. *FRAGSTATS: Spatial Pattern Analysis Program for Quantifying Landscape Structure*. USDA Forest Service General Technical Report PNW-351.

McNeely, J.A. and S. Scherr. 2003. *Ecoagriculture*. Island Press, Washington, DC.

Menon, S., R.G. Pontius, J. Rose, M.L. Khan, and K.S. Bawa. 2000. Identifying conservation priority areas in the tropics: a land-use change modeling approach. *Conservation Biology* 15:501–512.

O'Neill, R.V., J.R. Krummel, R.H. Gardner, G. Sugihara, B. Jackson, D.L. DeAngelis, B.T. Milne, M.G. Turner, B. Zygmunt, S.W. Christensen, V.H. Dale, and R.L. Graham. 1988. Indices of landscape pattern. *Landscape Ecology* 1:153–162.

Orr, B., B. Eiswerth, T. Finan, and L. Malembo. 1998. *Public Lands Utilization Study: Final Report*. Submitted to the government of the Republic of Malawi and USAID by the University of Arizona's Office of Arid Lands Studies and the Forestry Research Institute of Malawi. Ref. No. LRP/013.

Pontius, R.G., J. Cornell, and C.H. Hall. 2001. Modeling the spatial pattern of land-use change with GEOMOD2: application and validation for Costa Rica. *Agriculture, Ecosystems and Environment* 85(1–3):191–203.

Roshetko, J.M., M. Delaney, K. Hairiah, and P. Purnomosidhi. 2002. Carbon stocks in Indonesian homegarden systems: can smallholder systems be targeted for increased carbon storage? *American Journal of Alternative Agriculture* 17:138–148.

Serwan, M.J.B. and K. Wan-Yusof. 2003. Modelling optimum sites for locating reservoirs in tropical environments. *Water Resources Management* 17(1):1–17.

Slaymaker, D., H. Schultz, A. Hanson, E. Riseman, C. Holmes, M. Powell, and M. Delaney. 1999. Calculating forest biomass with small format aerial photography, videography and a profiling laser. In: Proceedings of the 17th Biennial Workshop on Color Photography and Videography in Resource Assessment, Reno, Nevada. American Society for Photogrammetry and Remote Sensing (ASPRS), Reno: 241–260.

Turner, M.G. 1990. Spatial and temporal analysis of landscape patterns. *Landscape Ecology* 4:21–30.

Turner, M.G. and R.H. Gardner (eds.). 1991. *Quantitative Methods in Landscape Ecology*. Springer-Verlag, New York.

Villa, F., L. Tunesi, and T. Agardy. 2002. Zoning marine protected areas through spatial multiple-criteria analysis: the case of the Asinara Island National Marine Reserve of Italy. *Conservation Biology* 16(2):515.

Woods, D.B. 1997. Siting artificial water holes at Mkuzi Game Reserve using GIS. Unpublished MSc dissertation, University of Natal, Pietermaritzburg, South Africa.

Part III

Institutional Foundations for Ecoagriculture

Overview

Achieving ecoagriculture landscapes will require innovations not only in production technology and conservation management but also in the institutions required to support stewardship by farming communities, collective landscape management, and enabling markets and policies. The eight chapters in this final section synthesize international experience on the institutional foundations of ecoagriculture.

A core finding of these reviews is that agricultural communities already play an important role in biodiversity conservation and ecosystem stewardship, and are at the center of ecoagriculture development. **Augusta Molnar, Sara J. Scherr, and Arvind Khare** review evidence of community-led forest conservation in ecoagriculture landscapes and conclude that communities protect—and finance—larger areas of forest devoted to conservation than are found in public protected areas.

Donato Bumacas, Delia C. Catacutan, Gladman Chibememe, and Claire Rhodes report on recommendations from grassroots communities themselves on how to develop and scale up ecoagriculture. They emphasize the role of knowledge processes, support services for community-led initiatives, secure tenure rights, and equitable access to resources, as well as policies that provide a clear and leading role for communities in setting priorities and implementation.

Experience shows that community action is necessary, but not sufficient, to achieve agricultural development, rural livelihood, and conservation objectives at farm and landscape scales. Ecoagriculture landscapes include a wide range of landowners and managers, conservation agencies, and other stakeholders who must achieve a degree of coordination in their activities. **William J. Jackson, Stewart Maginnis, and Sandeep Sengupta** discuss the challenges of ecoagriculture planning and implementation at the landscape scale, and some of the key lessons that have been learned about managing trade-offs and synergies, from experiences in forest landscape restoration.

Thomas P. Tomich, Dagmar Timmer, Sandra J. Velarde, Cheryl A. Palm, Meine van Noordwijk, and Andrew N. Gillison from the Alternatives for Slash-and-Burn (ASB) Partnership for the Tropical Forest Margins describe the participatory, multi-scale, integrated natural resource management approach used in ASB research to conserve globally significant wild biodiversity in forest habitats while at the same time addressing the chronic poverty of forest-dependent people. Their chapter illustrates how to forge effective research partnerships to address complex ecoagriculture challenges and describes a variety of research methods developed to tap into the strength of such partnerships to evaluate ecoagriculture options.

Minu Hemmati looks more closely at the values and processes required to establish multi-stakeholder partnerships and make them work effectively over the long term. She discusses the phases of partnership development and notes that the processes require a fundamental openness to learning, change, and the ability to cope with complexity and ambiguity. Such initiatives also need sufficient resources dedicated specifically to building and sustaining partnerships. The chapter illustrates these principles with ecoagriculture applications from the Seed Initiative.

Effective ecoagriculture approaches can exploit ecological or economic synergies, or minimize the trade-offs required among different actors in the landscape, in ways that improve local livelihoods. But the development and implementation of the technical and institutional innovations needed for ecoagriculture landscapes can involve significant financial costs. It is thus essential that markets provide financial incentives for key actors to invest in these changes. **Edward Millard** identifies the drivers—sustainability of supply and corporate responsibility—presently pushing the food industry to reward farmers for ecologically and socially beneficial production systems. He then describes, and illustrates with business examples, three major responses—new product development, certification, and individual company-sustainable sourcing initiatives.

In addition to food, medicines, and raw materials, ecoagriculture landscapes produce biodiversity services that provide financial or livelihood benefits to specific groups of consumers, governments, private companies, and environmentalists. **Sara J. Scherr, Jeffrey C. Milder, and Mira Inbar** assess a variety of ways such beneficiaries are beginning to reward farmers and communities for biodiversity stewardship, including payments for access to wildlife habitat, for biodiversity-conserving management, and for providing offsets for damage done to biodiversity elsewhere.

The concluding chapter by **Jeffrey A. McNeely and Sara J. Scherr** draws from findings in the rest of the book, and more broadly, to assess the state of science and practice of ecoagriculture and identify major gaps in knowledge and action. They outline steps that will be required to mobilize ecoagriculture initiatives on a scale that would meet global challenges for agricultural production, rural livelihoods and ecosystem management. Key aspects highlighted are the need to share knowledge, to build capacity across sectors, and to promote policies and institutions that support ecoagriculture.

Chapter 15

Community Stewardship of Biodiversity

Augusta Molnar, Sara J. Scherr, and Arvind Khare

Introduction

The current system of protected areas continues to be severely underfunded, while not including enough of the world's priority biodiversity and natural habitats. At the present coverage and quality of protection, biologists estimate that only 50 to 70% of the existing species will be conserved by the end of the century (Myers et al. 2000). Proposals for expanding protected areas in many of the developing countries continue to be made without a real understanding of the alternative choices or of the current financial, political, and social limitations for even maintaining the existing protected areas and related institutions. Significantly expanding the public protected areas system is not an option in most developing countries; neither is effective exclusion of people from many public protected areas viable nor affordable, particularly given the real costs of compensating for lost livelihoods or resettlement and the growing recognition of local rights.

The extent of human presence in the most biologically diverse regions is now well recognized. More than 1 billion people live on less than US$1 per day (and at least 25% of these are malnourished) in the 25 global biodiversity "hotspots" identified by Conservation International. Population growth in the world's last remaining wilderness areas is double the world average (Cincotta and Engleman 2000). The 2003 Durban Accord from the World Parks Congress endorses a more mainstream approach to conserving biodiversity that moves beyond protected areas and seeks to address root causes of biodiversity loss and promote biodiversity at a landscape scale. The Accord also recognizes the sovereignty of indigenous peoples and forest dwellers over forest areas considered part of the public domain and their potential role in determining categories of use and protection in a more flexible manner. The Accord was developed in line with the Millennium Development Goals, a global commitment to reducing poverty by 50% by the year 2015 and enhancing existing livelihoods.

A fundamental, far-reaching shift in approaches to biodiversity conservation is under way. Recent studies provide evidence that large forest areas under community ownership or management are being actively conserved by communities outside of formal protected areas systems, covering an area at least as large as that formally protected, and probably much more. A study by White and Martin (2002) looked at trends in community tenure in the world's forests and found that 420 million ha (11%) are legally owned or administered by communities, including 22% of the forests in the developing countries. With a modest level of financial and other support, community conservation efforts in the context of ecoagriculture landscapes could be increasingly effective and sustained, with a very high return to the planet.

The Extent of Community-Initiated and -Supported Conservation

The extent of community conservation has been expanding in recent decades with the recognition of indigenous and other community land rights and decentralization of government forest administration functions to community levels. This section examines the extent of community-driven conservation outside public protected area systems in community-owned and administered lands, and in some forests where communities practice active management without legal recognition. It relies upon the available case material to provide a reliable, minimum estimate of the global trend (see full data set in Molnar et al. 2004).

We first look at the main types of community-driven conservation, assess their benefits, and then estimate their potential scale in contributing to local and global conservation. The information on community conservation has been generated by several overlays, beginning with a global threat overlay created by First Nations Development Institute and Local Earth Observation to assess relationships among forests, tenure, biodiversity, global hotspots, and human presence. The community conservation data from case studies gathered by different research teams has been layered on this global assessment in Table 15.1 to provide a first, reasonable estimate of the scope of community conservation (data sources include Borrini-Feyerabend 2003; Chomitz et al. 2004; Cincotta and Engleman 2000; IIED 1994; McNeely and Scherr 2003). The World Conservation Union's (IUCN's) Commission on Economic, Environmental and

Table 15.1. Comparison of annual conservation investment by different actors

Government investment support to protected areas systems	*Overseas development assistance (ODA) and foundation support (forests and environment)*	*Community investment*
Stable	In decline	Growing
Total estimate is $3 billion per year; Developed countries spend $1000–3000/ha, whereas developing countries spend $12–200/ha LDC	Total ODA: $1.3 billion a year (1/3 of which is assigned to conservation, or $200 million a year)	Estimated $1.5–2.5 billion per year at a minimum

Social Policy (CEESP) and its Thematic Group on Indigenous and Local Communities and Equity in Protected Areas (TILCEPA) Web sites (in the IUCN knowledge portal) assemble a large number of community conservation examples inside protected areas. The data exclude human presence in public protected areas, although they include some buffer zones of biosphere reserve sites where communities have legal rights, such as the Maya Biosphere in Guatemala. Molnar et al. (2004) calculate that at least as much forest area is managed under community conservation (370 million ha) as in formal protected areas systems. The actual area could be double or triple this estimate if traditional agroforestry or agropastoral systems and other forest areas in Soviet Russia, Europe, and the Middle East are included.

Types of Community-Based Conservation

Community-conserved forest landscapes covered here fall into four main types based on forest use intensity, cultural relationship, and the length of time that the human population concerned has been managing that particular resource.

Type 1: Indigenous and Traditional Stewards of Large Areas of Natural Habitat Achieving Conservation Similar to That of Public Protected Areas

The most commonly identified category of community-based conservation is in indigenous and traditional peoples' lands and ancestral territories. People in such forests have sought to achieve cultural continuity and self-development on culturally relevant terms by managing their resource base to preserve traditional landscape biodiversity while supporting and improving local livelihoods. A significant segment of the population in this category would fall in the ILO 169 definition of *indigenous peoples*, whereas others would consider themselves "traditional people." The estimate excludes public forests demarcated as state land or protected areas where indigenous tenure or community responsibility is not currently recognized, such as high-cover state forest in tribal belts of central India. Examples include the following:

- Part of the 130 million ha of indigenous reserves or territorial lands in the Brazilian, Peruvian, and Bolivian Amazon (Nepstad et al. 2006)
- One million ha in the southern cone of Latin America (Oviedo 2002)
- Five million ha of forested areas of British Columbia, Ontario, Saskatchewan, and Quebec provinces in Canada, where indigenous peoples continue to have important use rights over extensive territories (Smith and Scherr 2002)
- Eight million ha of community-managed forest lands within the US Inter-Tribal Timber Council member territories (Brechin et al. 2003)
- At least 3 million ha of community or village forests recently devolved to traditional populations in 5000 African communities (Wily 2000)
- Forests in montane regions of the Andes, the Himalayas, China, and West Asia where traditional peoples have a high dependency on forests, yet forests cover significant landscapes of similar habitat and agropastoral systems are tightly linked to forestry (Khare et al. 2000; Poffenberger 2000; Poole 1995)

This category of community-conserved areas has important advantages for conservation, including large, nonfragmented areas able to support large species often protected by their religious value. A large portion of human languages are spoken by small numbers of people living in such traditional spaces—3400 of the world's 6000 languages are spoken by less than 8 million people, most living in forested landscapes. These language groups carry with them important cultural assets—generations of local adaptation and accumulation of knowledge, alternative cultural value systems, and alternative social governance institutions (Pretty 2002). Community-managed areas may support both resource and biodiversity conservation and local income and livelihoods; many livelihoods have been selected by communities for their long-term relationship with natural resources and adaptability to ecological changes.

Indigenous reserves in the Brazilian, Peruvian, and Bolivian Amazon are rapidly increasing in area due to the recognition of indigenous peoples' rights and due to a strong interest among these peoples in conserving their territories for long-term cultural survival and livelihood development. Such reserves can be highly effective in conserving biodiversity, in some cases even more so than the traditional protected areas established around them (Bamberger et al. 2003; Nepstad et al. 2006)

Type 2: Communities Managing Working Landscape Mosaics Compatible with or Favorable to Biodiversity Conservation

This second category of community-driven conservation is found in more intensively utilized spaces where people have a long-standing stewardship relationship to nature and their ecosystems and have developed extractive, cropping, grazing, and water and forest management practices over a long adaptive process. This includes privately owned land, community owned and administered forests, and lands with recognized usufruct. For example:

- At least 7 million ha of agroforests in Central, South, East, and West Africa (Adams and McShane 1992; Barrow et al. 2000; Neumann 1998)
- At least 7 million ha managed as commercially viable Community Forestry Enterprises in southern Mexico of the nation's 40 million ha of forest under *ejido* (a collective allocation of land to indigenous and non-indigenous farmers under land reform; communities are collective indigenous lands recognized as such through the land reform process) and community ownership (Bray et al. 2003)
- Three million ha of indigenous ecomanagement in Central America (Chapela 2000; Toledo 2002)
- Twenty million ha of complex agroforestry livelihood systems in South and Southeast Asia, including traditional and tribal peoples with successional forests (Colfer and Byron 2001; Poffenberger 2000)
- One million ha within the state-owned North American forests in the United States, which are traditionally a source of commercial and noncommercial nontimber forest products and which have active permit systems and, more recently, community contracts for extraction, such as the Appalachian or New Mexican forests (Rural Action and the Community Strategies Group 2002)
- Fourteen million ha of silvopastoral systems in Africa, the Himalayas, and Central Asia in and around savanna and montane forests (Barrow et al. 2000; Barry et al. 2003)
- One million ha of forest land used as pasturing systems for the Sami and Russian indigenous peoples in the boreal region (Sayer et al. 2005)
- Community forestry initiatives in at least 5 million ha of sub-Saharan Africa, which are expanding as forest management is decentralized to local levels and village forests are recognized as legal, local assets (USAID 2002; Wily 2000)
- More than 1 million ha of sacred groves each in India and Africa (Borrini-Fereyabend 2002; Pathak 2003)

In some of these situations peoples' management of nature is central to the composition and range of the present biodiversity, and local ecological knowledge and practice are crucial to that biodiversity's continuance. In some of these cases, communities have allocated portions of their forest resource for more strict conservation; in others, biodiversity values are conserved by complementary management of the resource for multiple purposes. The forest landscapes are fragmented but provide effective corridors as links to adjacent spaces.

Type 3: Community-Driven Conservation on the Agricultural Frontier

The third type of community conservation is found in large remaining patches of forests with natural habitat in and around land of agriculturalists and pastoralists. This category includes agricultural frontier zones where settlers are relatively recent arrivals in a region with important biodiversity values. They are adapting or willing to adapt their economic activities by securing adequate livelihood returns through sustainable management. Documented examples include:

- Extractive reserves in Brazil, which are now expanding as new groups of producers seek to form community concessions in the Amazon (Amaral and Amaral 2000; Sayer et al. 2005)
- Nontimber forest product collection areas in the western United States where long-term residents and foreign immigrants have strong livelihood ties and a growing interest in management (Rural Action and Community Strategies Group 2002)
- Forest concessions of communities in the Maya Biosphere Reserve, Guatemala (Sayer et al. 2005; Soza 2002; Sundberg 1998)
- Transmigration areas of the Indonesian and Malaysian archipelago where agricultural systems incorporate agroforestry and successional forests (Colfer and Byron 2001; Sardjono and Samsoedin 2001)
- Upland migrants who have maintained forested landscapes in some regions of the Philippines (Barry et al. 2003)

This category of community-driven conservation is perhaps least common because it has neither the scarcity-related incentives of the fourth category nor the local institutions and cultural norms present in the first two. Generally, the positive examples have emerged as a result of partnerships between settlers and nongovernmental organizations (NGOs) or government programs that let settlers organize themselves to protect their interests and find ways to adapt to the current policy and market environment. Some shifting cultivators are switching to perennial species of economic value and conserving secondary forests to reduce the use of fallows and fire. Researchers from the Center for International Forest Research (CIFOR) have documented that colonists in the rainforests of Brazil, Nicaragua, and Peru earn 10 to 20% of total income from a diverse set of forest products (Smith et al. 2001). In the Guatemalan Petén and Brazilian Amazon, immigrant farmers have adapted their agricultural systems to maintain forest ecosystems and aim to manage the resource base more sustainably (Amaral and Amaral 2000).

Type 4: Community-Driven Conservation and Restoration in Intensively Managed Landscapes

This category of community-driven conservation in intensively managed landscapes is perhaps the most widespread, but adequate information is not available to assess its real scope. Biodiversity is found in critical habitat niches that supply food and water sources, pollinator habitats, and other similar resources of value to local people. Some communities have organized land use to provide key connectivity among habitats.

- Organic and shade coffee cultivators of tropical forests in Latin America, many of whom are found in the humid cloud forest ecosystems (Soza 2002; Toledo 2002)
- A portion of the 150 million ha of community plantations and forests in agricultural villages in China (Miao et al. 2004)
- Ten million ha of agroforestry in South Asia with successional forests or restored forest landscapes where settled agricultural communities have reforested areas adjacent to their communities and protected them from grazing (Gilmore and Fisher 1995; Pretty 2002; McNeely and Scherr 2003)
- Silvopastoral communities in the forest savannas of sub-Saharan Africa where ecological balances were established between people and wild animals and forest products are still an important percent of local incomes (Sayer et al. 2003)
- Bushcare programs in Australia establishing biodiversity reserves in farmlands allocated for watershed rehabilitation (Garrity et al. 2000)
- Community windbreaks established in Costa Rica to protect crops and livestock that provide ecological connectivity between forest remnants.

Landscapes with high human populations or a large proportion of land under intensive management have limited capacity to retain or restore wild species that require large areas of contiguous habitat. But with adequate protection of critical habitat niches and networks and maintenance of more benign resource management practices, many wild species and ecological communities can be maintained with such landscapes.

Many submontane forested areas of Africa, Asia, and Latin America have landscape-scale evidence of extensive numbers of villages with high population densities where dense forests are stabilizing. Numerous researchers have compared land use from the 1950s to the present in such regions as Nepal, Central America, the Andes, Mexico, Vietnam, Thailand, and Laos to document little new deforestation in upper watershed forests since the late 1980s. Rather, forest cover in upper catchments has stabilized as land use in lower areas intensified to sustain upland forest systems.

Why and how are local people restoring habitats and biodiversity in this conservation category? In South Asia, Nepalese and Indian community forestry has become a key strategy for restoring degraded forests in and around human settlement, maintaining forest cover and a flow of forest products and watershed services in submontane and montane regions, and nontimber forest product (NTFP)-based economies in traditional tribal regions. Government-sponsored joint forest management in India covers 14 million ha with 63,000 user groups. Nepal's formally recognized community forestry user groups now total 12,000 in 900,000 ha, with many more informal ones. The Nepal–Australia forest project team (Gilmore and Fisher 1995) has documented significant biodiversity conservation in some of these forests, and new studies of tribal regions in India show important gains in forest cover and habitat diversity (Singh and Sinha 2004).

Existing and Potential Scale of Community Conservation

The preceding scoping analysis documents at least 370 million ha of community conservation on three continents. This is roughly equal to the forest area inside public protected areas. Population density, although determining the type of land use, is not a significant determinant of where community conservation systems concentrate. The cases include forested areas with relatively low population densities that are remote from market access, such as the indigenous lands in the Brazilian Amazon or the boreal forests in the Taiga. The majority of cases and greatest coverage, however, are in forested areas where rural population densities are medium to high and include indigenous communities in Mesoamerica, communities managing village forests in South Asia traditionally and through joint forest management, and village forests and village conservancies in sub-Saharan Africa and in North America.

Growing appreciation of the potential advantages of community-driven conservation is encouraging governments to recognize and support local efforts. Africa has numerous village woodlot and conservancy models, which evolved through the empowerment of local communities to manage forests. Tanzania has 400,000 ha under community management. Cameroon is testing participatory and community conservation of protected areas. Hunters and grazers in the savannas of Botswana, Kenya, Rwanda, and South Africa are seeking more integral rights and responsibilities in forests and protected areas (IIED 1994; Wily 2000). The evolution of these models in regions like the Congo Basin, where civil conflict is rife and historical relations between parks and local populations have been extremely bitter, is a surprising development.

Contribution of Communities to Financing Forest Conservation

Another key trend is the change in financing for conservation and the growing share of community investment in conservation compared to fiscal resources, overseas development finance, philanthropy, and private conservation by corporations and individuals.

Trends in Conservation Finance

The conservation community estimates a gap of US$27 to 30 billion annually in financing required for the management and expansion of the existing protected areas, if the full costs of infrastructure, research, land acquisition, outreach, and staffing requirements are taken into account. Current global trends, however, indicate that public expenditure and international financing for biodiversity are flat or declining, though private sector investment has possibly increased slightly. As a result, protected areas agencies and systems are likely to continue to suffer from limited budgets, lack of investment in building or maintaining infrastructure, limited resources for training and capacity building, and competition from other agencies for funds.

Public Funding of Protected Areas

Developed countries spend 80 to 100 times more on protected areas than the developing countries in terms of expenditure per hectare. A 1997 study of 123 conservation agencies in 108 developed and developing countries (representing 28% of protected areas) records US$3.2 billion in annual budgets or US$893/km^2 overall, but only US$10/km^2 in 13 of the developing countries studied and less than US$100/km^2 in 32 of the developing countries studied (Green and Paine 1997). The 60% of the sample parks that are in developing countries received only 10% of the total capital expenditure.

Overseas Development Assistance

Official Development Assistance (ODA) has been a major source of income to forest conservation. Bilateral flows were in the range of US$600 to 900 million in the late 1980s, reaching slightly more than a billion dollars in 1990–92 before declining to the previous range in the late 1990s. Multilateral flows in the late 1980s hovered around US$500 to 700 million, reaching a level of more than a billion dollars in 1990–92, and declining in the middle of the 1990s to a level lower than US$400 million.

International financing is key for particular countries. In Brazil, it constitutes 75% of the conservation funding and 50% in several megadiverse African countries. Private foundations are providing slightly more each year, but not more than US$150 million globally; and the private sector not more than US$20 to US$30 million (Khare et al. 2003). The resulting projection of ODA-protected areas expenditure is about US$1.5 per ha overall-and about US$6 per ha in the global "hotspots." Limited funds are therefore being dispersed among an ever larger number of hectares of protected areas.

Private Investment

An increasingly popular conservation model is the creation of private reserves, where governments can encourage permanent conservation by providing tax incentives, easements, or concessions to the private sector for conservation. Foreign conservationists have also purchased land for private conservation. There is certainly scope for future expansion of this model, but often these are not in the areas of highest priority for conservation. Some may also pose problems of elite land concentration, foreign land control or ownership, or disputed land claims.

The traditional private sector currently makes a negligible contribution to forest conservation. Global forest markets are also in flux, with new markets emerging for payments for ecosystem services and new biodiversity valuation, and with new markets for sustainably produced forest products linked to new models of corporate responsibility affecting both timber and nontimber products. As these new markets emerge, the private sector contribution may increase.

Community Financing of Conservation

Communities have been documented as spending significant amounts of time, labor, and financial resources on forest management and conservation activities, roughly estimated at US$1.2 billion to $2.6 billion per year in project reports from programs supporting community forestry (Khare et al. 2003). This is about the same as the annual budgetary allocation of the developing countries for their protected areas systems, and two to three times the annual allocation of all ODA for protected areas conservation worldwide. Table 15.1 shows the combined estimated financing from all sources, based on the earlier analysis.

The communities' contribution to the forest sector has increased most rapidly and reflects the positive benefits that emanate from recognition of their rights and decentralization of the forest management. Currently they constitute the largest single source of finance for forests, greater than ODA and public expenditure on forests.

Competing Political Claims Complicate Protected Area Designation

Along with limited financing sources a changing political environment for establishing new public protected areas or expanding existing areas has arisen. As many conservation analysts have observed, the establishment of protected areas in a particular space and at a particular time is as much a political as a scientific decision (Brechin et al. 2003). Historically, this decision was often taken by conservation interests, colonial authorities, and/or state governments with little voice from local people and little recognition of the biodiversity conservation value of traditional livelihood systems. Where population changes have created new pressures on the protected areas and surrounding resources, a historical legacy of distrust makes negotiation difficult. Where ecosystems are fragile and traditional livelihoods were characterized by a delicate human–nature balance, it is often difficult to devise alternative livelihoods once the original way of life and related ecological knowledge are lost (Colchester et al. 2001).

The conservation literature discusses many of the complicated pressures created by displacing traditional livelihoods. Masai pastoralists in Kenya, like many other African pastoral societies, responded to grazing restrictions in the Serengeti reserve by expanding agriculture, increasing competition over crops with a growing park population of elephants, and eliminating cattle–elephant balances that had earlier controlled wild mammal reproduction (Barrow et al. 2000). Strict control of Brazilian reserves forced Huarani populations to migrate illegally into Argentina and Peru, where there were no measures to accommodate them (Sayer et al. 2005).

In an environment of stable or declining finance, conservation authorities have underestimated—and continue to underestimate—the cost and livelihood impacts of restricting these areas on local people's assets and livelihoods, particularly in the event of weather or politically induced crises. Little attention is paid to the repercussions on settlements that become unwilling hosts of displaced populations (Geisler 2002). More demands are being made for fair compensation, not only for existing assets and the cost of resettlement but foregone rights and negative livelihood impacts.

Enabling Conditions for Community-Driven Conservation

Community conservation is clearly not a panacea for biodiversity conservation any more than are protected areas. This approach is only effective among certain communities and it carries risks related to organizational commitment; effectiveness of local strategies; and livelihood pressures resulting from natural disasters, civil conflicts, wars, and economic downturns that lead to resource degradation. Among the possible opportunities for long-term community conservation, some important enabling conditions make these models more likely to succeed and provide a more stable institutional base for the future. Without the minimum enabling conditions in place, community efforts will not succeed and commitments to conservation cannot be sustained in the face of the myriad counter pressures. Key enabling elements include:

1. Secure tenure rights and resource access, respecting indigenous peoples' movements
2. Adequate institutional, regulatory, and policy support and the flexibility to grow local community institutions
3. Access to markets, including green markets, that value community products and the multiple values that are produced along with these products
4. Finance channeled in a flexible way to complement local initiatives, rather than planning or designing models from outside or governing from above
5. Engagement of communities in conservation science and as research partners

The opportunities will differ for different types of community-driven conservation, and sustainable strategies to support local initiative must be tailored to local conditions. It is time for a serious and comprehensive reflection on tenure rights. Just as state control of production and protection forests is being questioned by those with historical claims and alternative models, so state designation of existing and new areas as officially protected requires rethinking. The appropriate tenure is that which respects rights and which provides the appropriate incentive structure for the desired management outcomes.

In this changing climate, policy and regulatory frameworks can have a sizable impact on the success of biodiversity conservation, so regulatory frameworks designed for a very different historical situation need to be reconsidered in light of new concepts and conditions (Scherr et al. 2004a). Regulatory reform can have dramatic results. An enterprise promotion program in Nepal doubled the price received by collectors of essential oils and bark for traditional paper in 30 villages simply by gaining approval for direct marketing and advertising market prices paid by intermediate buyers in neighboring India. The market price information generated by this small program became common knowledge, and producers throughout Nepal and northern India were able to gain higher prices for these products. In the program area, biodiversity increased because producers now had both incentives and income to invest in improving their resource base (Subedi 2002).

Forest product and service market opportunities can create financial incentives for forest conservation and sources of financing for local conservation initiatives. Indeed, while in many forest product and service markets low-income community producers may be at a disadvantage, in other markets they may have strong competitive advantages, including:

- Control of commercially valuable forest resources near domestic market demand
- Lower cost structure for some products
- Greater incentives for sustainable forest management and for maintaining landscape mosaics that retain biodiversity values
- Better monitoring and protection
- Branding in socially responsible markets (Scherr et al. 2004a)

Institutional rethinking and new conceptual models are needed at the level of policy and regulatory frameworks as well. The protected areas model was designed in a historical context of land-use zoning by centralized government administrations whereby protection and conservation were conceived as primarily a government or state responsibility to control overextraction. In the process of decentralizing and devolving state forests for production and multiple use to local actors and rights holders, new models of responsibility for conservation have emerged. Many of the forest policies, laws, and regulations designed to allocate public forests to private and state-mandated uses are being replaced by local governance systems, greater local rights, and the emergence of new markets, including those for ecosystem services.

Finally, communities are being enabled by new conservation science approaches that engage communities in research and monitoring of biodiversity and in developing strategies. Communities with local capacity for research often ask different questions and get different answers to research problems. Resulting recommendations favor local value systems, build on traditional knowledge, find locally acceptable solutions to overexploitation, and identify species that outsiders would miss.

Conclusions

Communities offer new institutional models for conservation in ecoagriculture landscapes that should be strengthened. Some traditional communities in large, intact, protected areas (type 1) require secure tenure rights and support for building local institutions and skills for better conservation outcomes. Others require stronger partnerships with government or private partners where their presence and control of boundaries are under threat from outsiders. Successful community managers in fragmented forest landscapes (type 2) have developed organizational structures and economic strategies that have competitive advantages which outside models too often seek to change, rather than replicate. Communities in and around newly settled agricultural or pastoral lands (type 3) tend to require more outside assistance to develop their management structures and seek viable enterprises, but can draw upon an existing or adapted knowledge base that is commonly underestimated. Communities that are actively restoring natural habitats in intensively managed agriculture–forest mosaics (type 4) often already have secure tenure rights. Yet policies often place formidable barriers and create disincentives for these communities to undertake conservation or economic activities that are compatible with their conservation goals.

Technical assistance and support are needed and should be provided on local terms (as further elaborated in Chapter 16, this volume). Finally, global conservation conventions and mechanisms need to foster and support community conservation. Exciting new markets for "ecosystem services" are emerging, but few of these are sensitive to equity issues or to the access of local communities to these markets and market players.

Large areas of the world's habitats are managed and, to varying degrees, conserved by rural communities. This presents both a unique opportunity and a unique challenge to governments, international organizations, the private sector, and civil society fostering more ecoagriculture landscapes. With global and rural populations increasing, it is timely—indeed urgent—to build on these positive community experiences.

References

Adams, J. and T. McShane. 1992. *The Myth of Wild Africa: Conservation without Illusion*. Norton Press, New York.

Amaral, P. and M. Amaral Neto. 2000. *Manejo forestal comunitario en la Amazonía brasileña: situación actual y perspectivas*. Instituto International de Educación del Brasil (IIEB), Brasilia.

Barrow, E., H. Gichohi, and M. Infield. 2000. *Rhetoric or Reality? A Review of Community Conservation Policy and Practice in East Africa*. Evaluating Eden Series no. 5. International Institute for Environment and Development (IIED), London.

Barry, D., J.Y. Campbell, J. Fahn, H. Mallee, and U. Pradhan. 2003. Achieving significant impact at scale: reflections on the challenge for global community forestry. CIFOR Conference on Rural Livelihoods, Forests, and Biodiversity, Bonn, Germany.

Borrini-Feyerabend, G. 2002. Indigenous and local communities and protected areas: rethinking the relationship. *Parks* 12(2):5–15.

Borrini-Feyerabend, G. 2003. *Community Conserved Areas and Comanaged Protected Areas: Towards Equitable and Effective Conservation in the Context of Global Change.* Report of the IUCN joint CEESP/WCPA Theme on Indigenous and Local Community, Equity and Protected Areas (TILCEPA) for the Ecosystem, Protected Areas and People (EPP) project. April. Draft. www.iucn.org/themes/ceesp/wkg_grp/TILCEPA/TILCEPA.htm.

Bray, D., L.Merino, P. Negreros, G. Segura, J.M. Torres, and H.F.M.Vester. 2003. Mexico's community managed forests as a global model for sustainable landscapes. *Conservation Biology* 17(3):672–677.

Brechin, S.R., P.R. Wilshusen, C.L. Fortwangler, and P.C. West (eds.). 2003. *Contested Nature: Promoting International Biodiversity with Social Justice in the Twenty-first Century.* State University of New York Press, Albany.

Chapela, F. 2000. "Indigenous peoples' community biodiversity management initiative." Consultation Workshop for Biodiversity Community Management. MBC Studies and Activities Central America Environment Projects. World Bank.

Chomitz, K.M., J. Robalino, and A. Nelson. 2004. A note on forest populations in Latin America and the Caribbean. Revised draft Working Paper. World Bank Group, Washington, DC.

Cincotta, R.P., and R. Engelman. 2000. *Nature's Place: Human Population and the Future of Biological Diversity.* Population Action International, Washington, DC.

Colchester, M., F. MacKay, et al. 2001. A survey of indigenous land tenure: a report for the land tenure service of the Food and Agriculture Organization. Forest Peoples Programme, Moreton-in-Marsh, England.

Colfer, C. and Y. Byron (eds.). 2001. *People Managing Forests: The Links between Managing Human Well-Being and Sustainability.* Resources for the Future and Center for International Forestry Research (CIFOR). Washington, DC.

Garrity, D.P., D. Catacutan, R. Alvarez, F.M. Mirasol. 2000. Replicating models of institutional innovation for devolved, participatory watershed management. In: *Choosing a Sustainable Future: SANREM CRSP 1999 Annual Report,* ed. K. Cason. Sustainable Agriculture and Natural Resource Management Collaborative Research Support Program, Watkinsville, GA.

Geisler, C. 2002. Endangered humans. *Foreign Policy* 130:80–81.

Gilmore, D.A. and R.J. Fisher. 1995. *Villagers, Forests and Foresters.* Sahogi Press, Ltd. Kathmandu, Nepal.

Green, M. and J. Paine. 1997. State of the world's protected areas at the end of the 20th century. Paper presented at IUCN World Commission on Protected Areas Symposium on Protected Areas in the 21st Century: From Islands to Networks, Albany, Australia, 24–29 November 1997.

IIED (International Institute for Environment and Development). 1994. *Whose Eden? An Overview of Community Approaches to Wildlife Management.* International Institute for Environment and Development, London.

Khare, A., M. Sarin, N.C.Saxena, S. Palit, S. Bathla, F. Vania, and M. Satyanarayana, 2000. *Joint Forest Management: Policy, Practice and Prospects.* Policy that works for forests and people series no. 3. World Wide Fund for Nature-India, New Delhi and International Institute for Environment and Development, London.

Khare, A. et al. 2003. *Funding Conservation: The Current Status of Conservation Financing in the Developing Countries.* Working Paper. Forest Trends, Washington, DC.

Maffi, L. (ed.). 2001. *On Biocultural Diversity: Linking Language, Knowledge, and the Environment.* Smithsonian Institution Press, Washington, DC.

McNeely, J.A. and S.J. Scherr. 2003. *Ecoagriculture: Strategies to Feed the World and Save Biodiversity.* Future Harvest and World Conservation Union (IUCN). Island Press, Washington, DC.

Miao, G., S. Zhou, K. Zhang, S. Gao, X. Huang, and J. Jiang. 2004. Collective forests in China. Report prepared for Forest Trends by the China National Forestry Economic and Development Research Center (FEDRC). Forest Trends, Washington, DC.

Molnar, A., S.J. Scherr, and A. Khare. 2004. *Who Conserves the World's Forests: Community Driven Strategies to Protect Forests and Respect Rights.* Forest Trends and Ecoagriculture Partners, Washington, DC.

Myers, N., R.A. Mittermeier, C.G. Mittermeier, G.A.B. da Fonseca, and J. Kent. 2000. Biodiversity hotspots for conservation priorities. *Nature* 403:853–858.

Nepstad, D.C., S. Schwartzman, B. Bamberger, M. Santilli, D. Ray, P. Schlesinger, P. Lefebvre, A. Alencar, E. Prinz, G. Fiske, and A. Rolla. 2006. Inhibition of Amazon deforestation and fire by parks and indigenous reserves. *Conservation Biology* 20(1):65–73.

Neumann, R. 1998. *Imposing Wilderness: Struggles over Livelihood and Nature Preservation in Africa.* University of California Press, Berkeley and Los Angeles.

Oviedo, G. 2002. Lessons learned in the establishment and management of protected areas by indigenous and local communities. World Conservation Union (IUCN), Gland, Switzerland. Mimeo.

Pathak, N. 2003. *Lessons Learnt in the Establishment and Management of Protected Areas by Indigenous and Local Communities in South Asia.* TILCEPA report, www.iucn.org/themes/ceesp/Wkg_grp/TILCEPA/community.htm#A and TILCEPA/CEESPWorking Paper, www.iucn.org/themes/ceesp/Publications/TILCEPA/CCA-NPathak.pdf.

Poffenberger, M., ed. 2000. *Communities and Forest Management in Southeast Asia. A Regional Profile of the Working Group on Community Involvement in Forest Management.* International Union for the Conservation of Nature (IUCN). Gland, Switzerland.

Poole, P. 1995. *Indigenous Peoples, Mapping and Biodiversity Conservation: An Analysis of Current Activities and Opportunities for Applying Geomatics Technologies.* Biodiversity Support Program.

Pretty, J.N. 2002. *Agriculture: Reconnecting People, Land and Nature.* Earthscan Publications, London.

Rural Action and the Community Strategies Group. 2002. *The Herb Basket of Appalachia: Community-Based Forestry and Sustainable Communities.* Forest Harvest Occasional Report 1. Community-Based Forestry Demonstration Program. Aspen Institute, Washington, DC.

Sardjono, M.A. and I. Samsoedin. 2001. Traditional knowledge and practice of biodiversity conservation: the Benuaq Dayak community of East Kalimantan, Indonesia. In:

People Managing Forests: The Link between Human Well-Being and Sustainability, ed. C. Colfer and Y. Byron. Center for International Forestry Research (CIFOR), Managua, Nicaraqua, and RFF, Washington, DC.

Sayer, J. (ed.). 2005. *The Earthscan reader in Forestry and Development*. Earthscan, Sterling, CA.

Sayer, J., C. Elliott, E. Barrow, S. Gretzinger, S. Maginnis, T. McShane, and G. Shepherd. 2005. The implications for biodiversity conservation of decentralized forest resources management. In: *The Politics of Decentralization: Forests, Power and People*, ed. C.J. Colfer and D. Capistrano. Earthscan, London.

Scherr, S.J., A. White, and D. Kaimowitz. 2004a. *A New Agenda for Forest Conservation and Poverty Alleviation: Making Markets Work for Low-Income Producers*. Forest Trends, Washington, DC. http://www.foresttrends.org.

Scherr, S., A. White, and A. Khare 2004b. *For Services Rendered: The Current Status and Future Potential of Markets for the Ecosystem Services Provided by Tropical Forests*. International Tropical Timber Organization (ITTO), Yokohama, Japan.

Singh, K.D. and B. Sinha. 2004. Findings from a study of CFM in Kandamahal District of Orissa. Presentation to the World Bank/WWF Alliance in Washington, DC, Government of Orissa, Bhubaneswar.

Smith, D. 2001. Can SL approaches learn from environmental mainstreaming? Background paper to a seminar on *Sustainable Livelihoods and Environment: Sharing approaches and principles.* November 1, 2001. Department for International Development. London.

Smith, J. and S.J. Scherr. 2002. *Forest Carbon and Local Livelihoods: Assessment of Opportunities and Policy Recommendations*. Occasional Paper no.37. Center for International Forestry Research, Bogor, Indonesia.

Soza, C. 2002. The process of forest certification in the Mayan Biosphere Reserve in Petan, Guatemala. Working Paper for A. Molnar et al. *Forest Certification and Communities: Looking Forward to the Next Decade*. http://www.forest-trends.org.

Subedi, B. 2002. Towards expanded property rights of local communities over forest resources in Nepal: lessons and strategies. Asia Network for Sustainable Agriculture and Bioresources (ANSAB), Nepal. Presentation to the Global Perspectives on Indigenous Peoples' Forestry: Linking Communities, Commerce and Conservation Conference, Vancouver, British Columbia, 2–6 June 2002. http://www.forest-trends.org.

Sundberg, J. 1998. Strategies for authenticity, space and place in the Maya Biosphere Reserve, Peten, Guatemala. In: *Yearbook*, Conference of Latin Americanist Geographers, 24:85–96.

Toledo, V.M. 2002. Ethnoecology: a conceptual framework for the study of indigenous knowledge of nature. In: *Ethnobiology and Biocultural Diversity*, ed. R. Stepp et al. University of Georgia Press, Athens: 224–235.

United States Agency for International Development (USAID). 2002. *Nature, Wealth and Power*, ed. Jon Anderson. Report prepared on Natural Resource Management in Africa by a joint institutional team: WRI, Cornell, Center for International Forestry Research (CIFOR), CLUSA, Winrock, and IRG. United States Agency for International Development, Washington, DC.

White, A. and A. Martin. 2002. *Who Owns the World's Forests?* Forest Trends, Washington, DC. www.forest-trends.org/whoweare/publications.

Wily, L. Alden. 2000. Forest law in eastern and southern Africa: moving towards a community-based forest future? *Unasylva* 203(4). United Nations Food and Agriculture Organization (FAO).

Chapter 16

Community Leadership in Ecoagriculture

Donato Bumacas, Delia C. Catacutan,
Gladman Chibememe, and Claire Rhodes

Introduction

Worldwide, local farming, herding, forest, and fishing producers and their communities demonstrate deep and diverse expertise in implementing ecoagriculture. Recognition of this expertise is growing, but the global community needs to take further action to appreciate and learn from the knowledge and innovation of grassroots practitioners worldwide. The international agriculture, rural development, and conservation communities respect local agents and indigenous peoples as both environmental stewards and rural producers, and enable them to play a central role in decision making about ecoagriculture strategies and actions.

This chapter reviews critical actions required at the local, national, and international levels to support, build upon, and mobilize community-based ecoagriculture expertise. It has four main sections. The first provides an overview of existing community-led ecoagriculture initiatives and challenges faced. The second addresses interlinked actions required to strengthen community-led ecoagriculture. The third section addresses opportunities and challenges for scaling up grassroots success, and the final section discusses policy actions needed to enable community-led ecoagriculture development. Recommendations are founded upon priorities articulated by grassroots and indigenous community representatives during the International Ecoagriculture Conference and Practitioners' Fair (Rhodes and Scherr 2005) and in preparation for the 2005 United Nations World Summit (Gillis and Southey 2005; United Nations 2005). In this chapter, the authors would like to acknowledge the contributions to our thinking provided by Odigha Odigha, Benson Venegas, all members of the Ecoagriculture Partners Community Facilitation team, United Nations Development Programme's (UNDP) Equator Initiative and all participants in the series of community-dialogue spaces, whose wisdom and recommendations form the foundation of this chapter.

Community Leadership in Ecoagriculture

Community-led ecoagriculture has been widely documented in recent years. Yet local communities remain marginalized in strategies to promote agricultural, rural development, and biodiversity conservation.

Evidence of Community-Led Ecoagriculture

Farming, pastoral, and forest communities worldwide conserve and manage millions of hectares of natural habitat within and beyond public protected areas. At least 22% of forests within developing countries are legally owned or administered by local communities, with more than 360 million ha of forest landscapes and forest–agriculture mosaics under community-led management within the Americas, Africa, and Asia (Molnar et al. 2004). Furthermore, over half of the world's 102,000 protected areas have been established on ancestral lands of indigenous and other traditional peoples (Borrini-Feyerabend et al. 2004).

Within and beyond these landscapes, diverse and innovative landscape-management approaches are employed by smallholder farmers and local communities to deliver positive, integrated outcomes for food security, rural livelihoods, and biodiversity conservation (Brookfield et al. 2002; McNeely and Scherr 2003). Such locally driven initiatives vary in their genesis, focus, and scope. Incentives for local communities to sustain or transition toward ecoagriculture approaches are diverse, variable, and highly context-specific, dependent on local socioeconomic, environmental, and political conditions but also on policy and market frameworks at the national and international level. Motivations may include the proven effectiveness of traditional management systems at meeting local needs and values, the need to maintain or restore essential resources and ecosystem services, or new enterprise development opportunities.

Community-driven "Landcare" groups are growing in a number of countries, including the Philippines, Australia, Uganda, and South Africa, to mobilize collective action by farmers and local communities concerned about land degradation and natural resource management challenges. In Rajasthan's Arvari Basin, a community-led watershed restoration program centered upon reinstating *johads*, an indigenous technology to collect water from uphill river tributaries, has significantly enhanced groundwater recharge and restored river flow, improving hillside forest productivity and water supplies for irrigation, wildlife, livestock, and domestic use. The Kalinga indigenous peoples of the Philippines are collectively managing their watersheds to naturally irrigate their rice terraces using proven traditional knowledge, integrating fish and vegetable production into rice terrace management. In Kenya's arid Marsabit region, the Pastoralist Integrated Support Program is working with over 11,000 pastoral people to protect dryland biodiversity from over-grazing by strategically managing the movement of herds around vulnerable water points. Diverse groups of local producers, particularly in Latin America, are self-organizing to collectively secure price premiums from fair trade, organic, and/or shade-grown certified systems. Of 400 community initiatives nominated worldwide for the 2002 Equator Prize for achievements in enhancing biodiversity while improving livelihoods, over 100 were ecoagriculture landscape initiatives (Isely and Scherr 2003).

Marginalization of Local Knowledge in Agricultural Development and Biodiversity Conservation Strategies

Neither the concept of community-based management nor the recognition of the need for integrated approaches to conservation and rural livelihoods development is new (Brookfield et al. 2002). Yet sector-specific public and NGO programs interventions continue to either address local symptoms while ignoring underlying policy constraints or to deal with macrolevel issues while ignoring local realities (McShane and Wells 2004).

Both conservation and rural development sectors commonly fail to reach, respond to, or meet the needs of the majority of the local communities worldwide. Conservation strategies driven by conservation values of urban-based environmentalists in the North have limited focus on species valued by local communities for food, medicines, or cultural significance. The centralized designation and management of many protected areas have at best ignored local people's dependence on crop, livestock, forest, and fishery production and at worst forced displacement, with local communities marginalized from the process and often the land itself (Mogelgaard 2003).

Meanwhile, strategies to promote agricultural development have also ignored inherent interdependencies between local livelihoods and sustaining the natural resource base, and have often failed to distinguish between enhancing agricultural productivity and achieving food security. Agricultural extension services have tended to encourage the replacement of diverse, traditional farming practices with more intensive production systems based upon a small number of new crop varieties. The lack of coordination between technical support offered by conservation, rural development, and agricultural actors has exacerbated the problem. Communities have often been the subject of technical assistance by these actors working in the same region, but each encouraging a different course of action (Sundberg 1998), based on its own interests rather than those of the community.

The need for local community involvement in natural resources management strategies became increasingly recognized during the 1990s, driven by unsatisfactory impacts from the "parks without people" model. Investment in community-based initiatives intentionally seeking to link biodiversity conservation and livelihood objectives was significantly upscaled, particularly through integrated conservation and development projects (ICDPs) and community-based natural resource management (CBNRM) approaches (Hughes and Fintan 2001; Newmark and Hough 2000; Wells et al. 1992; Worah 2000). Nonetheless, documented evidence to substantiate the effectiveness of such approaches and support the rationale for devolving management authority to local communities remains limited and poorly disseminated, leading critics to advocate a return to more traditional, protectionist approaches (Spinage 1998; Terborgh 1999).

Barriers to Community-Led Agricultural Development and Conservation

Local communities are not sufficiently valued and engaged as true, equitable partners in decision-making processes. Holistic management approaches that communities employ to sustain interlinked livelihood objectives and the institutional environment within which they are constrained to operate remain highly incompatible. It is not enough for projects to intentionally link conservation and development objectives. Rather, initiatives should emerge from and be driven by community representatives themselves, based on innovations that meet their inherently linked livelihood goals and objectives. Local land-use systems and decisions are often founded upon long-term accumulation of knowledge and experimentation about what will and will not work (Sayer and Campbell 2004). Such strategies must withstand unpredictability in climate, in socioeconomic circumstances, and also in external financing trends.

Current institutional environments are rarely conducive or receptive to understanding and supporting landscape management approaches or learning from challenges experienced by local communities and land managers. Communities do not operate in sector-specific, target-based, or timeline-defined environments. Nonetheless, a tendency remains to "shoehorn the complex, dynamic realities that shape community-led approaches into the constraints of time-bound tightly planned project frameworks" and "narrowly defined sectoral-based institutional support structures" (Sayer and Campbell 2004). Complex community- and landscape-scale dynamics are often underestimated, with projects and investments based on incomplete understanding or inaccurate assumptions about the relationships between livelihood and conservation objectives and needs within a community (Brookfield et al. 2002; Sayer and Campbell 2004). For example, to a community located within a protected area, losing an important grazing area to a community wildlife project that does not generate equal or more benefits to the same community ensures the project is doomed before it starts.

Time frames required to deliver tangible and sustainable outcomes are also underestimated, often falling well short of the period required to build trust, understanding, and thus the long-term processes and partnerships required to catalyze sustained change. Project financing does not allow for the delivery of long-term outcomes or the monitoring of long-term trends. For example, many CBNRM systems historically considered good practice are now experiencing challenges to sustain themselves under increased population pressures or changing market opportunities. Communities need to be actively adapting, innovating, and learning within dynamic circumstances, and successful partnerships must take this into account.

Strategies to Support Community-Led Ecoagriculture

As noted by the 2005 Millennium Ecosystem Assessment, a "significant constraint on developing effective management . . . [is the] failure of decision-making processes to adequately recognize and use information that does exist—particularly traditional and practitioners' knowledge and innovation—in support of management decisions."

To support and strengthen the potential of local communities to protect, develop, and adapt ecoagriculture landscapes requires a concerted effort to develop relevant institutions based on lessons learned from community-driven development. This section explores three sets of actions to create a more enabling environment:

- Build upon and strengthen the knowledge base and networks of local communities
- Support and finance community-led, needs-driven ecoagriculture initiatives
- Develop policies that enable and support community-led ecoagriculture

Building upon the Knowledge Base from the Bottom Up

Building the necessary knowledge base and local capacity for improving ecoagriculture systems involves documenting community knowledge, supporting community-led knowledge sharing, and strengthening community-driven research.

Documenting Community Knowledge and Expertise

Documenting community-based knowledge, management strategies, and lessons learned offers an important foundation upon which to catalyze knowledge-sharing among community-based ecoagriculture practitioners. A prerequisite to processes promoting knowledge and technology exchange is taking the time to identify and understand who knows what (Fairhead 1993). Such documentation can establish a baseline from which to identify knowledge, existing capacity and developments needs, and to demonstrate the value of community-based management to policymakers and researchers. A range of initiatives seeking to document local innovation and knowledge are underway, with varying purposes and target audiences. Internationally, the Convention on Biological Diversity (CBD) has encouraged traditional knowledge databases and registers in support of traditional knowledge and intellectual property protection (Kambu et al. 2004). A national-level example is India's National Innovation Foundation (NIF) and Honeybee Network, facilitating peer–peer knowledge exchange by documenting and disseminating grassroots technologies and traditional practices.

Within a community, individuals respond differently to changing ecological, sociological, and economic circumstances (Pinedo-Vasquez et al. 2002). The diversity of responses and collective knowledge offers considerable strength in terms of the community's resilience and responsiveness to change. However, documenting such intra-community expertise is challenging, especially understanding the cause–effect relationships between management practices and outcomes, and representing the depth of knowledge accumulated over many generations. Documenting ecoagriculture practices requires understanding the interdependencies between the conservation and farming systems, not just at farm and community levels but within the landscape mosaic.

Ensuring local ownership of documentation and monitoring processes is vital to enabling local communities to define the information and outcomes they wish to document, the purpose, and the target audience of the documentation process. As observed in the Talamanca community-based biomonitoring program, Costa Rica, "Communities learn to demand information, articulate their needs and directly link their management strategies with the outcomes they are delivering" (Benson Venegas, pers. comm., 2006) The information that communities may wish to document and the media they wish to use is likely to differ significantly from the information desired by other stakeholders, such as researchers or policymakers. An unthinking reliance on terminology and outcome indicators defined by those outside of the community may reveal little of use to communities themselves (Brookfield et al. 2002), or may fail to appropriately represent and balance traditional knowledge with more "science-based" perspectives.

Community-Led Processes for Knowledge Sharing

Ongoing initiatives substantiate the effectiveness of processes that enable communities to share their knowledge. These include individual farmers and community representatives demonstrating their practices to neighboring practitioners (as through farmer field schools), the facilitation of community–community learning exchanges, community-dialogue spaces, and intercommunity and farmer learning networks. The People, Land Management, and Ecosystem Conservation (PLEC) methodology, initiated in 1992, has supported community-based practitioners to share encouraging aspects of their management practice with contemporaries and take leadership in community discussions on capacity development, policy, and research needs (Brookfield et al. 2002). PLEC's approach has been effective. But lessons learned also caution against assuming that an individual expert in a particular practice will be willing to share knowledge or be most suited to take a leadership role. Practices demonstrated by individuals may not be appropriate and replicable by other community members when incentives, knowledge, and resources availability differ. Questions also arise around the incentives and motivations for individuals to share their knowledge, and around what knowledge can be shared, particularly if this might depreciate an individual's competitive advantage (Pinedo-Vasquez et al. 2002).

Co-learning and integrated knowledge management processes provide opportunities to move beyond individual leadership to recognize that diverse stakeholders within a landscape hold different knowledge in conservation, production, rural development. Together, this collective expertise is necessary to deliver landscape-scale ecoagriculture outcomes. Models are emerging to facilitate knowledge sharing and build a collective knowledge base, such as the Integrated Systems for Knowledge Management framework developed in New Zealand (Allen and Kilvington 2001). These frameworks enable a range of local land users, including communities, farmers, researchers, policymakers and other interest groups to share expertise and inform decision making on a range of resource management challenges.

The value of connecting individuals experiencing similar management challenges and opportunities extends beyond the landscape level, motivating a worldwide proliferation of community- and farmer-driven networks to facilitate peer–peer knowledge exchange and community-led advocacy. Notable examples include the aforementioned Landcare movements; the Great Limpopo Transfrontier Park Rural Communities' Network initiated to support information sharing on transboundary protected area management concerns (Box 16.3); Thailand's Local Wisdom farmer learning networks; and the Linking Local Learners programs established through Kenya, Uganda, and Tanzania to support learning and peer exchange among smallholder farmers. Even at an international level, when practitioners operate in very different environmental, socioeconomic, and political contexts, significant benefits are derived from enhanced access to innovative solutions and approaches being employed by other communities addressing shared challenges, particularly with respect to policies and markets. This has been demonstrated through the work of organizations such as GROOTS International and the United Nations Development Project (UNDP) Equator Initiative.

Community-Driven Research

Investment is also needed to strengthen and build upon communities' knowledge base. A significant disconnect remains between the demand for useful ecoagriculture-related knowledge and resources by grassroots practitioners and supply from research institutions, public agencies, and other service providers such as nongovernmental organizations (NGOs). Current research outputs are rarely accessible or appropriately adapted to meet the context-specific needs of grassroots practitioners. Inaccessibility relates to the media in which outputs are produced and disseminated, the terminology and language used, and their predominantly sector-specific focus. Research prioritization and investment remains predominantly shaped by the visions and desired outcomes of international and national research organizations; they need to instead be informed and shaped by community-based expertise and needs, building upon local knowledge of management challenges and potential solutions. Researchers can also support communities by "translating" community-generated knowledge into forms that resonate with researchers, policymakers, and landscape planners, and through research that informs landscape negotiation.

Increasing farmer participation in and influence over research and extension initiatives enables research toward outcomes relevant and responsive to local livelihoods (Hussein 2000). Participatory methodologies, including participatory rural appraisals, offer opportunity to improve understanding between local communities and researchers. Even so, leadership of process design and implementation remain primarily in the hands of researchers, with farmer and community representatives engaged in the work programs of researchers (Waters-Bayer et al. 2005).

However, promising mechanisms are evolving for local communities to drive the research agenda and to genuinely evaluate it. Local agricultural research committees (CIALs) within Latin America offer one model. Established as farmer-run research services, CIALs aim to directly respond to local community and smallholder needs (Ashby et al. 2000). Communities are responsible for electing the farmer-research committee. Research is conducted on priorities identified through community consultations, and research outcomes are reported back to the community. Each committee is provided with modest financing to offset research costs and risks associated with experimentation, and is offered the support of a trained facilitator if required (Ashby et al. 2000).

Currently CIALs are primarily focused on farm-level production challenges but could be adapted to increasingly address landscape-scale issues.

Investing in Community-Led Ecoagriculture

External support can play a valuable role in strengthening community-led ecoagriculture, particularly by providing needs-based support services, community-focused market initiatives, and financial investments.

Needs-Based Support Services

Effectively supporting communities to implement their ecoagriculture approaches requires the "retooling of service providers to offer advice, information and training on a much broader range of issues and processes" (Lightfoot et al. 2005). Interdisciplinary institutions are needed to provide "holistic" packages of services to support and respond to diverse management needs and motivations. These include not only technical support on agriculture and conservation management approaches but also business and financial management and market and enterprise development. Such institutions should enable communities to acquire advocacy and negotiation skills necessary to effectively participate in land-use decision making; negotiate and secure their rights to manage their resources; engage with private sector actors; mediate relationships with other stakeholders operating in the landscape (with whom they need to cooperate to achieve ecoagriculture); and engage in equitable dialogues with external actors on the appropriateness of recommended management strategies. Demand-driven information services are also required. Community-run Rural Knowledge Centres, such as those being established in India's Pondicherry region, offer one model for supplying timely, locally-specific information services to support crop, fisheries and livestock management through diverse media including radio, videos, self-help groups and community publications (M.S. Swaminathan Research Foundation 2005).

Producers' and other community-based organizations are emerging as key service providers to their constituencies (Hussein 2000), with a growing number of producer cooperatives transitioning from exclusively production-focused support toward a broader range of services. For example, the Fouta Djallon small-producer cooperative in Guinea encourages farmers to capitalize upon new production opportunities only after it can provide the necessary support services at each stage of the production chain (including credit, input supply, technical information, and marketing) (Hussein 2000). This is complemented by supportive action at higher institutional and policy levels to defend producer interests, particularly access to national markets. In Costa Rica, the Associación de Pequeños Productores de Talamanca (APPTA) offers a similar example. APPTA, a regional organic small farmer's cooperative, supports over 1500 small

Box 16.1. The role of smallholder producer organizations as service providers: an example from the Asociación de Pequeños Productores de Talamanca (APPTA)

APPTA, founded in 1987, is a regional organic cooperative supporting over 1500 smallholder farmers within La Amistad Biosphere Reserve to be successful in a competitive market, maximizing production and environmental benefits. La Amistad's 1,000,000+ ha reserve and World Heritage Site, primarily stretching along the Caribbean slope of Costa Rica and Panama, protects exceptionally diverse tropical ecosystems and is managed by local communities, nongovernmental organizations, and government representatives.

APPTA is committed to consolidating economically viable agroecosytems as an integral element of the region's biodiversity conservation strategy. This is achieved by promoting organic agroecosystem management; processing and marketing organic products from Talamanca's family farms, and providing training to the Talmancan people to encourage ways of life that are harmonious with indigenous culture, profitable, and environmentally ethical. The development of initial local processing infrastructures for organic cacao, bananas, and other fruits, quality-control checks, marketing strategies, and an organic certification program have enabled APPTA to become the largest-volume producer and exporter of organic products in Central America. In addition to creating new markets for some products, farmers receive an additional 15 to 60 percent revenue for their certified organic products.

These activities are a core element of the Talamanca Initiative, involving the collaboration of over 20 grassroots, community-based organizations, many small-scale producers, and the Costa Rican Ministry of the Environment, with leadership from the locally based organizations Asociación ANAI, APPTA, and Corredor Biologico Talamanca-Caribe (CBTC). These partners, each with its own specific objectives, share the common goal of improving quality of life in Talamanca through the preservation and environmentally ethical use of its unique biodiversity and ecosystems. A core belief is that the key to conservation and sustainable development is the successful management of these issues by the local people. Tangible environmental and economic benefits are realized by the rural poor through sustainable agriculture and forestry systems, locally owned ecotourism enterprises, and biodiversity monitoring and conservation. Beyond Talamanca, benefits have included organic and Fair Trade certification, and higher prices paid to over 2000 mostly indigenous farmers in Bocas del Toro, Panama, in collaboration with the Cooperativa de Cacao Bocatoreña (COCABO) (Venegas, pers. comm. 2006).

farmers managing small farm agroecosystems within and around the La Amistad Biosphere Reserve, Costa Rica/Panama, to be successful in a competitive market (Box 16.1).

Supporting community-based and producer organizations to play the role of holistic service providers will require significant investment to build upon existing capacity. For example, PROCYMAF, a program to support community forestry producers in southern Mexico, organized a roster of "providers for technical and professional services," who received training to enhance their expertise on relevant topics, including building social capital (Scherr et al. 2004). Capacity limitations mean that responsibility for service provision should be held by more than one locally based organization. The Namibian Association of CBNRM Support Organizations (NACSO) recognizes and responds to the reality that no "single institution is likely to house all of the skills, resources, and capacity to provide community organizations with the multidisciplinary assistance required." Instead NACSO aims to systematically strengthen coordination between various local-level stakeholders within the region, and draw upon the respective expertise of its 12 service providers (11 NGOs and the University of Namibia).

Market-Based Strategies to Support Community-Led Ecoagriculture

For many rural communities, ensuring long-term livelihood sustainability and independence from external financing will depend on their ability to access and benefit from a range of markets that value both products and ecosystem services from an ecoagriculture landscape. Diverse initiatives to enhance financial incentives for conservation within agricultural landscapes are emerging, including both product certification (Millard, Chapter 20, this volume) and ecosystem service payments (Scherr and Inbar, Chapter 21, this volume; Scherr et al 2007).

The extent to which existing incentives offer long-term livelihood opportunities for communities and smallholder producers demands further attention. Biodiversity payments are likely to hold most promise for communities located in areas supporting globally or nationally significant biodiversity, for which there is a willingness to pay for its conservation and/or where communities are able to produce products or services with high national or international demand. Yet barriers to communities accessing and benefiting from these incentives are considerable. Financial premiums from production certification and ecosystem service payments are often only attainable once high compliance verification costs have been met. Rather than offering long-term financing for resource stewardship, certification and payment schemes may unwittingly continue the trend of imposing external standards and management models on communities, or foster dependency on financial and technical support to meet standards (van Dam 2002). Molnar (2003) highlights the risks of certification bodies and technical support becoming preoccupied with compliance standards and blinded to the real objectives of community management.

The challenges faced by communities in the development of a certifiable enterprise include the need to assess prospective buyers, market opportunities, and their capacity to sustain supply—in both volume and quality—within unpredictable, competitive product and service markets. Coordinating diverse smallholder farmers to collectively supply "bundles" of products and services is often necessary, yet highly complex. Communities also need the capacity to negotiate and enforce business contracts with private- and public-sector buyers. The majority of current certification and technical support services are not equipped to appropriately advise communities on these critical issues (Molnar 2003). Thus communities themselves need to help drive such market-based initiatives and inform their development, based on their knowledge of their short- and long-term potentials to supply and capacity to manage marketing processes. Currently, small-scale producers are self-organizing to develop their processing and marketing capacities to access national and international markets for "green" products. Examples include APPTA in Costa Rica and Café de la Selva in Mexico.

Financing Community-Led Ecoagriculture

Financing frameworks for rural development and conservation are currently not conducive to supporting community-driven innovation. The reluctance of governments and donors to invest in long-term relationship-building and learning processes leads to piecemeal investment in short-term projects, with no commitment to long-term follow-up or to applying the knowledge acquired and lessons learned from past projects. Unwillingness to fund such learning and relationship-building processes fosters "islands of success/best practice," which are readily drawn upon and cited, yet remain isolated (Roe et al. 2000) and unable to realize broader-level impacts (Binswanger and Aiyar 2003; Pretty 1998).

Logical frameworks outlining project resource allocation leave little freedom for local-level innovation, or time for newly acquired knowledge to be evaluated, adapted, and applied. Results and deliverables are often determined before the project even starts, without any clear assurance that these outcomes are those desired by the project's intended beneficiaries (Sundberg 1998). Local communities and producers are afforded little engagement in decision making on desired deliverables, how resources are allocated, timeframes for delivery, or indicators designed to evaluate "success."

Opportunities are being explored to support community-led innovation more effectively, through the direct delivery of financing. Piloting of local innovation support funds (LISFs) is under way in Cambodia, South Africa, Sudan,

Uganda, Ethiopia, and Nepal (Waters-Bayer et al. 2005). LISFs place finances directly in the control of local community-based organizations (LBL 2002), aiming to enable small-scale farmers the flexibility and independence to undertake their own initiatives to address local problems. Principles associated with LISFs include transparency in funding application and disbursement, minimum paperwork, and rapid decision-making. Funds offer communities the flexibility to experiment and learn, but also to fail, by covering risks associated with trying "creative ideas without knowing for sure what the results will be"—a freedom often precluded from conventional financing frameworks.

Concerns regarding the financial management capacity of community-based organizations are commonly raised by donors in response to calls for decentralized funding mechanisms. Community-based organizations are themselves recognizing and responding to this constraint. For example, the Kalinga Mission for Indigenous Communities and Youth Development, Inc. offers financial management systems training to communities (Box 16.2). As recommended by community representatives presenting to the 2005 Millennium Summit (Community Commons Declaration 2005), the establishment of financing mechanisms, such as an international "global community learning fund," are required to support the replication and upscaling of innovative, community-based resource management strategies. Such a fund should be designed and primarily governed by community representatives themselves, with funds disbursed directly to community-based organizations.

Support for Scaling Up Grassroots Success: Lessons Learned from Landcare

Landcare movements began emerging in the mid-1980s as an approach for mobilizing collective action by local farming and ranching communities concerned about land degradation and natural resource management challenges. The approach centers on forming community Landcare groups, supported to varying degrees through partnerships with government and nongovernment agencies (Cramb and Culasero 2003). Groups with a common agenda work together to identify how problems can be solved and mobilize resources to solve them, based on the principles of volunteerism, genuine participation, responding to local demand, and building partnerships and support from the local level. Groups engage in varying activities, including total farm care, catchment care, vegetation management, coastal management, and property planning. Landcare approaches thus aim to build the necessary partnerships between farmer, catchment, regional approaches, and government policy to deliver broader landscape change, employing facilitators and coordinators to provide an interface between government agencies and Landcare groups.

The growth of Australia's Landcare movement has been explosive. There are now over 4000 Landcare groups, with approximately one-third of Australian

Box 16.2. Community-led capacity development in financial management and governance: an example from the Kalinga Mission for Indigenous Communities and Youth Development, the Philippines

For many centuries, the Kalinga Indigenous Peoples in the Philippines have sought to sustain their livelihoods while conserving their mountain biodiversity through an integrated, landscape-level approach. Local communities manage their watersheds to ensure a continual supply of rainwater to communal irrigation systems. Fish and vegetable production is integrated into irrigated rice terrace management. Within Kalinga, locally-led sustainable community development initiatives are supported by Kalinga Mission for Indigenous Communities and Youth Development, Inc. (KAMICYDI). KAMICYDI recognized and responded to the need for community-based organisations (CBOs) to institutionalize good governance and financial management systems as an essential prerequisite to developing, managing and scaling-up effective ecoagriculture initiatives. Such systems are designed to complement and strengthen the existing CBO's traditional organizational knowledge and management systems. To date, approximately fifty locally-based organizations within Kalinga have been empowered to improve their governance. Training in financial accounting, management, and reporting have enabled these CBOs to build the confidence to negotiate and build relationships with potential donors. This simple yet highly effective local capacity development support is now in high demand within the region and at a national level.

farming families engaged (AFFA 2003). A Landcare council has been established to provide ministerial advice and create a platform to discuss and present views to the national government, while a National Landcare Facilitator Program supports the training needs of local Landcare facilitators and coordinators. Within the Philippines, Landcare developed independently as a grassroots initiative based on a three-way partnership of farmers, local government units, and the World Agroforestry Center to support the adoption of soil and water conservation technologies in the uplands in response to soil erosion and low-productivity challenges. The rapid proliferation of Landcare groups; their diversification into a range of other activities, including participation in municipal natural resource management planning; and the development of municipal Landcare federations influencing watershed and protected area management has sparked widespread interest regionally, nationally, and globally.

Landcare experiences within different local and national contexts illustrate the challenges of moving to scale. Relationships relevant at one scale, in one particular context, do not easily transcend scales or contexts (Lovell et al. 2003). Scaling up Landcare practices requires flexibility to use different models, depending on locally specific conditions. Successful upscaling does not depend on replicating the program itself but on adapting (and creating) the conditions under which the program can work (Berman and Nelson 1997; Schorr et al. 1999). Experiences demonstrate the need to balance community-initiated change, partnerships with local governments, and promotion of technological and institutional innovations by external actors. For example, it was easier to facilitate technology adoption than to initiate institutional change—yet institutional changes were essential to establish a foundation from which to move the process forward.

Landcare groups have flourished in situations where locally adapted technologies have emerged, local authorities are supportive of grassroots initiatives and have the desire to work with farmers and other agencies, and a long-term research and extension presence is provided by supportive NGOs. Catacutan and Cramb (2006) highlight conditions that influence potentials for scaling up, their relative importance depending on local realities:

- Successful initiatives offered a set of widely adoptable management options, providing farmers the flexibility to develop more sustainable management approaches according to their needs, resources, and preferences. Some built on existing local technologies or knowledge systems, others adapted "modern" technologies. Farmers need to own the decision-making process and adopt the approach most appropriate to their needs.
- Landcare flourished in areas where conservation efforts were promoted and supported, and farmers were not affected by rapid economic change such as the growth of large-scale agribusiness or nonagricultural employment. Landcare had better prospects where local politics were stable and Landcare leaders were able to establish good relationships with local government officials.
- Effective training, communication, and facilitation were central to Landcare's farmer-based extension approach, and also essential for scaling up.
- Even when local organizations are strong, grassroots initiatives significantly benefit from the concerted support of local and national governments and other nongovernmental actors to mobilize resources. Within Australia, Landcare could not have grown so rapidly without national government support, which included US$340 million of funding over 10 years (Campbell 1994; Lockie and Vanclay 1997). Unlike Australia, the Philippines national government had insufficient resources to respond at the scale necessary to deliver meaningful improvements to livelihoods and natural resource management. But with a growing network of like-minded communities, local government representatives, NGOs, and project partners, it is possible to mobilize a critical mass of actors, from the bottom up, to influence the broader policy agenda.

Policies That Recognize and Support Community-Led Ecoagriculture

Enabling community-based ecoagriculture requires coordinated responses at multiple scales and across multiple sectors. Locally driven ecoagriculture approaches will have limited impact unless complemented by supportive policy frameworks. Yet there remains a disconnect between the sector-based policy targets determined by policymakers at the international and national level, and the individuals on the ground who must play a central role in operationalizing them. Two broad sets of actions are urgently required: processes that effectively engage local communities in policy formulation, and legislation and policy frameworks that support community-led ecoagriculture.

Strengthen Community Representation in Policymaking Processes

Mechanisms to facilitate community representation in major international policy process are improving. The establishment of the UN Permanent Forum on Indigenous Issues in 2000 promotes the integration and representation of indigenous issues within the United Nations systems. Community dialogue spaces have been integrated into a number of international policy processes to facilitate dialogue between local stakeholders and international decision makers (Gillis and Southey 2005). Such mechanisms should be more systematically integrated into the full range of international policy processes that impact communities, including rural development, production, trade, and climate change. For example, community recommendations to the 2005 UN World Summit (Community Commons Declaration 2005) highlight the need for strengthened community representation on global- and national-level task forces established to review national development strategies, particularly ongoing Millennium Development Goal–based poverty reduction strategies. The potential to drive and inform policy processes from the bottom up, through consistently ensuring community representation in decision-making bodies at the local, regional, and national level, is exemplified by the approach of the Chibememe Earth Healing Association (Box 16.3).

Effective mechanisms to facilitate community engagement in decision-making processes will require that policymakers and community representatives relate to and understand one another. Campbell (1994) highlights the profound value of high-profile, committed political leaders that understand and are committed to supporting grassroots initiatives. Genuine dialogue and partnership have to be based on equitability, trust, and mutual respect, in forms and languages that resonate with all partners. In turn, community representatives may require support to relate their expertise to the broader policy context in order to strengthen their advocacy and negotiation capacity.

Box 16.3. Community-driven advocacy to inform policy from the bottom up in Zimbabwe

Since 1999, the Chibememe Earth Healing Association (CHIEHA) of Chiredzi, Zimbabwe, has played an influential role in ensuring that local and national polices enable community engagement in the management of the Great Limpopo Transfrontier Park (GLTP). This 35,000 km^2 conservation area spans Kruger National Park in South Africa, Gonarezhou National Park in Zimbabwe, and Limpopo National Park in Mozambique. A series of major awareness-raising activities has been pivotal to CHIEHA's considerable policy impact. For example, a high-profile 990 km cycle ride from Zimbabwe to the Durban World Parks Congress, South Africa, consulting with local communities en route, was undertaken to focus attention on the need for policies that allow communities to benefit from protected areas, and to lobby for enhanced community participation in the planning and management of the GLTP. Through such processes, CHIEHA has reached out to over 50,000 stakeholders in and around the GLTP to date—acting as a significant driver of enhanced community participation in national environmental policy development, particularly Zimbabwe's National Environmental and Wildlife Based Land Reform Policy.

Major policy impacts include recognition of community conservation efforts and initiatives as a pillar of biodiversity conservation and sustainable use in and around protected areas, particularly the critical role of local communities in transboundary protected areas management. Community conserved areas and the value of indigenous and traditional knowledge, innovations, and practices. have been legitimized with CHIEHA representing on the National Expert Working Group on Access and Benefit Sharing (ABS). They are particularly involved in designing ABS comanagement models that engage the private sector, nongovernmental organizations, communities, and government in policy development to protect traditional Indigenous Knowledge Systems and Technologies. These ensure that prior informed consent is secured, and apply the Convention on Biological Diversity's Code of Conduct for undertaking research in local and indigenous communities. A key principle underlying CHIEHA's work is empowering communities to understand the policies they are both influenced by and seeking to influence. Community-driven capacity development processes include the Transfrontier Parks and Protected Areas Rural Communities' Network—a regional network facilitating knowledge exchange between South African, Mozambican, and Zimbabwean GLTP communities—and the development of community-led training programs on ABS and community enterprise development.

Legislative and Policy Frameworks That Enable Community-Led Ecoagriculture

Without legislation and policy frameworks that enable local communities and small-scale producers to own and derive benefits from their resources, their willingness and capacity to implement ecoagriculture is severely compromised. As a foundation, policies are required that recognize the resource and intellectual property rights of communities and afford them both decision-making and management authority over their resources, whether for agricultural production or for conservation. The last two decades have seen significant progress in establishing international precedents. Article 8(j) of the Convention on Biological Diversity calls for parties to respect, preserve, and maintain knowledge, innovations, and practices of indigenous and local communities, promote their wider applications with the approval of knowledge holders, and encourage equitable sharing of benefits that arise. The Durban Accord of the 2003 World Parks Congress (WCPA 2003) emphasizes the need for local communities to share both the costs and the benefits of living close to protected areas, and for governments to institute participatory mechanisms that engage communities in identifying, creating, and managing protected areas. The United Nations Declaration on the Rights of Indigenous Peoples (UN Human Rights Council 2006) emphasizes the need to "respect indigenous knowledge, cultures and traditional practices, and their contribution to sustainable and equitable development and proper environmental management."

The effectiveness of these international precedents remains contingent on the political will to translate and enforce them within national-level legislation. Many international policies supporting community action in natural resource management are "soft" policies—optional, not obligatory. Even leadership in setting national-level conservation precedents is insufficient if contravened by contradictory trade, development, and production policies. A frequent obstacle has been the failure of national legislation to recognize customary tenure systems, including community-conserved areas, or local or indigenous communities as legal entities (Borrini-Feyerabend et al. 2004). Legal recognition and secure ownership rights are required with respect to customary laws and traditions, and ancestral lands, domains, and resources including air, water, and seed supplies, within and beyond designated community conserved areas. Legislation must also protect intellectual property rights, along with mechanisms to ensure the free prior informed consent of communities is obtained before development initiatives are undertaken on their land and to guarantee them equitable access and benefit sharing from resources utilized.

Political will is also required to ensure that policy- and decision making is decentralized to the most appropriate level and coordinated with the allocation of funding (Borrini-Feyerabend et al. 2004). Too often, communities receive project financing but lack authority to manage it, or vice versa. In the Philippines, national legislation set promising precedents through the 1997 Indigenous Peoples Rights Act—the first law within Asia to comprehensively recognize indigenous peoples' rights with respect to education, and traditions and customary laws, ancestral lands, domains, and resources, including air, water and land. Yet the experience of the Philippines' Landcare movement (Catacutan and Cramb 2006) highlights the challenges of operationalizing such policies. Although government recognition of rights and devolution of decision-making authority support grassroots initiatives, the government was unable to provide the necessary resources to support the mobilization of grassroots initiatives on a scale that would generate nationally significant improvements in rural livelihoods or natural resource management practices.

Conclusions

Collective action by local communities and farmers is essential to sustain or develop ecoagriculture landscapes that jointly deliver conservation and livelihood benefits. Such action will be fundamental to achieving sustainable rural development, realizing the Millennium Development Goals and implementing multilateral environmental conventions. By explicitly supporting grassroots ecoagriculture initiatives, decision makers at international, national, and district levels can reconcile sectoral goals with the integrated strategies needed to deliver them on the ground. Those strategies must be flexible enough to support different models of community-led change in different local contexts, fully valuing and equitably engaging farmers and community organizations as partners in decision- and policymaking processes.

References

AFFA. 2003. *Report of the Review of the National Landcare Program.* Australian Government Department of Agriculture Fisheries and Forestry. www.daffa.gov.au/__data/assets/pdf_file/29190/nlp_review_report_final.pdf.

Agrawal, A. and C.C. Gibson (eds.). 2001. *Communities and the Environment: Ethnicity, Gender and the State in Community-Based Conservation.* Rutgers University Press, Piscataway, NJ.

Allen, W. and M. Kilvington. 2001. *An Outline of a Participatory Approach to Environmental Research and Development Initiatives.* www.landcareresearch.co.nz/sal/iskm.asp.

Ashby, J.A., A.R. Braun, T. Gracia, M.P. Guerrero, L.A. Hernandez, C.A. Quiros, and J.A. Roa. 2000. *Investing in Farmers as Researchers: Experiences with Local Agricultural Research Committees in Latin America.* International Center for Tropical Agriculture (CIAT), Cali, Colombia. www.ciat.cgiar.org/downloads/pdf/Investing_farmers.pdf.

Bebbington, A. and J. Farrington. 1993. Governments, NGOs and agricultural development: perspectives on changing inter-organisational relationships. *Journal of Development Studies* 29(2):199–219.

Berman, P. and B. Nelson. 1997. Replication: adapt or fail. In: *Innovations in American Government*, ed. A. Altshuler and R. Behn. Brookings Institute Press, Washington, DC: 319–31.

Binswanger, H.P. and S. Aiyar. 2003. *Scaling Up Community Driven Development: Theoretical Underpinnings and Program Design Implications*. World Bank, Washington, DC.

Borrini-Feyerabend, G., A. Kothari, and G. Oviedo. 2004. *Indigenous and Local Communities and Protected Areas: Towards Equity and Enhanced Conservation*. World Conservation Union (IUCN), Gland, Switzerland, and Cambridge, UK.

Brookfield, H., C. Padoch, H. Parsons, and M. Stocking (eds.). 2002. *Cultivating Biodiversity: Understanding, Analysing and Using Agricultural Diversity*. Intermediate Technology Development Group (ITDG) Publishing, London.

Campbell, A. 1994. *Landcare: Communities Shaping the Land and the Future*. Allen and Unwin, St Leonards, Australia.

Catacutan, Delia and R.A. Cramb. Forthcoming. Mobilizing communities for ecoagriculture: Lessons from landcare in the Philippines and Australia. Draft paper, forthcoming.

Community Commons Declaration. 2005. www.undp.org/equatorinitiative/secondary/events/CommunityCommons/CommunityCommons_Declarations.htm.

Convention on Biological Diversity (CBD). 1992. www.biodiv.org.

Cramb, R.A. and Z. Culasero. 2003. Landcare and livelihoods: the promotion and adoption of conservation farming systems in the Philippine uplands. *International Journal of Agricultural Sustainability* 1(2):141–54.

Fairhead, J. 1993. Representing knowledge: the "new farmer" in research fashions. In: *Practising development. Social science perspectives*, ed. J. Pottier. Routledge, New York: 187–204.

Gillis, N. and S. Southey. 2005. *New Strategies for Development: A Community Dialogue for Meeting the Millennium Development Goals*. Fordham University Press, Bronx.

Hughes, R. and F. Fintan. 2001. *Integrating Conservation and Development Experience: A Review and Bibliography of the ICDP Literature*. International Institute for Environment and Development, London.

Hussein, K. 2000. *Farmers' Organizations and Agricultural Technology: Institutions That Give Farmers a Voice*. Overseas Development Institute, London. www.livelihoods.org/pip/pip/refo1.html.

Isely, C. and S.J. Scherr. 2003. *Community Based Ecoagriculture Initiatives: Findings from the 2002 UNDP Equator Prize Nominations*. Ecoagriculture Partners and Equator Initiative, Washington, DC. www.ecoagriculturepartners.org/documents/reports/CIreportdraft12-23%5B1%5D.pdf

Kambu, A., M. Alexander, K. Chamundeeswari, M. Ruiz, and B. Tobin. 2004. *The Role of Registers and Databases in the Protection of Traditional Knowledge: A Comparative Analysis*. United Nations University Institute of Advanced Studies (UNU-IAS), Yokohama, Japan. www.ias.unu.edu/binaries/UNUIAS_TKRegistersReport.pdf.

LBL (Landwirtschaftliche Beratungszentrale Lindau). 2002. *Innovative Approaches to Fi-*

nancing Extension for Agriculture and Natural Resource Management. Landwirtschaftliche Beratungszentrale Lindau.

Lightfoot, C., K. Gallagher, U. Scheuermeier, and J. Nymand. 2005. *Insights for the Emergence of Demand Driven Services.* Linking Local Learners Briefing Note no. 8. www.linkinglearners.net/downloads/brief8.pdf.

Lockie, S. and F. Vanclay. 1997. *Critical Landcare.* Center for Rural Social Research, Charles Sturt University, New South Wales, Australia.

Lovell, C.J., A. Mandondo, and P. Moriarty. 2003. The question of scale in integrated natural resource management. In: *Integrated Natural Resource Management: Linking Productivity, the Environment and Development,* ed. B.M. Campbell and J.A Sayer. CABI Publishing in Association with the Centre for International Forestry Research, Bogor, Indonesia.

McNeely, J.A. and S.J. Scherr. 2003. *Ecoagriculture: Strategies to Feed the World and Save Biodiversity.* Island Press, Washington, DC.

McShane, T.O. and M.P. Wells. 2004. *Getting Biodiversity Projects to Work: Towards More Effective Conservation and Development.* Columbia University Press, New York.

Millennium Ecosystem Assessment. 2006. *Millennium Ecosystem Assessment Synthesis Report.* The Millennium Assessment, Washington, DC. www.millenniumassessment .org.

Mogelgaard, K. 2003. *Helping People, Saving Biodiversity: An Overview of Integrated Approaches to Conservation and Development.* Occasional Paper, March 2003. Population Action International, Washington, DC.

Molnar, A. 2003. *Forest Certification and Communities: Looking Forward to the Next Decade.* Forest Trends, Washington, DC.

Molnar, A., S.J. Scherr, and A. Khare. 2004. *Who Conserves the World's Forest?* Forest Trends and Ecoagriculture Partners, Washington, DC.

M.S. Swaminathan Research Foundation. 2005. *Toolkit for Setting Up Rural Knowledge Centers (RKC).* www.mssrf.org/tsunami/rkc_toolkit.pdf.

Newmark, W.D and J.L. Hough. 2000. Conserving wildlife in Africa: integrated conservation and development projects and beyond. *BioScience* 50(7):585–592.

Pinedo-Vasquez, M., E. Gyasi, and K. Coffey. 2002. PLEC demonstration activities: a review of procedures and experiences. In: *Cultivating Biodiversity: Understanding, Analysing and Using Agricultural Diversity,* ed. H. Brookfield, C. Padoch, H. Parsons and M. Stocking. Intermediate Technology Development Group (ITDG) Publishing, London: 116–122.

Pretty, J. 1998. Supportive policies and practice for scaling up sustainable agriculture. In: *Facilitating Sustainable Agriculture,* ed. N.G. Roling and M.A.E. Wagemakers, Cambridge University Press, Cambridge: 23–46.

Rhodes, C. and S.J. Scherr (eds.). 2005. *Developing Ecoagriculture to Improve Livelihoods, Biodiversity Conservation and Sustainable Production at a Landscape Scale: Assessment and Recommendations from the First International Ecoagriculture Conference and Practitioners' Fair, Sept. 25–Oct. 1, 2004.* Ecoagriculture Partners, Washington, DC. www .ecoagriculturepartners.org.

Roe, D., J. Mayers, M. Grieg-Gran, A. Kothari, C. Fabricius, and R. Hughes. 2000. *Evaluating Eden: Exploring the Myths and Realities of Community-Based Wildlife Management.*

Evaluating Eden Series no 8. International Institute for Environment and Development (IIED), London, UK.

Sayer, J. and B. Campbell. 2004. *The Science of Sustainable Development: Local Livelihoods and the Global Environment.* Cambridge University Press, Cambridge.

Scherr, S.J. J.C. Milder, L. Lipper, and M. Zurek. Forthcoming. *Poverty and Payments for Ecosystem Services.* Ecoagriculture Partners and FAO. Washington DC.

Scherr, S.J., A. White, and D. Kaimowitz. 2004. *A New Agenda for Forest Conservation and Poverty Reduction: Making Markets Work for Low-Income Producers.* Forest Trends, Washington, DC, Center for International Forestry Research (CIFOR), Managua Nicaraqua, and World Conservation Union (IUCN), Gland, Switzerland.

Schorr, L., K. Sylvester and M. Dunkle. 1999. Strategies to Achieve a Common Purpose: Tools for Turning Good Ideas into Good Policies. Special Report No. 12. Technical Policy Exchange: Institute for Educational Leadership.

Spinage, C. 1998. Social change and conservation misrepresentation in Africa. *Oryx* 32(4):265–276.

Sundberg, J. 1998. Strategies for authenticity, space, and place in the Maya Biosphere Reserve, Petén, Guatemala. *Conference of Latin Americanist Geographers* 24:85–96.

Terborgh, J. 1999. *Requiem for Nature.* Island Press, Washington, DC.

UN Human Rights Council. 2006. United Nations Declaration on the Rights of Indigenous Peoples. www.ohchr.org/english/issues/indigenous/docs/declaration.doc

United Nations. 2005. *World Summit Outcome (15 September 2005): Final Document.* http://daccessdds.un.org/doc/UNDOC/LTD/N05/511/30/PDF/N0551130.pdf?OpenElement.

Van Dam, C. 2002. Certificación forestal, equidad y participación. Paper prepared for the electronic conference of the Participation Network CODERSA-ECLNV, August 5– September 1. www.red-participation.com.

Waters-Bayer, A., L. van Veldhuizen, M. Wongtschowski, and S. Killough. 2005. *Innovation Support Funds for Farmer-Led Research and Development.* IK Notes no. 85, October 2005. www.prolinnova.net/Downloadable_files/isf.pdf.

WCPA (World Commission on Protected Areas). 2003. Durban Accord. www.iucn.org/themes/wcpa/wpc2003/pdfs/english/Proceedings/durbanaccord.pdf.

Wells, M., K. Brandon, and L. Hannah. 1992. *People and Parks: Linking Protected Areas with Local Communities.* The World Bank, World Wildlife Fund and US Agency for International Development (USAID). World Bank, Washington, DC.

Worah, S. 2000. International history of ICDPs. In: Proceedings of Integrated Conservation and Development Projects Lessons Learned Workshop, 12–13 June 2000, Hanoi, Vietnam. ed. United Nations Development Program (UNDP), World Bank, and World Wide Fund for Nature (WWF), Hanoi, Vietnam.

Chapter 17

Planning at a Landscape Scale

William J. Jackson, Stewart Maginnis, and Sandeep Sengupta

Introduction

Land-use planning is critical to the successful design and implementation of ecoagricultural landscapes, whether in developed temperate zones or in undeveloped tropical landscapes. Because they contain the greatest level of the earth's remaining biodiversity and are the most at risk, this chapter focuses on the latter. Moist tropical forests alone contain about 60% and possibly even as much as 90% of all terrestrial species found on Earth (UNEP 2001). Consistent high rates of forest loss over the last several decades, together with extensive and ongoing fragmentation and simplification of forest ecosystems, means that global biodiversity continues to remain under serious threat. This is despite the fact that about 12.4% of the world's forests currently enjoy protected area status (FAO 2001). Between 1990 and 2000 the world lost an estimated 94 million ha of forests—an area equivalent to 2.4% of the world's total forests (FAO 2001). Habitat loss and degradation, especially in lowland and mountain tropical rainforests, has been identified as a primary cause of biodiversity loss, affecting 89% of all threatened birds, 83% of threatened mammals, and 91% of the threatened plants assessed (IUCN 2003).

Agricultural expansion has, by far, been the most prominent cause (Geist and Lambin 2002). Nearly 70% of the total area that was deforested in the 1990s was converted to agriculture, predominantly under permanent rather than shifting systems (UNEP 2002). However, it also remains a fact that in many biodiversity-rich developing countries with high populations of rural poor, the conversion of forests to agricultural lands has resulted in impressive growth in food production, which in turn has led to better livelihood security and economic development for the local people. With global population projected to increase by 3 billion to a total of 9 billion by 2050, more forest land is predicted to be cleared for food production in the future, especially in developing countries (FAO 2003; Poore 2003). This will make biodiversity conservation an even more challenging task.

Ecoagriculture landscapes are designed both to increase agricultural production and farmers' income and to conserve wild biodiversity in the agricultural landscapes that surround protected areas. The quest for an integrated land-use management system that meets the diverse needs of economic development, food production, livelihood security, and conservation of biodiversity is not a new one. Approaches to achieve integrated natural resource management, particularly in developing countries where the clashes between diverging agendas of "environment" and "development" are most prominent, have included integrated rural development programs in the 1960s and 1970s, integrated conservation and development projects (ICDPs) in the 1970s and 1980s, and more recent ecoregional approaches to conservation, integrated soil and water management projects, integrated watershed/catchment management, and so on. Although all these different approaches have been driven by a common desire to achieve integrated land-use management, the inability to translate theories into practical achievements has resulted in widespread frustration and disillusionment (Sayer and Campbell 2004). Moreover, these efforts have mostly been "sectorally driven," stressing promotion and implementation of a particular sector's form of integrated planning and management rather than using a commonly agreed upon strategy for meeting the needs of agriculture, forestry, irrigation, and rural development.

Why have these approaches failed, and how can the planning and management of integrated land-use systems change in order to achieve effective implementation in the future? This chapter first looks at what is wrong with conventional land-use planning. Using our experiences from forest landscape restoration, essentially a framework for the restoration of multiple forest goods and services in degraded and deforested landscapes, we then outline some of the lessons we have learned about achieving this hitherto elusive goal of integrated land-use management.

The Problem with Conventional Land-Use Planning

Conventional land-use planning and practice has typically followed an approach described by Charland (1996) as the problem-isolation paradigm. According to this paradigm, a complex problem—such as how to deliver a socially and environmentally optimal mix of land-use services—is dealt with by breaking it down into smaller, easily understood elements. Each of the different elements is tackled independently through generally standardized responses, in the hope that when they are all put back together, the end product will be an optimal one. Thus, in the context of different land-use requirements, food and fiber production needs are met through intensive agriculture involving forest clearance; timber needs are met through industrial plantations or through manipulating and simplifying natural forest stands; and conservation needs are met by corralling biodiversity into islands designated as protected areas (Maginnis et al. 2004). However, such compartmentalized planning approaches have failed to realize that the biophysical and social relationships and pressures that shape natural production systems in the real world are complex, interdependent, and cannot be easily predicted or reduced to discrete components.

What may seem like a good solution for satisfying society's need for one particular ecosystem service, such as the steady and reliable supply of industrial roundwood, can end up creating a problem for the delivery of another. Likewise, whatever the extent of the spatial unit selected for land-use planning, it can seldom be treated as a closed system: what happens "outside the box" often influences what happens inside it. For example, even in the Yellowstone National Park, for many a model of how protected areas should be set up and run, rapid population growth in the private lands surrounding it is now beginning to threaten the integrity of the core wilderness area, making conservation an increasingly controversial issue in the region (Stohlgren 1996).

In addition, land-use planning has often been a centralized, top-down, prescriptive, and expert-driven process with "outsiders" making land-use decisions, to the almost complete exclusion of local stakeholders. Invariably it is "[g]overnment bureaucrats [who] choose the location for a plantation to resolve a timber shortage problem, a hydroelectric scheme to resolve an energy problem, and an irrigation scheme to solve a foreign exchange problem, with little if any consideration for the needs of the communities that will be most affected" (Maginnis et al. 2004).

With respect to biodiversity conservation planning, choices such as what to save, how much to save, and where to save it have typically been shaped by a small, elite group, usually comprising central and provincial government authorities or civil servants, conservationists (often representing distinctly Western views and values about conservation), and scientists. Rarely are the local people who live in the affected areas, and whose livelihoods are directly affected by these decisions, consulted about them in any meaningful manner. The ways in which many protected areas have been created in the past are a good case in point. Several national parks across India, for example, were established on the whim of Lady Curzon (wife of a British colonial), and in the United States, even as late as 1940, 4000 people were forced to leave their homes in order to establish the Great Smokey Mountains National Park (Maginnis et al. 2004).

The modern history of land-use planning is full of examples of dispossessing many to suit the desires of a few, but these plans also failed on their own declared objectives as a result of ignoring or not sufficiently factoring in local needs and value systems (Guha 1997). Even though in recent years greater efforts have been made to enhance the participation of local people in land-use and conservation planning, such as in ICDPs or in comanagement systems like joint forest management, a common criticism is that in many of these cases the nature of peoples' participation is superficial and there is little real and meaningful change in the way in which land-use decisions are planned or made (Larson 2004).

Developing a People-Centered Landscape Perspective to Land-Use Planning

Critiquing conventional land-use planning is not to question the value of establishing protected areas or plantation forests, or imply that local needs and values must always take precedence over wider interests or global public goods. Indeed, the logical step-by-step approach of the problem-isolation paradigm in disaggregating complex large-scale land requirements into smaller and more manageable units can facilitate better understanding among stakeholders of the various land-use problems. However, for satisfactory outcomes to be realized on the ground, planners and decision makers should not promote a "one-size-fits-all" solution. Nor should experts assume that the understanding they have gained about the nature of biophysical, social, and economic interactions at one location will play out in the same way at all locations. Optimizing integrated land-use management will require land-use planning and decision making to become a much more democratic process that considers the knowledge and needs of all local stakeholders. Even if one ignores the basic ethical arguments against excluding local stakeholders from land-use decision making, the practical problem remains that, unless local communities and stakeholders are at the center of land-use decision making and conservation planning, plans that are imposed by experts from the top will face a serious risk of being undermined by realities on the ground.

Tropical agroecosystems represent a typical challenge as to reconciling the needs of local stakeholders with the desires of remote but influential national and global players. Commonly, one will encounter a matrix of different land uses within a larger landscape, with each land holding being managed in a slightly different way so as to deliver a specific function or a suite of specialized functions that its owner values or desires. Whether the net impact of these individual land holdings is a wide or narrow range of goods and services at the landscape scale depends on three interrelated factors: the individual decisions taken on-farm; the position of the farm with the landscape; and the sum of like-minded decisions across the landscape. A landscape can be defined as "a contiguous area, intermediate in size between an 'ecoregion' and a 'site,' with a specific set of ecological, cultural, and socioeconomic characteristics distinct from its neighbours." However, in practice all landscapes are social constructs and the definition of a landscape lies largely in the eye of the beholder (Maginnis et al. 2004). When discussing land-use planning it is important to realize that many of those individual decisions will be driven more by market and policy signals and local agreements than by lines drawn by experts on maps.

Building a landscape perspective into land-use planning and land-use policy depends on identifying those conditions and incentives under which local stakeholders will be prepared to make land-use decisions at the site level that are conducive to the delivery of a wide and balanced range of goods and services within the larger landscape, and vice versa. For example, a typical tropical agroecosystem landscape includes not only farms, grazing areas, and settlements, but often natural and seminatural areas that have high biodiversity values. Whether this biodiversity will continue to persist in the landscape will depend not only on agreements and rules to maintain the actual remnant natural areas but also the agreements and rules that govern productive on-farm activities. For example, an agreement to reduce the density of on-farm trees may reduce genetic flows across the landscape and particularly between those sites of high natural value. Equally, annual burning of crop residues may cause long-term deterioration in the quality of the forest fringe and increase the likelihood of a catastrophic forest fire.

However, a landscape perspective is not just about imposing ever more rules on farmers in order to ensure the delivery of goods and services that they do not particularly value. Rather, it provides a framework within which stakeholders can negotiate their individual land-use trade-offs and agree on a mosaic of different land uses that meets their needs as well as delivering national and global values such as biodiversity. Critically, by focusing on the delivery of goods and services rather than particular land-use configurations, a landscape perspective can actually be used to lift constraints on local farmers.

A good example of this is the role of secondary forest as an on-farm resource. When conservationists see secondary forest regenerate on farmland the temptation is to view it as a new forest resource and therefore protect it so that it delivers those values that conservationists tend to prefer. Farmers, on the other hand, may see secondary forest regrowth not as a forest resource but as an on-farm resource, such as a ready supply of fencing poles, which perhaps will be cut down at some later stage and converted back to agriculture. Regulations that prevent farmers from using secondary forests in this way discourage farmers from allowing any forest to regenerate anywhere on their land. They are thus deprived of fencing poles and the opportunity to rest some of their land and recondition their soil, and conservationists are deprived of having a permanent net increase of forest habitat at a landscape level, though not necessarily always at the same site. A landscape perspective might therefore draw the conclusion that rules preventing the harvest of secondary forest regrowth ought to be relaxed, and farmers should be granted greater latitude in deciding how to use this on-farm resource.

Therefore, a landscape perspective differs from conventional large-scale land-use planning in that it does not depend on a top-down approach driven by experts, but rather a bottom-up approach, informed by expert opinion, that involves stakeholders in negotiating land uses across a landscape (Fisher et al. 2005). This requires creating appropriate and effective institutional frameworks that allow all stakeholders to undertake negotiations on an equitable footing.

The Experience of Forest Landscape Restoration

Recognizing the drawbacks of conventional approaches to the restoration and rehabilitation of degraded and deforested landscapes, since 2001 the World Conservation Union (IUCN) and the World Wide Fund for Nature (WWF), in collaboration with a number of governments and nongovernmental organizations (NGOs), have promoted the concept of forest landscape restoration as a "process that aims to regain ecological integrity and enhance human well-being in degraded or deforested forest landscapes." Over the ensuing years, this concept has evolved into an established community of practice operating as the Global Partnership on Forest Landscape Restoration, a network of governments, organizations, communities, and individuals who recognize the importance of forest landscape restoration and want to be part of a coordinated global effort to promote it. The partnership currently comprises several leading international organizations, governments, state agencies, and NGOs.

The tropical forest landscape is no longer an unending stretch of undisturbed forest, but often a fragmented mix of primary forests, managed forests, secondary forests, plantations, and degraded forest lands, interspersed with extensive (and often expanding) areas of other nonforest land uses, predominantly agriculture. By some estimates, agricultural and forest resources in these fragmented tropical forest landscapes today support around 500 million, largely poor, people (Maginnis and Jackson 2000).

Unlike conventional planning, which has generally tended to view forest landscapes narrowly, either as biodiversity reserves or as timber-producing estates, forest landscape restoration recognizes the multiple functions that fragmented landscapes provide to multiple stakeholders—farmers, indigenous peoples, pastoralists, and timber companies, among others—who depend on them for pursuing their different, and sometimes conflicting, land-use objectives. In many tropical landscapes it is unrealistic to expect that deforested landscapes can be restored to their original state. Thus forest landscape restoration pragmatically focuses not on how to restore forest cover per se, but on how to restore forest *functionality* (i.e., the full range of goods, services, and ecological processes that forests provide, including biodiversity conservation) at the overall landscape level.

Forest landscape restoration encourages individual stakeholders to make site-level decisions while keeping in mind the site's linkages with the broader landscape (for example, the effect of site-level decisions on hydrological functions). Forest landscape restoration emphasizes site-level interventions, such as ecological restoration, natural forest management, regeneration of secondary forests, reforestation of degraded forest land and rangelands, as well as agroforestry and urban and periurban forests, which can be applied to effectively meet both the production and the conservation needs across a landscape.

Forest landscape restoration seeks to involve local communities and stakeholders in the land-use planning and decision-making process, thereby placing local people's needs at the center of all forest land-use decision making while also seeking to incorporate sustainable forest conservation and management objectives and practices.

Forest landscape restoration is a promising approach to achieving ecologically sound landscapes. McNeely and Scherr (2003) document the development of farm windbreaks and silvipastures in Costa Rica and the agroforests of Sumatra, among others, as examples of ecoagriculture that are using unfarmed areas and forest mosaics to develop habitat networks.

In the Shinyanga region of northern Tanzania, Sukuma agropastoralists have restored 500,000 ha of degraded landscapes on farmlands and rangelands since 1985 through developing *ngitili* fodder enclosures, which are now providing them with fodder, firewood, thatching grass, and increased water supply (Barrow et al. 2003). In Senegal, women in the Popenguine village have successfully restored 1000 ha of the Popenguine nature reserve, which has resulted in the return of local biodiversity, and have also established nurseries on their village lands to meet their local fuelwood needs. The same women have subsequently formed a larger women's network across eight villages to restore 10,000 ha of the degraded Ker Cupaam Community Space. Similarly, efforts made by local farming communities have restored large tracts of degraded lands in the Middle Hills of Nepal (Pye-Smith and Saint-Laurent 2003). In all these cases, the process of restoration has also entailed a significant increase in the occurrence of on-farm trees on countless hectares of productive agricultural land.

Managing Trade-Offs at the Landscape Scale

Ecoagriculture strategies seek to achieve enhanced outcome for both agricultural production and biodiversity at the landscape scale. In some cases, intervention at a site can achieve both (see cases in McNeely and Scherr 2003). More often, especially in the short term, trade-offs exist at a site that must be resolved at a higher scale. For example, ICDPs have been criticized for overstating the extent to which they can be vehicles for delivering win–win conservation-development outcomes, often ignoring the fact that final outcomes on the ground are generally likely to involve a large measure of compromise. Recognizing this, practitioners have started to qualify the win–win optimism of some of the early approaches of integrated resource management and have begun to look at more realistic trade-off based options for integrating conservation and development. Tomich et al. (2001), for instance, outline a process of trade-off assessment among six principal stakeholder groups (indigenous peoples, small farmers, large estates, midsize absentee landowners, public institutions, and the international community) with respect to various land-use systems in Sumatra, Indonesia.

Forest landscape restoration accepts that win–win outcomes may be difficult to achieve, and it is sometimes necessary to accept trade-offs between one set of priorities and another, particularly at the site level. Often, in those landscapes where different stakeholders compete for the same parcel of land, trade-offs will occur independently of whether a formal planning process exists. For example, small farmers will make their trade-offs with respect to forest clearance based on returns to land and labor. If no action is taken to help direct these trade-offs, say, by changing the incentive basis or compensating farmers fairly for the provision of ecosystem goods and services, then forest loss will be the most likely outcome.

In some cases it might be necessary to forgo the achievement of all the biodiversity goals of some areas to support economic development, while doing the opposite in other areas to avoid a complete collapse of biodiversity across the entire landscape. For example, in Mount Elgon National Park, the Ugandan Wildlife Authority negotiated comanagement and access rights to certain areas of the protected area with the local communities. Even though this decision affected biodiversity within the comanaged areas, it helped to secure the integrity of the core of the protected area. From the communities' point of view, trading off access to the remote mountain area allowed them to gather forest products legally from the comanaged zone (Maginnis et al. 2004).

Likewise, in Sabah, Malaysia, oil palm plantation managers along the Kinabatangan River realized that flooding was damaging their oil palm crops, which occupy almost 300,000 ha of once-forested land. They stopped their usual practice of planting oil palm right up to the edge of the river, and instead encouraged the regeneration of secondary and planted forests along the river banks in collaboration with WWF and local communities. This expanded species habitat and enhanced landscape connectivity for threatened species such as orangutans and elephants (Pye-Smith and Saint-Laurent 2003).

Nevertheless, basing conservation or development strategies on trade-offs poses a certain degree of risk to both goals. Trade-offs are generally easier to achieve with increased scale, because "dominant-use" can be applied for site-level activities, whereas "multiple-functionality" can be achieved at the landscape scale (Maginnis and Jackson 2000). In other words, it is more likely that one can accommodate a broader range of uses at a larger scale than in a smaller area, where the chances are that one specific activity, or land use, will dominate. Attempting to balance multiple land uses at a relatively small scale—for example, to expect one small area of forest to fulfill all possible functions ranging from conservation to timber production—precludes the benefits of site specialization and often results in suboptimal conservation and human development results. Conversely, allowing specialization of land uses over large areas can produce uniform, sterile landscapes. This means that trade-offs between livelihood needs and conservation needs will need to be made at the site level so as to ensure ecological integrity and human well-being at the landscape level (Brown 2004). In other words, although ecological integrity and human well-being can be traded off at the site level, they must not be traded off at the landscape level. Figure 17.1 illustrates the extent to which goods and services can be traded off against each other at different scales without undermining either biodiversity or other aspects of sustainable development.

Unit of Scale	Acceptability of net losses of environmental functions (goods, services, future options, ecological processes) at a particular unit of scale.	Trends in land-use specialisation
Regional / national (10^5 km^2)	Ethically unacceptable - even if possible and economically rational. Examples include Lawrence Summer's proposal of shipping and storing industrial countries' toxic waste to developing countries, or phosphate mining in Nauru. Would inevitably lead to greater social inequity and species extinction.	Multiple use
Provincial / Eco-regional (10^4 km^2)	Strongly discourage in virtually all circumstances. Examples include conversion of most peninsular Malaysian lowland forest to estate crops. Directing one provincial administration to pursue industrial development and the other to pursue conservation will raise practical governance issues that will be virtually impossible to resolve democratically. Likelihood of social inequity and species loss high.	↑
District / Landscape (10^3 km^2)	Discouraged in the majority of circumstances. If trade-offs are to be made at this scale the strong justification is needed along with robust mechanisms to balance loss functions elsewhere. Governance issues likely to be a problem and ensuring social equity is properly addressed will be a major challenge.	↓
Stand / Site (10^1 km^2)	Acceptable in the majority of cases as long as good management practices are pursued and there is reasonable likelihood of social equity issues being adequately addressed.	Dominant use

Figure 17.1. Acceptability of trade-offs at different spatial scales (adapted from Maginnis et al. 2004)

Negotiating Landscapes: Risks, Roles, and Responsibilities

Given that trade-offs and synergies through coordination should be negotiated between multiple stakeholders and managed in a balanced manner at a landscape scale, it is important to address questions such as how these are decided, who makes trade-off decisions, whether balanced trade-offs are even possible, and, if they are not, who chooses the winners and losers. To do so we draw on some of the recent work that IUCN has carried out on conservation for poverty reduction (see Fisher et al. 2005).

Creating a "Level Playing Field"

Planning land use at the landscape scale requires that the interests of different stakeholders are understood and reconciled, and that the final land-use configuration reflects a negotiated settlement among the different stakeholder groups. Different stakeholders often have different capacities for negotiation.

As is often seen in "participatory" village-level planning processes that typify recent ICDP and other comanagement initiatives, local communities find it extremely hard to have any real say in the final land-use decisions. The opportunity costs for participating in such negotiations are invariably much higher for local communities than they are for large commercial interests or government bureaucracies. Pluralist stakeholder negotiations need to be structured to address the needs of the less powerful (Fisher et al. 2005; Wollenberg et al. 2001). Thus, if integrated land-use management approaches such as ecoagriculture or forest landscape restoration are to succeed in having a people-centered perspective in decision making, they will first need to focus on strengthening the institutional and policy arrangements that exist for undertaking landscape-scale negotiations, and put in place equitable compensatory mechanisms where trade-offs cannot be balanced.

As Brown (2003) notes, it is not easy to design institutional arrangements that can accommodate diverse groups of interests, especially at a landscape-scale where actors and stakeholders come from remote locations and varying levels of power. It is, nevertheless, essential. It is also important that stakeholder negotiations recognize the diversity and power imbalances that exist within local community groups themselves, and that the institutional and policy arrangements also enable the meaningful participation of weaker sections within the community, such as women and the landless.

Strategies for Conflict Management

Only when the institutional arrangements necessary to ensure the equitable participation of stakeholders are in place and the playing field has been leveled should the actual negotiations regarding land-use configurations begin. Particular care must be taken to ensure that all stakeholders have a common understanding of the larger conservation landscape and also of each others' cultural landscapes and livelihood needs before negotiating trade-offs. This will require development of negotiation support systems, such as the World Conservation Union's (IUCN's) Wellbeing-of-Forests and other scenario-building tools (2006).

When win–win consensus solutions cannot be found, efforts must be made to achieve a compromise "win-more–lose-less" outcome by making site-level land-use trade-offs that maintain ecological integrity and human well-being at the landscape scale. As Fisher et al. (2005) point out, a key lesson from conflict management theory is the idea that parties in negotiations should focus on interests and not argue over positions, because negotiations will work best if focused on interests and outcomes rather than the preordained positions of the various stakeholders (Fisher et al. 1981). However, there will also be circumstances when land-use conflicts cannot be resolved at the landscape level, either because key groups hold extremely polarized opinions or, more likely, because the key factors that influence land-use practices are outside the control of local players (Fisher et al. 2005; Maginnis et al. 2004). When this happens, other incentive mechanisms and solutions will have to be sought at a higher scale, for example, for payments through ecosystem services by beneficiaries outside the landscape.

Redefining the Role of Experts

As we have stressed, integrated land-use management is possible only if local communities and stakeholders have a central role in deciding what their landscapes should look like and what land should be used for what purpose. However, this does not mean that "experts" or "science" should not have a role in the land-use planning and management process. Indeed, the setting of overall conservation goals, the determination of what constitutes a conservation landscape, and the identification of "natural thresholds" beyond which trade-offs should not be permitted, will require technical and scientific expertise. However, these inputs should be used to *inform* all stakeholders as they negotiate site-level land-use trade-offs, rather than to *enforce* decisions that are not likely to be implemented in the longer term. That is, land-use planning should ultimately be a matter of societal choice, but a choice that is well informed by the best available knowledge and free from coercion.

The challenge is to ensure that government oversight is maintained without undermining the local decision-making process. The minimum standards approach proposed by Ribot (2002), which suggests that governments should provide a list of the few things that *cannot* be done, rather than a long list of things that *must* be done, is one way of resolving this problem.

Finding a Neutral Facilitator

To be successful, negotiations generally need facilitation. Ensuring that a neutral facilitator is available to manage, guide, and, when necessary, push the various stakeholders and interest groups involved in negotiating land-use trade-offs, is a key factor, and certainly one of the major challenges that face landscape-scale negotiation and decision-making. The facilitator could be an individual or an organization. But as Fisher et al. (2005) point out, it is essential to ensure that the facilitator is trusted by all parties. The roles of facilitator and convenor (if these are two separate individuals or organizations) should not be seen as having a strong vested interest in particular outcomes.

In some cases conservation NGOs have attempted to fulfill the role of facilitator on the basis that they have a fuller understanding of both the conservation as well as the cultural landscapes of the various stakeholder groups. However, because they are an interested party and are not always successful in remaining unbiased, the outcomes have not always been acceptable to stakeholders. In most cases an outsider with no connections to other stakeholders who can act as a mediator or honest broker will be more effective.

The Importance of Lifting Constraints

Managing trade-offs and identifying synergies alone will not ensure improved human well-being and biodiversity conservation. Key factors that impede local players in following balanced land-use practices include constraints that are externally imposed either by government policies or through market forces. Often poverty is not due to the lack of assets or resources, but the fact that these assets cannot be legally accessed, collected, or sold. Other constraints can include inadequate marketing systems and people-unfriendly policies (Fisher et al. 2005). By shifting from conservation through incentives to conservation by removing constraints, initiatives may be better able to achieve poverty reduction and environmental conservation objectives. In addition to strengthening institutions to enable effective landscape-scale negotiations, and providing appropriate economic and policy incentives, efforts have to also be made to lift constraints that prevent landscapes from being managed in a balanced and integrated manner.

Conclusion

If conservation is to succeed in the 21st century, then action has to be taken in the wider, expanding agricultural and degraded and deforested landscape, particularly in those biodiversity-rich developing countries where the pressures on forests and protected areas are often highest. However, these are also the countries where the goals of poverty reduction and human development are pressing societal challenges. Ethics and reality dictate that one cannot be achieved at the cost of the other. Ecoagriculture, in attempting to create spaces for conserving wild biodiversity in agricultural landscapes in a way that enhances both agricultural productivity and the well-being of poor farmers, is both a relevant and a timely concept.

Using our experiences in developing forest landscape restoration, we have tried to show how the landscape perspective of ecoagriculture can be practically achieved. Interventions should not be imposed in a top-down, prescriptive manner by experts or central authorities, but should ensure the meaningful participation of all stakeholders whose interests are affected by such interventions. This will require the creation of effective institutional arrangements that can provide a level playing field for negotiating site-level trade-offs, the lifting of existing constraints, and a process of continuous learning and adaptive management.

References

Barrow, E. B. Kaale, and W. Mlenge. 2003. *Forest Landscape Restoration in Shinyanga, Tanzania.* Presented at the 12th World Forestry Congress, Québec City, Canada.

Brown, K. 2003. Integrating conservation and development: a case of institutional misfit. *Frontiers in Ecology and the Environment* 1(9):479–487.

Brown, K. 2004. Trade-off analysis for integrated conservation and development. In: *Getting Biodiversity Projects to Work: Towards More Effective Conservation and Development*, ed. T.O. McShane and M.P. Wells. Columbia University Press, New York.

Charland, J.W. 1996. The "problem–isolation paradigm" in natural resource management. *Journal of Forestry* 94(5):6-9.

FAO (Food and Agriculture Organization). 2001. *Global Forest Resources Assessment 2000: Main Report.* FAO Forestry Paper no. 140. Food and Agriculture Organization, Rome. www.fao.org/forestry/fo/fra/main/index.jsp).

FAO (Food and Agriculture Organization). 2003. *The State of the World's Forests 2003.* Food and Agriculture Organization, Rome. http://www.fao.org/forestry/foris/webview/forestry2/index.jsp?siteId=3321&langId=1.

Fisher, R. and W. Ury with B. Patton (eds.). 1981. *Getting to Yes: Negotiating Agreement without Giving In.* Houghton Mifflin, Boston.

Fisher, R.J., S. Jeanrenaud, S. Maginnis, W.J. Jackson, and E. Barrow. 2005 (forthcoming). *Poverty and Conservation: Landscapes, People and Power.* The World Conservation Union (IUCN), Gland, Switzerland,

Geist, H.J. and E.F. Lambin. 2002. Proximate causes and underlying driving forces of tropical deforestation. *BioScience* 52(2):143–150.

Guha, R. 1997. The environmentalism of the poor. In: *Between Resistance and Revolution: Cultural Politics and Social Protest*, ed. R. Fox and O. Starn. Rutgers University Press, Piscataway, NJ: 17–39.

Howard, S. and J. Stead. 2001. *The Forest Industry in the 21st Century.* World Wide Fund for Nature (WWF), Godalming, UK.

IUCN. 2003. *IUCN Red List of Threatened Species.* IUCN Species Survival Commission, Gland, Switzerland. www.iucn.org and www.redlist.org.

Larson, A.M. 2004. *Democratic Decentralization in the Forestry Sector: Lessons Learned from Africa, Asia and Latin America.* Center for International Forestry Research (CIFOR), Managua, Nicaragua. http://www.cifor.cgiar.org/publications/pdf_files/interlaken/Anne_Larson.pdf.

Maginnis, S. and W.J. Jackson. 2000. *Restoring Forest Landscapes.* IUCN Forest Conservation Programme. http://www.iucn.org/themes/fcp/publications/files/restoring_forest_landscapes.pdf.

Maginnis, S., W.J. Jackson, and N. Dudley. 2004. Conservation landscapes: whose landscapes? Whose trade-offs? In: *Getting Biodiversity Projects to Work: Towards More Effective Conservation and Development*, ed. T.O. McShane and M.P. Wells. Columbia University Press, New York.

McNeely, J.A. and S.J. Scherr. 2003. *Ecoagriculture: Strategies to Feed the World and Save Wild Biodiversity.* Island Press, Washington, DC.

Poore, D. 2003. *Changing Landscapes: The Development of the International Tropical Timber Organization and Its Influence on Tropical Forest Management.* Earthscan Publications, London.

Pye-Smith, C. and C. Saint-Laurent. 2003. *Global Partnership in Forest Landscape Restoration: Investing in People and Nature.* Demonstration Portfolio, IUCN–The World Conservation Union, World Wildlife Fund International, Forestry Commission of Great Britain. http://www.iucn.org/themes/fcp/publications/files/global_partnership_demo_portfolio.pdf.

Ribot, J.C. 2002. *Democratic Decentralization of Natural Resources: Institutionalizing Popular Participation.* World Resources Institute, Washington, DC.

Sayer, J. and B. Campbell. 2004. *The Science of Sustainable Development: Local Livelihoods and the Global Environment.* Cambridge University Press, Cambridge.

Stohlgren, T. 1996. *The Rocky Mountains.* National Biological Service, Washington, DC.

Tomich, T.P., M. van Noordwijk, S. Budidarsono, A. Gillison, T. Kusamanto, D. Murdiyarso, F. Stolle, and A.M. Fagi. 2001. Agricultural intensification, deforestation, and the environment: assessing tradeoffs in Sumatra, Indonesia. In: *Tradeoffs or Synergies? Agricultural Intensification, Economic Development, and the Environment*, ed. D.R. Lee and C.B. Barrett. CABI, Oxford: 221- 224.

UNEP (United Nations Environment Program). 2001. *The Global Biodiversity Outlook Report.* Convention on Biological Diversity (CBD)/UNEP. http://www.biodiv.org/gbo/.

UNEP (United Nations Environment Program). 2002. *Global Environmental Outlook 3: Past, Present and Future Perspectives.* Earthscan, London, Sterling, VA.

Wellbeing Scores Software. 2005. World Conservation Union. http://www.iucn.org/places/canada/prog/TBFP/WOF.htm.

Wollenberg, E., J. Anderson, and D. Edmunds. 2001. Pluralism and the less powerful: accommodating multiple interests in local forest management. *International Journal of Agriculture, Resources, Governance and Ecology (IJARGE)*, Special Issue, 1(3–4):199–222.

Chapter 18

Research Partnerships

Thomas P. Tomich, Dagmar Timmer, Sandra J. Velarde, Cheryl A. Palm, Meine van Noordwijk, and Andrew N. Gillison

Introduction

The development of ecoagriculture landscapes is essential to addressing the dual challenge of enhancing wild biodiversity in critical habitats and improving rural livelihoods of agriculture-dependent people. As illustrated by the many examples in this volume, there is considerable scope for local innovations through collective action and local trial and error. But in many landscapes, there is no clear pathway to ecoagriculture solutions, and developing them will require integrating the efforts of scientists with a range of disciplinary expertise, interacting closely with land users. There remain major gaps in knowledge to overcome major tradeoffs and achieve real synergies between biodiversity and agricultural livelihoods.

There is thus a need for large-scale research initiatives in major ecozones to understand how agroecosystems function and to identify opportunities and constraints for ecoagriculture under diverse conditions. But conventional research approaches are inadequate. Few existing research methods are adapted for application in complex and dynamic landscape mosaics. The agricultural and conservation research communities are weakly linked, and their research frameworks are often incompatible. Disciplinary research priorities often fail to address the major constraints of field practitioners (Rhodes and Scherr 2005).

The challenge for ecoagriculture is particularly evident in the practice of fallow-based (slash-and-burn) agriculture now practiced in the tropical forest margins, where over a billion chronically poor people depend on forest resources, and globally important biodiversity is threatened. These systems are found across the humid tropics. Working in collaboration since 1994, a large group of research partners have developed and tested an institutional model for research in the tropical forest margins using a collaborative, participatory, multiscale, integrated natural resource management approach and research methods.

This organizational model for collaborative research on integrated natural resource management is the ASB Partnership for the Tropical Forest Margins (formerly known as the Alternatives to Slash-and-Burn program). The partnership focuses on landscape mosaics (comprising both forests and agriculture) where global environmental problems and local poverty converge at the margins of remaining tropical forests. This chapter describes the ASB partnership, its approach to assessment of ecosystem services in the tropical forest margins and the trade-offs and synergies among these services, and the tools ASB partners developed to support landscape-scale assessment and action. Through participatory action research, ASB partners have identified promising ways to achieve an equitable balance between biodiversity conservation and agricultural livelihood options for poor, rural communities. This organizational model potentially has broad applicability for the development of ecoagriculture landscapes in other biomes.

Key Features of the ASB Partnership for the Tropical Forest Margins

The ASB Partnership for the Tropical Forest Margins is a systemwide program of the Consultative Group on International Agricultural Research (CGIAR) that was born out of recommendations agreed upon at the 1992 Rio Earth Summit (*Agenda 21*, Chapter 11, on Combating Deforestation) (Sanchez et al. 2005). It is a global consortium of more than 80 local, national, and international partners, including research institutes, nongovernmental organizations (NGOs), universities, community organizations, and farmers' groups. Key features of the partnership include an ecoregional focus, a growing emphasis on landscape mosaics within its multiscale design, an integrated natural resource management approach, and a multisite, multidisciplinary, multipartner network.

Ecoregional Focus

ASB benchmark sites have been established in the Amazon of Brazil and Peru, the Congo Basin forest of Cameroon, the island of Sumatra in Indonesia, the mountains of Northern Thailand, and the island of Mindanao in the Philippines. These are landscapes of roughly 100 to 1000 square kilometers for long-term study and engagement by ASB partners with households, communities, and policymakers at various levels.

All the ASB benchmark sites are in the humid tropical and subtropical broadleaf forest biome as mapped by the World Wide Fund for Nature (WWF). The most biologically diverse terrestrial biome by far, conversion of these forests leads to the greatest species loss per unit area of any land cover change. Current estimates by ASB indicate that of the more than 1.8 billion people within this tropical forest biome, 1.2 billion are rural. Most are poor households that depend directly on forest resources and agriculture for their livelihoods. Other poor households suffer indirectly from activities such as commercial logging that can lead to resource wastage and environmental degradation. Tropical deforestation has no single cause but is the outcome of a complex web of factors whose mix varies greatly in time and space. Understanding the nexus of these factors in a given situation is a crucial first step if policymakers are to introduce effective measures to curb deforestation, and to do so in ways that reduce poverty (http://www.asb.cgiar.org/PDFwebdocs/PolicyBrief6.pdf). Across the benchmark sites, deforestation is often blamed on the slash-and-burn practices of migratory smallholders, millions of whom do clear and cultivate small areas of forest by this method. However, plantation owners, other medium- and large-scale farmers, ranchers, loggers, and state-run enterprises and projects clear much larger areas, leading to conflict, often with smallholders, who in many cases have longstanding prior claims that are rarely officially recognized.

Growing Emphasis on Landscape Mosaics

Although clearing forests for pasture development is readily visible through remote sensing, much of the change at the forest margin is more gradual and defies accurate classification by current remote sensing techniques. The combined land-use systems, with portions of fields, farms, and landscapes in annual crops, pastures, agroforestry, and forests are classified as agriculture and forest mosaics and are now recognized as quite extensive in area and importance in terms of their impact on ecosystem functions. However, the mapping legends used in studies of land cover change do not in general do justice to these landscape mosaics. Furthermore, some of the processes by which these complex landscapes function in terms of providing ecosystem services are not yet understood. These questions have become the focus of ASB's ongoing work.

Integrated Natural Resource Management Approach

ASB operates as a multidisciplinary research and development consortium that aims to implement integrated natural resource management (iNRM) through long-term engagement with local communities and policymakers at various

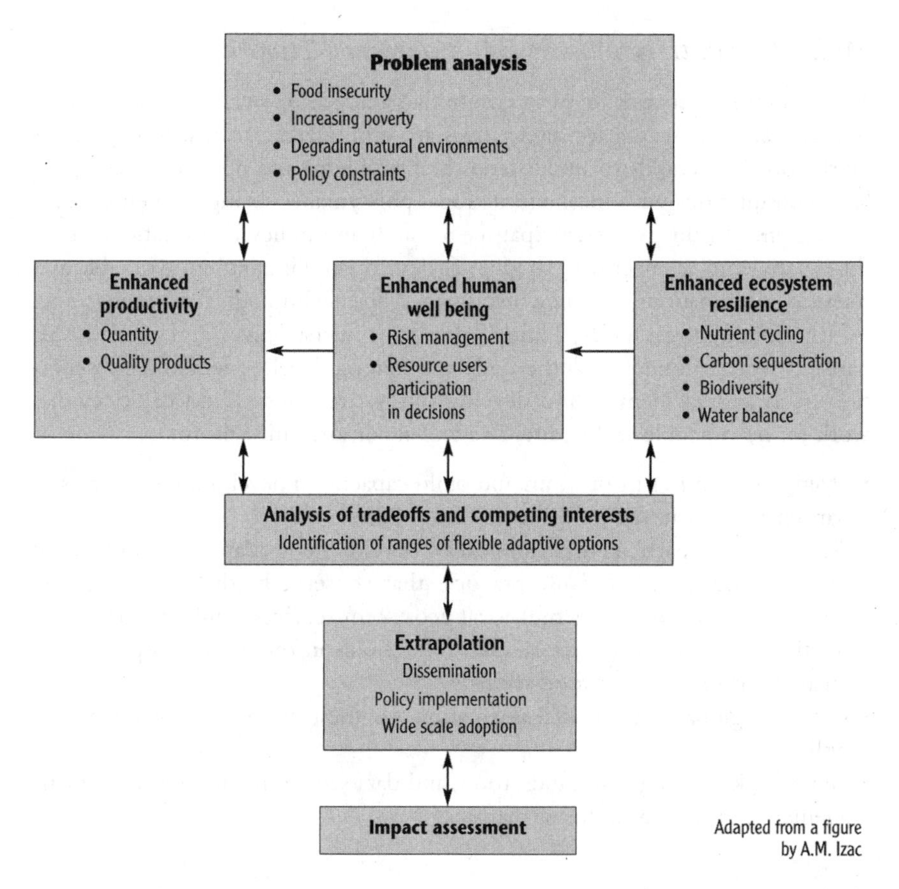

Figure 18.1. The integrated applied natural resource management (iNRM) paradigm (adapted from a figure by A.M. Izac)

levels. As shown in Figure 18.1, iNRM begins with problem analysis. An integrated analysis of a broad range of land-use alternatives is used to quantify the local, national, and global benefits they entail as well as the institutional realities that may favor or hinder their further development in three distinct dimensions: (1) enhanced productivity, (2) enhanced ecosystem integrity and resilience, and (3) enhanced human well-being. These three elements correspond closely to the three legs of the ecoagriculture "stool": agricultural productivity, biodiversity conservation, and contributions to human well-being (livelihood support, poverty reduction, and rural vitality) (Buck et al. 2004; McNeely and Scherr 2003).

Multisite, Multidisciplinary, Multipartner Network

ASB's multisite network helps to ensure that analyses of local and national perspectives and the search for alternatives are grounded in reality. ASB partners work with households to understand their problems and opportunities. Similarly, consultations with local and national policymakers bring in their distinctive insights. In this way, participatory research and policy consultations guide the iterative process necessary to identify, develop, and implement workable and relevant combinations of policy, institutional, and technological options.

By bringing together local knowledge, policy perspectives, and science, ASB partners work to understand the trade-offs among development and conservation goals and to identify and develop innovative policies and practices that work for both people and nature. The global consortium seeks to:

- Strengthen partner institutions and build capacity in developing countries to participate fully in the global search for solutions
- Accelerate the participatory assessment, development, adaptation, and spread of technologies and land-use practices that conserve biodiversity, store carbon, and maintain (or restore) local ecosystem services while providing attractive opportunities for poor rural households in the humid tropics to increase their income and food security
- Facilitate global sharing of information on these innovative practices and policies
- Generate knowledge, methods, tools, and data sets on natural resource management in the tropical forest margins

ASB Approach to Assessment of Ecosystem Goods and Services

This section covers how the ways to provision, regulate, and support ecosystem services are studied and incorporated within the ASB approach.

Assessment Approach

Ecosystem services provided by tropical forests and forest-derived land-use systems range from the plot to the global level, making cross-scale assessment a vital necessity. As part of the Millennium Ecosystem Assessment (MA), ASB partners have undertaken a multiscale assessment of forest and agroecosystem tradeoffs in the humid tropics, attempting to synthesize findings across ASB sites. Over the past decade, ASB partners have undertaken research, evaluated development experience, and built capacity at the benchmark sites and elsewhere. These have resulted in proven methods and new databases for plot-level indicators. Methods for landscape and watershed scale assessment are under development, as well as a pantropic analysis of the nexus among tropical hydrology, biodiversity, and poverty. Stakeholder consultations to identify user needs have provided the basis for developing questions to guide assessment teams for specific topics.

For each of the assessment topics, indicators reflect user needs and concerns regarding specific outcomes, focusing on land use, land cover change, and resource management. These include protocols for the following:

- Benchmark site characterization and multiscale assessment (Palm et al. 2000)
- Assessment of global environmental services, specifically carbon stocks and greenhouse gas fluxes. The ASB report can be viewed at http://www.asb.cgiar.org/publications/wgreports/wg_climatechange
- Assessment of below-ground biodiversity (Bignell et al. 2005) and agronomic sustainability at the plot level (Hairiah et al. 2005). The ASB report can be viewed at http://www.asb.cgiar.org/publications/wgreports/wg_biodiversity.asp
- Assessment of social and economic issues, including smallholder farmers' concerns and national policymakers' concerns. The ASB report can be viewed at http://www.asb.cgiar.org/publications/wgreports/wg_socioecon.asp

These include measures of output and profitability, valued at both private and socially adjusted costs and benefits; labor requirements, including the establishment and operational phases for various land uses; means of meeting household food security; and institutional capacities, such as existence and functioning of markets for purchased inputs and outputs, labor, and capital; access to technological information; property rights and resource access; equity issues; and degree of social cooperation.

These plot-level indicators have been measured for locally significant land uses at ASB benchmark sites (http://www.asb.cgiar.org/publications/countryreports/). The underlying data on plot-level indicators for the specific land uses are compiled in a format referred to as the ASB matrix (Table 18.1), available at http://www.asb.cgiar.org/gallery/ASB_matrix.ppt, which facilitates assessment of trade-offs across land uses (Tomich et al. 1999, 2005b).

To conduct cross-site (and intercontinental) comparisons within the biome, land uses for benchmark site assessments were also selected with reference to a set of "meta land uses" http://www.asb.cgiar.org/gallery/ASBmetalanduse.ppt. Similar indicators have been developed for landscape mosaics in Northern Thailand (Thomas et al. 2005). Indicators for output of ecosystem goods (provisioning services) underpin the profitability indicators described earlier and will be derived from those existing databases and other secondary data for

Table 18.1. The alternatives to slash-and-burn (ASB) matrix

Meta land uses	*Global environmental concerns*	*Agronomic sustainability*	*Smallholders' socioeconomic concerns*	*Policy and institutional issues*
Natural forest				
Forest extraction				
Complex, multistrata agroforestry systems				
Simple tree-crop systems				
Crop/fallow systems				
Continuous annual cropping systems				
Grasslands/pasture				

Note: completed matrices for the forest margins of Cameroon, Northern Thailand, the Peruvian Amazon, Sumatra in Indonesia, the western Amazon of Brazil are available at http://www.asb.cgiar.org/data/dataset/

which work is ongoing (see later discussion). In 2002 and 2003, assessment activities were extended to include hydrological functions and their coincidence with biologically significant habitats at various scales, from the local/landscape scale to medium-sized river basins (the Mekong) to the pantropic scale.

Natural or undisturbed forests were considered as the reference or baseline condition against which other tropical forest–derived systems could be compared. ASB assessed subsets of conditions and trends at the global, benchmark, landscape, and land-use (plot) level. The ASB assessment "priority list" includes ecosystem goods (of which food supply received particular emphasis from some users); the regulating services of carbon sequestration and air quality; water supply; and nutrient supply; and the resource base and supporting services such as soil, soil formation and biological diversity. The full discussion is available in Tomich et al. (2005a). Two categories of clear importance—regulation of pests and diseases (of humans, plants, and animals) and cultural and spiritual values—were not covered in detail because of limited capacity and data within the consortium and limited time and other resources to develop new partnerships in these areas.

"Provisioning" Ecosystem Services

Building on the synthesis of users' needs, a protocol was developed to assess food, fiber (including timber and fuelwood), feed for livestock, and a host of other major products that are obtained from forests and the diverse forest-derived land-use systems found at the tropical forest margins. The main products differ greatly by land-use system, between and within regions, and also with changes over time (e.g., during droughts). But regardless of their location, rural people in these regions typically rank these goods, especially food and timber, as their primary concerns among all ecosystem services. After synthesizing data collected on users' needs, the ASB-MA ecosystem goods (provisioning services) assessment intends to "scale up" the goods analyses from benchmark sites to states/provinces and other larger units.

"Regulating" Ecosystem Services

The ASB paradigm works in conjunction with ecosystem services, supporting their functions while attempting to maximize production in a sustainable way. Measures to quantify the effectiveness of management approaches allows land stewards to most effectively utilize natural controls on climate change, air quality, water, nutrient supply, crop pests, and disease.

Carbon Sequestration and Climate Change Mitigation

Carbon stored in tropical forests is vital to maintaining and reducing the level of greenhouse gases in the atmosphere. There is considerable uncertainty in the carbon dioxide flux and storage from the tropics due to inadequate estimates for rates of different land-use transitions, the biomass of the vegetation that is cleared, the rates of regrowth, and levels of biomass recovery of the subsequent land-use systems. In particular, little information is available on the carbon stored and the potential to sequester carbon in many of the land-use systems of the humid tropics, other than for continuous cropping and pasture systems, both of which have low carbon storage potential. However, significant tree cover on deforested, agricultural, and abandoned land in the rain-fed, or humid, tropics could provide a potentially large sink for carbon. Carbon storage in soils is essential for maintaining many ecosystem services, including nutrient availability, water infiltration, and detoxification of certain minerals and chemicals. A database developed by ASB from the carbon stock measurements taken in each of the major land-use types at each site was used for assessing the condition of carbon sequestration in forests and agroecosystems. (See the ASB Climate Change working group report at http://www.asb.cgiar.org/publications/wgreports/wg_climatechange.asp.) Because many of the land-use systems (agroforests and tree plantations) in the humid forest margins are cyclic, the carbon stored in these systems at the end of a rotation cycle is the maximum carbon stored and is thus an overestimate of the carbon stock of those systems over their full rotation period. To account for the changes in carbon stocks from the harvest, clearing, and regrowth phases, a time-averaged carbon index was calculated to indicate the average carbon stocks stored in each land-use system over the rotation time of the system (Palm et al. 2005b).

Air Quality and Mitigation of Smoke Pollution

Smoke pollution from burning of forest, grassland, and agricultural fields is a serious public health problem and disrupts livelihoods in large areas of the humid tropics. ASB research has emphasized options to manage smoke from land clearing activities (Byron 2004; Tomich et al. 2004). Fires are not intrinsically harmful and if well managed can be a legitimate, low-cost technique for clearing unwanted vegetation. The challenge for policymakers is to minimize the adverse effects of fire and smoke, not to stamp out the use of fire (http://www.asb.cgiar.org/PDFwebdocs/ASBPolicyBriefs4.pdf).

Water Supply and Buffering Lowland Flooding

Hydrological integrity is extremely important in providing sustainable water yields, reducing flood hazard, and controlling erosion. Habitat conservation and integrated watershed management are intimately connected to the water cycle of humid tropical ecosystems. A recent assessment of available evidence for the humid tropics (Bruijnzeel 2004) indicates that the role of natural forest cover in providing reliable water supplies to humans and their agroecosystems may be significantly overstated in this biome.

Nutrient Supply

The continued productivity of land-use systems depends on the supply of nutrients through either external inputs or internal cycling, the latter being an ecosystem service upon which many tropical agroecosystems depend. Depletion of nutrient stocks through the repeated harvest and removal of goods without replenishment of those nutrients will also result in a decline in the provisioning services of the agroecosystem. Two indicators were developed to assess whether the nutrient balance was (or could potentially be) maintained in the different land-use systems: (1) net nutrient export (which equals nutrient input minus nutrients harvested), and (2) relative nutrient replacement cost (which equals the costs, in the form of chemical fertilizer, of replacing nutrients "exported" from the agroecosystem in harvested products).

Regulation of Crop Pests and Diseases

Many pests and diseases of domesticated crops can be controlled "biologically" through natural parasites and predators that occur within the same agroecosystem or within patches of vegetation in the landscape. As agroecosystems are intensified they are frequently simplified with accompanying losses of biodiversity. In such cases pests and diseases can occur at levels that negatively affect the provisioning service of the system (crop and tree productivity and yields). Continued production of the same crops without rotation may also lead to increased pest problems. Similarly, as landscapes become more homogeneous with forest clearance and land-use intensification, forest refugia for natural predators disappear, and pest and disease outbreaks may become more common. Overall, there are few immediate answers for the practical questions regarding biodiversity function at the landscape scale (Swift et al. 2004; Tomich et al. 2004). One obvious priority for further work is to examine whether the risk of pests and diseases increases as biodiversity richness declines within these changing landscapes (Naylor and Ehrlich 1997). Although farmers in the humid tropics typically rank crop pests and diseases (including weeds) as their paramount resource management concern, with few exceptions (collective action for pig hunting in Sumatra, locust control, synchrony in rice planting to reduce opportunities for rats), interventions beyond the plot/household scale seem rare.

Resource Base and "Supporting" Ecosystem Services

This section deals with key resources underpinning ecosystem services: soil and biodiversity above and below ground.

Soils

The physical, chemical, and biological properties of soils determine their capacity for overall supporting services, and are affected by deforestation and land-use change. The role of soils and the interaction with above- and below-ground biodiversity is a central theme to the sustainability of land-use systems in the tropics and has been investigated at several scales. The work by Ziegler et al. (2004) illustrates what can be accomplished through informed scientific efforts to measure lateral flows. They provide evidence that unpaved roads produce as much sediment as agricultural land in an upper catchment in Northern Thailand, despite the fact that these roads occupy less than one-tenth of the area occupied by agriculture.

Biological Diversity, Above and Below Ground

Although biological diversity is not necessarily an ecosystem service per se, it provides a reservoir of biota that can and does provide a regulatory service, such as regulation of pests and diseases. Considerable ecological debate surrounds the issue of whether (and how) high biological diversity enhances ecosystem resilience and stability, although most empirical information suggests this is the case in tropical moist broadleaf forests.

It is clear that in addition to cultural, aesthetic, and spiritual values, natural biological diversity is vital for providing many local livelihood needs in terms of foods, medicines, and cash. Agrobiodiversity (in its broadest interpretation) also can help regulate pests and diseases and provides a diverse set of crops to reduce risk to climatic or economic fluctuations and to provide nutritional diversity.

ASB researchers have developed a highly cost-effective method for rapid appraisal of vascular plant biodiversity, which was tested at all benchmark sites. The method was originally designed to quantify changes in species composition, functional types, and vegetation structure along land-use intensity gradients from primary forest to degraded cropland in ASB sites. Multidisciplinary surveys in Indonesia and Thailand (Gillison 2000; Gillison et al. 2003) and the Brazilian Amazon (Gillison et al. 2003) have also shown that the method provides a robust statistical basis for using vegetation features as indicators of above-ground carbon, certain faunal groups, and potential agricultural productivity. These ASB rapid survey methods have led to major progress on biodiversity conservation in Indonesia, most tangibly playing a key role in designation of a new national park in Sumatra, currently covering 33,000 ha and planned to be expanded by another 120,000 ha. The methods were also influential in declaration of Tesso Nilo as a conservation priority area and national park in October 2003.

A global database currently held by the Center for Biodiversity Management contains uniformly collected data from more than 1600 sites, including all ASB biodiversity sample sites. The rapid survey protocol developed in association with ASB makes possible rapid assessment and comparison of data across all global scales, and this is facilitated by the hierarchical nature of the variables sampled, from local species to generic plant functional types (PFTs) and vegetation structure. Data collected using this approach in Sumatra and Thailand (Gillison and Liswanti 2004; Gillison et al. 2004) have shown useful predictive correlates between land use and biodiversity at a landscape scale.

ASB researchers have also been pioneers in the study of the functional values of below-ground biological diversity (BGBD) to better assess the potential uses of soil biodiversity in ecosystem management. A project coordinated by an ASB partner, the Tropical Soil Biology and Fertility Institute of the International Center for Tropical Agriculture (CIAT), is developing internationally accepted standards for characterizing and evaluating BGBD. These include indicators of BGBD loss, inventories of BGBD at sites representing a broad range of globally significant ecosystems and land-use types, and a global network to exchange that information.

Ecological Knowledge

The research program included assessment and comparison of local, public/policy, and scientists'/modelers' ecological knowledge. Initial findings suggest that understanding the differences among these "ways of knowing" can help to reduce conflict and find practical solutions. Local ecological knowledge on, for example, watershed functions is process-based and well articulated for observable phenomena such as overland flow, erosion, sedimentation, and filter effects. It does not depend on strict land-use categories. By contrast, policy knowledge is based on established categories, for example, the attributes that are supposed to go along with forest and nonforest land cover. Science can potentially help bridge between these process- and category-based types of understanding and can usefully interact in both arenas. Progress in actual stakeholder negotiations can come from developing a shared articulation of the underlying cause–effect relations and the criteria and indicators that reflect the various concerns. Breaking through existing categories at the policy level, and especially recognizing the intermediate systems and forest mosaics as the focus of interest in natural resource management, requires a change to evidence-based discourse (Joshi et al. 2004).

Analyses of Tradeoffs and Synergies

Poverty reduction in most of the tropics depends on finding ways to raise productivity of labor and land through intensification of smallholder production systems. Although there may be opportunities to alleviate poverty while conserving tropical rainforests, it is naive to expect that productivity increases necessarily slow forest conversion or improve the environment.

Balancing Biodiversity, Carbon Storage, and Production

Provisioning ecosystem services are often produced at the expense of some regulatory or support services. Yet some land-use combinations appear to minimize these trade-offs or even achieve synergies, producing both additional globally valued ecosystem services and provisions for local livelihoods and national economic development. For example, complex agroforestry systems provide food, including nutritionally diverse fruits, products sold locally for cash, timber, and international commodity crops such as rubber, cacao, and coffee. These systems can also maintain 25 to 50% of the carbon of the natural forest, often contain high levels of biodiversity, and also effectively sustain the main hydrological functions and soil-supporting functions. These tree-based systems at current levels of production appear to be sustainable in terms of their provisioning and regulating services. But not all agroforestry or tree-based systems maintain carbon and biodiversity. As an example from Cameroon, most of the tree-based systems contain similar amounts of carbon but have very different levels of biodiversity. And efforts to intensify annual food crop production in the rainfed uplands of the humid tropics frequently are unable to maintain sufficient inputs to sustain production.

Given that forest-derived farming systems cannot match the biological diversity and carbon storage of old-growth forests, it can be difficult to strike a balance between the legitimate interests of pro-poor development, on the one hand, and concerns over the environmental consequences of tropical deforestation, on the other. However, ASB research in Southeast Asia and West Africa has found that agroforests (complex, multistrata systems) that harbor high levels of biodiversity often represent the next best option to natural forest for conserving biodiversity and storing atmospheric carbon, while also providing attractive livelihood opportunities for poor people.

Moreover, it is often the direction of change (degradation vs. restoration) that determines whether the environment benefits. If farmers replace unsustainable cassava production, for example, with an improved rubber agroforest, they help restore habitats and carbon stocks while raising their incomes. Tree-based restoration activities (agroforestry), in which both the environment and poor people can benefit from a change in land use, constitute some of the major win–win opportunities in the humid tropics. Such tree-based restoration activities can increase carbon storage and be profitable (privately and socially). But planting trees does not necessarily quickly restore all the ecosystem services of old-growth natural forests. In particular, the hydrological values of such restoration remain uncertain, and we know very little about ecological functions in tropical landscape mosaics.

Landscape-Level Responses to Balance Tradeoffs

The forest margins of the humid tropics tend to be relatively isolated and also lie at the margins of formal administrative influence. Human activities are generally the result of private initiative, spatially dispersed and uncoordinated by central administrative authorities. Attempts to impose coordinated responses (e.g., government interventions such as land-use planning or zoning) will face difficult prospects and risk perverse results if these interventions ignore the interests "on the ground," which produce a vast number of private responses that dominate decisions about land use and land cover change. Ecosystem assessment and management in the tropics need to evolve from a focus on simplistic management categories to an acknowledgment of the chaotic reality, where a large number of actors make their own decisions, disregarding official plans, maps, rules, and "academic" categories and typologies. Often, the best contribution outside actors (research and development agencies) can make will be to facilitate and support a process of negotiation among these stakeholders, who usually have conflicting interests among various ecosystem goods and services (van Noordwijk et al. 2001). Assessment of responses and response options needs to include both the imbalanced political economy and the weak organizational capacity of public institutions.

Tools for Landscape Assessment and Action

Despite these challenges, ASB experience shows it is possible to achieve a better balance between people's needs and nature conservation. The following paragraphs describe tools developed through research partnerships to support landscape assessment and action, and "success stories" associated with their application.

Standardized methodologies, co-located measurements, and a set of sites and land uses that represent the extremes of land use at the humid forest margins provide ASB with unique yet comparable datasets. These have been used successfully to investigate the site specificity or generalities in the trade-offs and synergies among services.

Participatory Scenarios

Building scenarios about the future can help people making decisions about dynamic landscapes to gain a better understanding of the potential impacts of land-use change. National and regional partners of the ASB consortium have underlined their need for such decision-making tools to deal with the uncertainties they face. In response, ASB, the World Agroforestry Centre (ICRAF), and the Millenium Ecosystem Assessment (MA) conducted a global training workshop on scenarios development for facilitators who, subsequently, have both applied these skills themselves and trained others in their use (Lopez et al. 2006; Prieto et al. 2005; Rao and Velarde 2005; Thongbai et al. 2006; Ugarte et al. 2005). Participatory scenario formulation and use are key elements of ASB's efforts to develop consensus and manage conflicts at the local, national, and regional levels.

Simulation Models

In Cameroon, the ASB team built models to simulate land use decisions at the tropical forest margins. Participatory social and agroecological mapping exercises were undertaken at a sample village to identify indicators of changes at the village landscape level. Using global positioning system (GPS) tools as well as supplementary socioeconomic information, models of land-use change were developed and calibrated for each mosaic of land uses. ASB modeling efforts in Brazil helped Embrapa (Brazil's national agricultural research organization) analyze the potential contributions of increased rice yields to reducing deforestation, which ran well below their previous expectations.

Strategic Stakeholder Analysis

ASB developed questions and categories that frame strategic research on social and political dynamics associated with forest margin areas. At a national level, applying this protocol will increase the practical understanding of challenges at specific sites. Engaging in dialogue with local and national policymakers, smallholder farmers, and private sector and other stakeholders to explore their perceptions of information, technical, and policy needs helps ensure that researchers are addressing issues of real concern to the major stakeholder groups. Such dialogue also facilitates participation by key local and national stakeholders in seeking solutions when there are conflicting interests regarding forests, land, and ecosystem services.

Innovative Technologies and Resource Management Practices

ASB research (particularly in Indonesia and Cameroon) has revealed the feasibility of a "middle path" of development involving smallholder agroforests and community forest management for timber and other products. Such a path could deliver an attractive balance between environmental benefits and equitable economic growth. "Could" is the operative word, however, because whether this balance is struck in practice will depend on these countries' ability to deliver the necessary policy and institutional innovations. Indeed, one of the main lessons of ASB experience is that workable iNRM interventions involve *combinations* of technological, institutional, and policy innovations. A number of site-specific alternatives to slash-and-burn agriculture, such as sustainable forest management, smallholder agroforestry, improved pastures, and *Imperata cylindrica* grassland reclamation, are assessed in depth by Palm et al. (2005a).

Recognition of Indigenous Systems

ASB results have improved official understanding and supported government decisions to promote rubber agroforests in Indonesia. ASB research showed that it was beneficial to maintain rubber agroforests, as well as highlighting opportunities to increase productivity and smallholder income through improvements in rubber germplasm. As a result, the government implemented a national program on smallholder rubber agroforests, called Integrated Rubber Forest Management, and reformed trade and marketing policies for rubber wood. The program has the potential to directly improve livelihoods of many of the estimated 7 million people in Sumatra and Kalimantan who make their living from rubber agroforests.

A major outcome of ASB activities has been a contribution to policy dialogues at the local and national level on the ways forest functions can be maintained in the context of development. For example, official recognition of the valuable role of agroforests and other sustainable land-use systems at a national and local level provides a first step toward empowering the farmers that manage these systems.

Empowerment through Measurement

In ongoing work in the Mae Chaem watershed in Northern Thailand, scientists are working with communities to develop and validate methods that the communities themselves then use to monitor their watershed functions and the environment (http://www.asb.cgiar.org/PDFwebdocs/PolicyBrief7.pdf). Paying attention to local environmental knowledge also improves scientists' understanding of environmental problems and widens the range of local livelihood options and appropriate interventions considered by researchers. Early experiences with monitoring suggest that scientists can greatly strengthen communities' positions in negotiations to resolve environmental problems with neighboring communities and with the Thai Royal Forest Department. One of the communities won a national award for its environmental efforts.

These communities in Thailand are working to scale up the knowledge acquired through watershed monitoring by establishing watershed management networks. The importance and potential role of such networks is being recognized by local leaders elsewhere in Mae Chaem, as well as by high-priority national efforts coordinated by Thailand's new Ministry of Natural Resources and Environment to develop river basin and subbasin management approaches. Subcatchment management networks are seen as the basic building block for larger-scale watershed management. Approaches developed by villagers in pilot subcatchments of Mae Chaem are expected to be a major example for other parts of the larger Ping River Basin. Moreover, efforts in the Ping basin are serving as a pilot project for further efforts in the other 24 designated river basins in Thailand.

Recognition of Local Rights

The Krui case in southwest Sumatra shows how scientists helped win the argument for community management by documenting the environmental and social benefits of the Krui agroforests. The Krui agroforests were threatened with conversion to large oil palm plantations, and the people were threatened with eviction from their land and the loss of their livelihoods. This work produced immediate benefits from increased security for at least 7000 families in the 32,000 ha of reclassified Krui lands. These communities won the national environmental award, Kalpataru. This principle of local management could be extended to benefit hundreds of thousands of rural Indonesians in similar areas. Indonesian NGOs have identified at least 50 other communities across the archipelago that have developed production systems comparable to the Krui case (http://www.asb.cgiar.org/PDFwebdocs/PolicyBrief2.pdf).

Conflict Management

In hopes of securing rights to contested land in Sumber Jaya, Lampung Province, Indonesia, groups have formed to apply for stewardship contracts through the community forestry program (HKM). ASB researchers are working with several of these groups, local government, and the forestry department to facilitate negotiation for HKM status. The overarching goal is to develop a process by which the government can meet its environmental objectives to protect watersheds and park boundaries, while also enabling established settlers to make a living by managing their coffee systems in ways that are environmentally sound.

Rewards for Ecosystem Services

ASB research has shown that certain land uses that follow forest conversion maintain *some* key ecosystem services. The analysis of tradeoffs between the profitability and ecosystem services showed that these systems are really very useful from a broad social perspective, yet they are under threat because options like oil palm may be slightly more profitable from a private perspective. Mechanisms designed to reward the upland poor for these ecosystem services may be able to tip the balance in favor of the broader social and environmental concerns.

Evolving Research Partnerships

Together with WWF, the World Conservation Union (IUCN), and the Center for International Forestry Research (CIFOR), ASB is a founding partner in the Rainforest Challenge Partnership (RCP). RCP adds a protected area component lacking in ASB and involves a broader range of conservation, development, and research partners to share and adapt concrete strategies for poverty reduction with conservation of unique tropical forest ecosystems (Box 18.1).

It is envisioned that RCP will integrate ASB activities into a much broader network of ongoing conservation and development initiatives, taking advantage

Box 18.1. A vision of what could be achieved

- Diverse landscapes with more "wild" species
- Poor people empowered through tenure reform
- Poor people have secure access to resources
- Livelihood options expand; people move out of poverty
- Voice for disadvantaged groups in land-use decisions
- Rural poor rewarded for nurturing their land and forests
- Young people have skills and options to choose a better future
- Science and technology harnessed for sustainable development
- Major conservation, development, and private sector organizations change how they operate

of complementarities among sites to tackle conservation and development issues of both local and global significance. This proposed global network of "learning landscapes" is expected to extend lessons learned so far about developing ecoagriculture in the tropical forest margins, and accelerate the learning process (ASB et al. 2003; http://www.asb.cgiar.org/about_us/future.asp).

The ecoregion- and landscape-focused partnership model of ASB offers a promising institutional approach for research on ecoagriculture systems in other ecosystems. Suitable combinations of agricultural and natural area management systems to achieve joint production, biodiversity, and livelihood goals will likely be landscape specific. But strategic, cross-site action research, in close collaboration with farmers and conservationists, can be a cost-effective means to develop, learn from successful practice, and disseminate ecoagriculture components and principles relevant to many landscapes.

References

Alternatives to Slash and Burn Programme (ASB), Center for International Forestry Research (CIFOR), World Agroforestry Centre (ICRAF), World Conservation Union (IUCN), World Wide Fund for Nature (WWF). 2003. *Forests as Resources for the Poor: The Rainforest Challenge.* Nairobi, Kenya, Bogor, Indonesia, and Gland, Switzerland.

Bignell, D.E., J. Tondoh, K.L. Dibog, S. Pin Huang, F. Moreira, D. Nwaga, B. Pashanasi, E. Guimarães Pereira, F.-X. Susilo, and M.J. Swift. 2005. Below-ground biodiversity assessment: developing a key functional group approach in best-bet alternatives to slash and burn. In: *Slash-and-Burn Agriculture: The Search for Alternatives*, ed. C.A. Palm, S.A. Vosti, P.A. Sanchez, and P.J. Ericksen. Columbia University Press, New York.

Bruijnzeel, L.A. 2004. Hydrological functions of tropical forests: not seeing the soil for the trees? In: *Environmental Services and Land Use Change: Bridging the Gap between Policy and Research in Southeast Asia*, ed. T.P. Tomich, M. van Noordwijk and D.E. Thomas. *Agriculture, Ecosystems and Environment* Special Issue 104(1):185–228. http://www.asb.cgiar.org/pdfwebdocs/AGEE_special_Bruijnzeel_Hydrological_functions.pdf.

Buck, L.E., T.A. Gavin, D.R. Lee, and N.T. Uphoff. 2004. *Ecoagriculture: A Review and Assessment of Its Scientific Foundations*. Cornell University, Ithaca, New York.

Byron, N. 2004. Managing smoke: bridging the gap between policy and research. In: *Environmental Services and Land Use Change: Bridging the Gap between Policy and Research in Southeast Asia*, ed. T.P. Tomich, M. van Noordwijk, and D.E. Thomas. *Agriculture, Ecosystems and Environment* Special Issue 104(1):185–228. http://www.asb.cgiar.org/pdfwebdocs/AGEE_special_Byron_Managing_Smoke.pdf.

Gillison, A.N. 2000. Rapid vegetation survey. In: *Above-Ground Biodiversity Assessment Working Group Summary Report 1996–99: Impact of Different Land Uses on Biodiversity*, A.N. Gillison (coord.). Alternatives to Slash and Burn project. World Agroforestry Centre (ICRAF), Nairobi: 25–38.

Gillison, A.N., D.T. Jones, F.-X. Susilo, and D.E. Bignell. 2003. Vegetation indicates diversity of soil macroinvertebrates: a case study with termites along a land-use intensification gradient in lowland Sumatra. *Organisms, Diversity and Evolution* 3:111–126.

Gillison, A.N. and N. Liswanti. 2004. Assessing biodiversity at landscape level in Northern Thailand and Sumatra (Indonesia): the importance of environmental context. In: *Environmental Services and Land Use Change: Bridging the Gap between Policy and Research in Southeast Asia*, ed. T.P. Tomich, M. van Noordwijk, and D.E. Thomas. *Agriculture, Ecosystems and Environment* Special Issue 104(1):75–86. http://www.asb.cgiar.org/PDFwebdocs/AGEE_special_ANGillison_Assessing%20biodiversity.pdf.

Gillison, A.N., N. Liswanti, S. Budidarsono, M. van Noordwijk, and T.P. Tomich. 2004. Impact of cropping methods on biodiversity in coffee agroecosystems in Sumatra, Indonesia. *Ecology and Society* 9(2):7. http://www.ecologyandsociety.org/vol9/iss2/art7.

Hairiah, K., M. van Noordwijk, and S. Weise. 2005. Sustainability of tropical land use systems after forest conversion. In: *Slash-and-Burn Agriculture: The Search for Alternatives*, ed. C.A. Palm, S.A. Vosti, P.A. Sanchez and P.J. Ericksen. Columbia University Press, New York.

Joshi, L., W. Schalenbourg, L. Johansson, N. Khasanah, E. Stefanus, M.H. Fagerström, and M. van Noordwijk. 2004. Soil and water movement: combining local ecological knowledge with that of modelers when scaling up from plot to landscape level. In: *Belowground Interactions in Tropical Agroecosystems*, ed. M. van Noordwijk, G. Cadisch, and C.K. Ong. CAB International, Wallingford, UK.

Lopez, M., R. Prieto, S.J. Velarde. 2006. Construyendo el Futuro de Chalaco: Reporte del Taller de Escenarios, 20 y 21 de Mayo de 2005. Municipalidad Distrital de Chalaco, Colegio Secundario "San Fernando," Chalaco. ASB, Universidad Nacional Agraria La Molina (UNALM), World Agroforestry Centre and Millennium Ecosystem Assessment, Piura, Peru. http://www.asb.cgiar.org/PDFwebdocs/Lopez-et-al-2006-Reporte-Taller-Escenarios-Chalaco-Piura.pdf.

McNeely, J.A. and S.J. Scherr. 2003. *Ecoagriculture: Strategies to Feed the World and Save Wild Biodiversity*. Island Press, Washington, DC.

Naylor, R.L. and P.R. Ehrlich. 1997. Natural pest control services and agriculture. In: *Nature's Services: Societal Dependence on Natural Ecosystems*, ed. G.C. Daily. Island Press, Washington, DC: 151–174.

Palm, C.A., A.-M.N. Izac, and S.A. Vosti. 2000. Procedural guidelines for characterization. Technical Report. ICRAF, Nairobi: 31. http://www.asb.cgiar.org/PDFwebdocs/Characterization_and_Guidelines_ASB.pdf.

Palm, C.A., M. van Noordwijk, P.L. Woomer, J.C. Alegre, L. Arévalo, C.E. Castilla, D.G. Cordeiro, K. Hairiah, J. Kotto-Same, A. Moukam, W.J. Parton, A. Ricse, V. Rodrigues and S.M. Sitompol. 2005b. Carbon losses and sequestration after land use change in the humid tropics. In: *Slash-and-Burn Agriculture: The Search for Alternatives*, ed. C.A. Palm, S.A. Vosti, P.A. Sanchez, and P.J. Ericksen. Columbia University Press, New York.

Palm, C.A., S.A. Vosti, P.A. Sanchez, and P.J. Ericksen (eds.). 2005a. *Slash-and-Burn Agriculture: The Search for Alternatives*. Columbia University Press, New York.

Prieto, R.P., F. Patiño, J. Ugarte, S.J. Velarde, C. Rivadeneyra. 2005. Explorando el Futuro: Madre de Dios. Reporte del taller Escenarios, 28 y 29 de Mayo del 2005, Universidad Nacional de Madre de Dios. ASB, Universidad Nacional Agraria La Molina (UNALM), World Agroforestry Centre and Millennium Ecosystem Assessment, Puerto Maldonado, Peru. http://www.asb.cgiar.org/PDFwebdocs/Prieto-et-al-2005-Reporte-Taller-Escenarios-Puerto-Maldonado.pdf.

Rao, S. and S.J. Velarde. 2005. ASB Global Scenarios Training Course. 17th November–23rd November 2004, Chiang Mai, Thailand. Training Course Report. World Agroforestry Centre, ICRAF, Nairobi, Kenya. http://www.asb.cgiar.org/PDFwebdocs/Rao-Velarde-2005-Training-course-report-ASB-global-scenarios.pdf.

Rhodes, C. and S.J. Scherr (eds.). 2005. *Developing Ecoagriculture to Improve Livelihoods, Biodiversity Conservation and Sustainable Production at a Landscape Scale: Assessment and Recommendations from the First International Ecoagriculture Conference and Practitioners' Fair, Sept. 25–Oct. 1, 2004*. Ecoagriculture Partners, Washington, DC. www.ecoagriculturepartners.org.

Sanchez, P.A., C.A. Palm, S.A. Vosti, T.P. Tomich, and J. Kasyoki. 2005. Alternatives to slash and burn: challenge and approaches of an international consortium. In: *Slash-and-Burn Agriculture: The Search for Alternatives*, ed. C.A. Palm, S.A. Vosti, P.A. Sanchez, and P.J. Ericksen. Columbia University Press, New York.

Swift, M.J., A.-M.N. Izac, and M. van Noordwijk. 2004. Biodiversity and ecosystem services in agricultural landscapes: are we asking the right questions? In: *Environmental Services and Land Use Change: Bridging the Gap between Policy and Research in Southeast Asia*, ed. T.P. Tomich, M. van Noordwijk, and D.E. Thomas (eds.). *Agriculture, Ecosystems and Environment*, Special Issue 104(1):113–134. http://www.asb.cgiar.org/pdfwebdocs/AGEE_special_Swift_Biodiversity_and_ecosystem.pdf.

Thomas, D.E., P. Saipothong, P. Preechapanya, N. Badenoch, P. Wangpakapattawong, C. Hutacharoen, and B. Ekasingh (eds.). 2005. *Landscape Agroforestry in Northern*

Thailand: Impacts of Changing Land Use in an Upper Tributary Watershed of Montane Mainland Southeast Asia. ASB-Thailand Synthesis Report. Alternatives to Slash-and-Burn (ASB)-Thailand and World Agroforesty Centre (ICRAF) Chiang Mai, Chiang Mai.

Thongbai, P., P. Pipattawattanakul, P. Preechapanya, and K. Manassrisuksi. 2006. *Participatory Scenarios for Sustainable Management of an ASB Benchmark Site in Thailand: The Case of Mae Kong Kha Sub-watershed of Mae Chaem Watershed.* Paper presented at International Symposium, Towards Sustainable Livelihoods and Ecosystems in Mountainous Regions, 7 – 9 March 2006, Chiang Mai, Thailand.

Tomich, T.P., A. Cattaneo, S. Chater, H.J. Geist, J. Gockowski, D. Kaimowitz, E.F. Lambin, J. Lewis, O. Ndoye, C.A. Palm, F. Stolle, W.D. Sunderlin, J.F. Valentim, M. van Noordwijk, and S.A. Vosti. 2005b. Balancing agricultural development and environmental objectives: assessing tradeoffs in the humid tropics. In: *Slash-and-Burn Agriculture: The Search for Alternatives*, ed. C.A. Palm, S.A. Vosti, P.A. Sanchez and P.J. Ericksen. Columbia University Press, New York. http://www.asb.cgiar.org/gallery/ASB_matrix.ppt.

Tomich, T.P., C.A. Palm, S.J. Velarde, H. Geist, A.N. Gillison, L. Lebel, M. Locatelli, W. Mala, M. van Noordwijk, K. Sebastian, D. Timmer, and D. White. 2005a. *Forest and Agroecosystem Tradeoffs in the Humid Tropics: A Crosscutting Assessment by the Alternatives to Slash-and-Burn Consortium Conducted as a Sub-global Component of the Millennium Ecosystem Assessment.* Alternatives to Slash-and-Burn Programme, Nairobi, Kenya. http://www.asb.cgiar.org/ma/ASB-MA_statusreport_ver5.0.pdf.

Tomich, T.P., D.E. Thomas, and M. van Noordwijk. 1999. Research Abstracts and Key Policy Questions. *Environmental Services and Land Use Change: Bridging the Gap between Policy and Research in Southeast Asia.* ASB-Indonesia Report Number 10. Alternatives to Slash-and-Burn (ASB)-Indonesia and World Agroforestry Centre (ICRAF) SE Asia, Bogor, Indonesia.

Tomich, T.P., D.E. Thomas, and M. van Noordwijk. 2004. Environmental services and land use change in Southeast Asia: from recognition to regulation or reward? In: *Environmental Services and Land Use Change: Bridging the Gap between Policy and Research in Southeast Asia*, ed. T.P. Tomich, M. van Noordwijk, and D.E. Thomas. *Agriculture, Ecosystems and Environment* Special Issue 104(1):229–244. http://www.asb.cgiar.org/pdfwebdocs/AGEE_special_Tomich_Environmental_Services.pdf.

Tomich, T.P., M. van Noordwijk, S.A. Vosti, and J. Witcover. 1998. Agricultural development with rainforest conservation: methods for seeking best bet alternatives to slash-and-burn, with applications to Brazil and Indonesia. *Agricultural Economics* 19:159–174.

Ugarte, J., R.P. Prieto, M. Lopez, S.J. Velarde, C. Rivadeneyra. 2005. *Explorando el Futuro: Ucayali.* Reporte del taller Escenarios, 10 de Junio del 2005, Sala de Conferencias del Hotel Sol del Oriente, Pucallpa. ASB, Universidad Nacional Agraria La Molina (UNALM), World Agroforestry Centre and Millennium Ecosystem Assessment, Ucayali, Peru. http://www.asb.cgiar.org/PDFwebdocs/Ugarte-et-al-2005-Reporte-Taller-Escenarios-Ucayali.pdf.

Van Noordwijk, M., T.P. Tomich, and B. Verbist. 2001. Negotiation support models for integrated natural resource management in tropical forest margins. *Conservation Ecology* 5(2). http://www.consecol.org/vol5/iss2/art21.

Ziegler, A.D., T.W. Giambelluca, R.A. Sutherland, Y. Pongpayack, S. Yarnasarn, M.A. Nullet, J. Pintong, T. Vana, S. Jaiaree, and S. Boonchee. 2004. Toward understanding the cumulative impacts of roads in agricultural watersheds of montane mainland Southeast Asia. In: *Environmental Services and Land Use Change: Bridging the Gap between Policy and Research in Southeast Asia* ed. T.P. Tomich, M. van Noordwijk and D.E. Thomas. *Agriculture, Ecosystems and Environment* Special Issue 104(1):145–158. http://www.asb.cgiar.org/pdfwebdocs/AGEE_special_Ziegler_Towards_understanding.pdf.

Chapter 19

Multistakeholder Partnerships

Minu Hemmati

Introduction

In many ways, communication and collaboration between people lie at the heart of ecoagriculture. As other chapters in this book demonstrate, increasing our understanding and practice of ecoagricultural concepts requires building new connections between issues and relevant areas of knowledge, skills, and power, as well as between people and organizations who carry such capacities. It requires replicating, adapting, and scaling up successful programs in new settings, and new initiatives that enable people to work in partnership. In other words, ecoagriculture is about developing continuous *dialogue* between different stakeholders, and about building partnerships and inspiring *collaborative action*.

"People and process" significantly impact the outcomes: how we communicate, make decisions, and work together can enable or obstruct success. Ideally, dialogue and partnership serve to create new spaces of social learning and collective leadership, based on mutual trust and an emerging common vision. This chapter outlines how collaborative partnership approaches can be applied to ecoagriculture initiatives, integrating the interests of diverst stakeholders to achieve mutually beneficial outcomes.

Some Concepts and Definitions

Stakeholders are those who have an interest in a particular decision or program, either as individuals or as representatives of a group. This includes people and their organizations that can influence it as well as those affected by it (Hemmati 2002).

Stakeholder groups are actors, and their organizations, from different sectors in society, such as public authorities; international agencies; businesses; community-based organizations; nongovernmental organizations (NGOs); labor unions; women's groups; indigenous peoples organizations; and research and academic institutions and foundations.

These are broad definitions of stakeholders and stakeholder groups, outside of the traditional box of a "tripartite approach" (government, private sector, civil society). Stakeholder analysis begins with looking at the issue at hand to examine who has the power and capacity to affect it, who is affected by it, who is needed to address it, who could be helpful, and who could be obstructive. In that sense, what ecoagriculture is depends on who is taking a stake in it.

Multistakeholder processes (MSPs) are those that aim to bring together all major stakeholders in a new form of communication, decision-finding (and possibly decision-making) structure on a particular issue. They recognize the importance of achieving equity and accountability in communication between stakeholders, with equitable representation of different groups and their views. They are based on democratic principles of transparency and participation and aim to develop partnerships and strengthened networks between and among stakeholders. MSPs cover a wide spectrum of structures and levels of engagement. They can comprise dialogues or grow into processes encompassing consensus building, decision making, and collaborative implementation (i.e., partnerships).

Objectives of MSPs include promoting better decisions by means of wider input, integrating diverse viewpoints where possible and appropriate, bringing together the principal actors, creating trust through honoring each participant as contributing a necessary component of the bigger picture, searching for structures and mutual benefits (win–win rather than win–lose situations), developing shared power, creating commitment through participants identifying with the outcome and thus increasing the likelihood of successful implementation.

MSPs can be found at all levels, from local to global. They have emerged because there is a need for a more inclusive, effective process. They require not only new institutional structures, meeting styles, and knowledge management but also fundamentally challenge the culture of politics. While acknowledging diverse interests of actors, they seek to overcome competition. They challenge all participants to identify common ground and ways forward that are based on mutual respect and consensus. They require transparency and active listening instead of fighting for speaking time, understanding and role-taking instead of convincing or coercing. They have the potential to create new and unlikely coalitions and thus more powerful forces for change.

Box 19.1. Who is a stakeholder in ecoagriculture?

Principally, the following groups and types of organizations (listed alphabetically) should be considered as potential stakeholders in an ecoagricultural undertaking. Who should be included in any particular case depends on the specific conditions at a community, landscape, national, regional, or global scale, respectively. It is also important to note that none of the following stakeholder groups is homogeneous but that each in itself represents a range of perspectives and interests.

- *Business*: from large food industry to small and large organic producers: depend on production quality and quantity, and largely determine agricultural practices
- *Communities*: contribute to the knowledge base for ecoagriculture and need to be engaged in, if not leading, land-use planning in their areas
- *Environmental organizations*: campaign for conservation goals, and often work closely with natural scientists and governments
- *Farmers*: small and large, and their associations: can determine agricultural practices and generate significant knowledge relevant to ecoagriculture
- *Foundations*: provide resources for research, advocacy, participation, and other crucial components of moving ecoagriculture forward
- *Governments:* at local, provincial, national, and international levels: provide legal, institutional, and financial frameworks, which facilitate, or hinder, ecoagriculture
- *Indigenous peoples*: hold and use traditional knowledge about their land and make decisions about land use
- *International organizations*: impact directions of research and policy and convene many relevant stakeholders
- *Researchers*: and research institutions and networks: provide a knowledge base for integrating agriculture and biodiversity conservation needs
- *Trade unions*: represent the interests of agricultural workers and have become increasingly active on environmental and sustainable development issues
- *Women*: are responsible for food production in many developing countries but are also often less empowered than men to voice their interests and secure their rights

MSPs ideally develop a manner of communication that facilitates social learning, and hence potentially leads to changes in individuals' perspectives and the group's approach and actions. This notion of dialogue is different from debate, discussion, or negotiation and is closely related to achieving a dialogic process of societal change (see Buber 1997 for a careful analysis of the concept; also Bawden 1997; Künkel 2004, 2005; Sampson 1993; Waddell 2005; Woodhill 2003 applying this same concept in different fields).

This is a description of MSPs guided by some basic values, ideals, and principles that are largely not culturally specific. Rather, I would argue that aiming to practice such principles increases the likelihood of success—meaningful dialogue, effective collaboration—because of the way people function individually and in groups (see Hemmati 2002 for an analysis of values and principles, and of 20 case examples of MSPs).

Why Engage in Multistakeholder Processes?

In recent years, multistakeholder partnerships have attracted increasing interest, often in a rather ideologically charged context. Examples range from increasing efforts to establish partnerships between the United Nations (UN) and the private sector, as in the United Nations Global Compact, to the international sustainable development debate and the Johannesburg World Summit on Sustainable Development (2002). In the political context of fierce debates over economic liberalization, privatization of services, and connected loan conditionality, for example, the focus on partnerships is, unfortunately, all too often misused in a rather cynical manner, and hence criticized for disguising a lopsided privatization agenda. In this context, it is useful to review pragmatically the benefits of collaborative approaches. Potential benefits of MSPs can be summarized broadly under the following points: quality, credibility, equity, and likelihood of implementation.

Quality

Drawing together expertise and perspectives from a number of disciplines and areas of work, such as biodiversity research, participatory impact assessment, and profitable business investment, can yield improved understanding of the challenges and better solutions than when we rely on one area of expertise. Often, a multistakeholder process can tackle problems that simply cannot be addressed by one organization or stakeholder alone. In a well-designed process, diversity is conducive to human creativity. In ecoagriculture, for example, this can mean better integrating the environmental, social, and economic concerns relating to food production and nature conservation, and implementing approaches that better satisfy the need to address competing goals of communities and environmental conservation at the same time.

Credibility

Broad and meaningful participation is increasingly understood as a basic democratic requirement. In many countries, and at the international level, public authorities and others are experimenting with participatory approaches beyond traditional parties or regular elections. Including stakeholders in consultation and decision making lends credibility to any decision because it (at least) appears to be based on the widest possible range of inputs and balancing various interests. People know that collaborating across interest groups is difficult, the result of identifying common ground, building trust, and, often, compromise. This can create results that are more likely to be seen as legitimate than efforts that are undertaken by one group alone. Such credibility, however, is lost when stakeholders are merely "heard" and their inputs have no predictable bearing on decisions taken.

Box 19.2. Ecoagricultural parternships

The Seed Initiative—Supporting Entrepreneurs for Environment and Development—is a global network for action on sustainable development partnerships that also identifies outstanding partnership initiatives through a biennial award scheme. Two partnerships that won the Seed Awards 2004 illustrate the kind of partnerships that are possible in ecoagriculture:

The Global Marketing Partnership for SRI Indigenous Rice brings together rice farmer organizations in Cambodia, Madagascar, and Sri Lanka and Cornell University (USA) to promote production and marketing of traditional rice varieties grown using the System of Rice Intensification (SRI). SRI includes a range of environmentally friendly methods for growing rice that reduce water consumption, seed, and agrochemical input while significantly increasing yield. The approach relies on farmers' knowledge of their land, the whole ecosystem, and their seed varieties, as well as scientific results from across the world and in situ experimentation with different SRI methods. The global marketing of SRI rice will induce other stakeholders to join the partnership effort, such as marketing specialists, international food companies, and restaurant chains.

Harvesting Seabuckthorn at the Top of the World is a partnership committed to combating land degradation and to generating income for the local population in Mustang, Nepal. A nongovernmental organization, a hospital, and a Nepali university have joined forces with local communities to collect and grow the seabuckthorn berry, to research seabuckthorn's medicinal and cosmetic purposes, and to sell seabuckthorn products to national and international markets. This partnership has established additional nurseries and is training local communities in harvesting and processing. Other Nepali initiatives are replicating the partnership's model to scale up their own efforts.

Equity

Participation also needs to be equitable; otherwise people will get the impression that resourceful groups are gaining undue influence and recognition of their specific interests. Equity depends on equitable recognition, but also on equitable capacities. In many processes, investment in capacity development will benefit the outcomes. Designing capacity development, again, should be demand-driven and conducted collaboratively; otherwise it will not work effectively.

Likelihood of Implementation

Both better quality and improved credibility enhance the likelihood that a solution or decision actually gets implemented. Solutions developed in a multistakeholder fashion are likely to be more workable at a substantive level. Participation breeds people's commitment to decisions and increases their dedication to implementation.

Typical Phases of Multistakeholder Partnerships

Multistakeholder dialogues and collaborative actions go through different phases over time (e.g., Tennyson 2003). Typically, they first involve a *development phase*, an initial idea that develops into a plan of activities. Plans change as thinking evolves and more partners are sought and engaged. This period is often marked by initial skepticism (toward ideas and toward each other) and a building of trust—often a veritable enthusiastic "honeymoon" ensues. More detailed action planning leads to agreement on who should be doing what, when, and with whom. It is also often the phase of acquiring financial resources because most partnership programs require additional resources to be implemented. The need for a fiduciary agent also often prompts the development of contractual arrangements among the partners. By then, the process is beginning to go into its *implementation phase*, when joint actions are undertaken. The transition from development to implementation is of course crucial, and often the point where the process gets stuck. This can have a number of reasons, such as "diffusion of responsibility" (when people implicitly assume that the respective other(s) were supposed to take this or the other action), or delays due to initial lack of resources or the need for intraorganizational clearance. Also, increasing experience with the partnership often leads to restructuring and reassignment of roles and responsibilities. All of these challenges can cause frustration and negative group dynamics, and need to be managed carefully. The next phase(s) can include further, permanent *institutionalizing* of the partnership, *monitoring and evaluation*, and/or *closure* of the joint activity. Again, transitions pose any number of potential pitfalls and need to be carefully managed.

A few examples may illustrate these phases and transitions: When moving from the development of the ideas to the implementation of an action plan, there may be agreement on basic ideas but when developing an action plan, differences of understanding of the same terms. For example, differences of perspectives and ideas about roles and responsibilities commonly arise. It also commonly takes much more time and effort to move from ideas to concrete action plans than expected. In other cases, all partners may want the program to be implemented, but nobody has the resources to work on it, manage it, or raise funds for it.

Some Key Lessons Learned

The following summarizes some key lessons learned through research, practical experience, and exchange among the networks of practitioners of dialogue and multistakeholder processes.

Clear Principles and Flexible Practice

Key principles of effective multistakeholder communication and collaboration include transparency, equity, and integration of perspectives (rather than domination of one perspective); inclusion; trust and trustworthiness; joint learning; shared ownership; and good governance.

Principles can be practiced in very different ways, and the groups not only have to agree on principles but also learn how to apply them in their particular case. It is crucial to explicitly agree on basic principles, *and* to work out in practice, together, as a group, how these principles shall be enacted. "Walking the principled path with practical feet" is a social learning experience that builds commitment and a fundament for success.

For example, people agree that a well-governed partnership is based on principles of transparency, fairness, and inclusion. However, no one governance structure can be applied to all kinds of collaboration. How transparency, for example, is being realized depends on the number of partners involved, their geographical distance, their access to information and communication technology, the amount of time they can put into the partnership, and so on. In addition, practical requirements are likely to change over time. For example, a loose network structure may have to create a formal institution in order to receive funds, and hence will have to adapt governance structures such as executive offices and boards, and draw up written agreements.

Communicating about the way one communicates and collaborates is called *metacommunication*. Sufficient and well-functioning metacommunication is a crucial fundament of successful collaboration. Although one should not allow endless, unproductive "navel-gazing," such reflective periods should be seen as an investment rather than a waste of time and resources.

A good example of the need for metacommunication is the fact that many multistakeholder groups include partners who are not equally powerful. While striving for equity between partners, the common lack of appropriate tools to deal with power gaps leaves many partnerships at risk to be run inequitably and thus fail to benefit from the full contributions of all partners. The same applies to different bases of power; for example, financial resources (often business) versus social consciousness (often nongovernmental organizations) and community based organizations (CBOs). This poses challenges that partners need to address within their process. A widespread tendency to consider such difficult questions as "taboo" is understandable but not helpful. Creating a reflective setting and allowing the group to step back and look at their structure and process, possibly through role-play, can help to further an open and fruitful conversation about power relations within the group.

Control and Chaos

Wollenberg et al. (2004) called their article on MSPs and social learning "Muddling Towards Cooperation." Kader Asmal (2000), the former South African minister for water and chair of the World Commission on Dams, talked about the Dams Commission not as an "organized, neat concerto" but as a "cacophony" of melodies and sounds. Both authors refer to the fact that multistakeholder dialogues and partnerships involve very diverse groups of people, perspectives, and interests. Facilitating such processes, and participating in them, is often compared to juggling. Keeping several "balls" of multipronged, formal and informal, multiscaled strands of communication "in the air" requires skill, concentration, and a level of confidence. It is often necessary to employ diverse creative mechanisms to further communication, intergroup problem solving, and collaboration.

Psychologists talk about the "tolerance toward ambiguity" that is required to cope with the fact that things are (seemingly) contradictory and difficult to predict, with many parts of a complex mosaic developing while adapting the larger picture without losing its key components. Plans are being continuously adapted as the process moves along. Operating successfully in such complex and ambiguous situations requires the individual to put forward positive expectations and trust and a strong positive vision. It does not allow for as much control over what is going on as many of us are used to in our work.

Not surprisingly, tolerance toward ambiguity is positively linked to

Box 19.3. The Nairobi Declaration process

At the International Ecoagriculture Conference, held in Nairobi, Kenya in October 2004, a Declaration was finalized that formed one part of the conference outcomes. Initially, it was planned that a small group of people, who had produced a draft Declaration before the conference, would receive feedback, comments, and amendments in written form from participants and would incorporate them by the end of the conference. A box for submitting comments was installed in the main conference room. However, some participants felt very strongly that the process should be more open and transparent, and that the drafting group should include more diverse perspectives. The conference organizers then suggested a second feedback loop, with a second draft of the Declaration being distributed on the middle day of the conference. Although some participants were content with that, others felt a need for an open face-to-face discussion. This was convened one evening, with the whole drafting group present. This meeting discussed the Declaration and also agreed what its status would be. The Declaration was redrafted, and presented on the final day of the conference, making sure that its agreed status was made clear to all participants as well as the media (Nairobi Declaration 2004).

One lesson that can be drawn from this process is that there is not an objective way to assess a procedure's openness, transparency, balance, and so forth. In fact, although all people share the desire for procedural justice, it can imply very different requirements of time, feedback loops, and kinds of desired interactions (also see LeResche 2005). Hence, when facilitating a process, it is important to keep in mind that not only the *content* but also the *process* will require inputs from different stakeholders so that everyone feels comfortable and able to commit.

intelligence, or the ability to perceive and understand complex environments, and to communicate effectively in diverse groups. It is negatively correlated to authoritarianism, which involves an extremely strong need to control.

Learning

However convenient it would be, we cannot "inject" others' knowledge into our own heads. Neither can we expect people to behave straightforwardly on the basis of their knowledge. As much as human behavior is codetermined by a number of factors (knowledge being one of them), so is behavior change. Even if we "know" it all (theoretically) in advance, we might still have to go through a learning and habituation process in the group and as a group.

Constant learning is necessary in situations of complexity and ambiguity that develop over time. Ecoagriculture is a complex concept, and its implementation is a complex undertaking that requires contributions from all stakeholders. People learn from documents, from individual experience and reflection, from interaction with others, and from working with others. Dialogue and collaboration in the context of ecoagriculture have to become interactive, dynamic learning environments. Again, for the individual, embracing learning and change requires a certain level of humility, curiosity, and confidence. It requires knowing that we don't hold "the truth," but that we all hold assumptions and knowledge that are *true to us*.

Benefits of Facilitation

Many partnerships develop from an idea that one individual had, or two or three people had, over the course of a conversation. Continuing their exchange, they may start working together, add additional partners, and get on with developing and implementing the partnership (no matter if they call it one or not). As the process moves along, and as the group of core or associated partners grows, they will often find that their meetings and telephone conferences benefit from somebody facilitating the conversation. More often than not, this will actually be a member of the group, a person in a leadership role, often one of the initiators. Sometimes, people rotate that role consciously, but sometimes the "leader" facilitating will create inequities that bother the others over time.

Partnerships benefit from independent facilitation, provided by someone who has expertise and experience in working with diverse groups. While working with the partners and individual members on a regular basis, the facilitator maintains a somewhat "outside" perspective as someone who keeps an eye on people and process and does not get caught up in developing the ideas or implementing agreed-upon actions. Such independent facilitation helps build communication and good working relationships within the partners' group by facilitating equitable, constructive, and outcome-oriented communication; enforcing ground rules; and providing observations that help the group reflect on how they work together. It helps to develop leadership within the group through observations, feedback, and mentoring. It also helps to identify problems, roadblocks, and opportunities that partners, focusing on the substance rather than on the process, may not identify themselves. Facilitators can also help build knowledge management procedures, and generally ensure that essential steps are taken throughout development and implementation.

Leadership Challenges

Leadership, as defined from a social psychological perspective, is a "process in which group members are permitted to influence and motivate others to help attain group goals" (Smith and Mackie 2000). It is interesting to note that this definition focuses on the people being "led" as actors (i.e., *permitting*). This view makes it clear that leadership depends on being recognized and allowed rather than a leader emerging out of a desire to lead. Although this is certainly an interactive process of mutual influence and permission, it is an important point in the context of MSPs.

In contrast, one of the difficulties in thinking about leadership is that our usual perception is that leadership is what leaders *do*—leaders lead and followers follow. Traditional (autocratic, paternalistic) and Machiavellian (manipulative) and authoritarian modes of leadership, which are found in all parts of the world, tend to disempower the ones they "lead." Such leaders exercise control by overcentralizing decision making, thereby coercing others into agreement. The effects might be overt agreement—and internal withdrawal, or open opposition—and a failure of the group process toward shared goals and commitment.

The emergence of "servant leadership" has contributed to a shift in orientation, namely, an orientation to leaders as *serving* the needs of "followers" so that the followers become in fact the leaders. And *visionary leadership* tends to shift our concept of leadership away from leaders and toward *shared purpose and vision*. Such leadership will express itself in service to and empowerment of others. It will foster collective decision making and collective action. Multistakeholder settings represent a model where new forms of leadership can be explored and developed.

We can think about multistakeholder partnerships as leaders *themselves*, further embracing a concept of *collective leadership*. In a functioning partnership, all participants lead, in their respective roles, and the group as a whole leads the partnership program. Thus the group not only reaps the benefits of increased quality, innovation, credibility, and better implementation, it also creates a new form of leadership toward the outside. A group of partners demonstrates collaboration, integration, and learning.

Challenges of leadership in multistakeholder settings include dealing with the diversity of cultures, interests, and goals, and also with the initial lack of trust between people from different backgrounds and power bases. The multistakeholder approach is relatively new and hence requires some patience and experimentation. This quality attracts quite a bit of public interest, which increases the pressure to deliver quick results.

Although initiating a dialogue or partnership process benefits from thoughtful leadership, well-articulated vision, and thorough process design, it is essential that leading persons—ideally, the whole of the group—succeed in constantly balancing some contradictory or complementary requirements. These include establishing a clearly defined but open space; listening and encouraging group members but also articulating for the group; and providing input and vision but also challenging each other to lead collectively. This requires individuals to cultivate strong principles as well as flexibility toward their practice, and clear goals as well as flexibility as regards developing objectives. In other words, leading individuals—and groups—need to know that they are continually learning and not the sole purveyors of "truths."

Conclusions and Outlook

Better partnerships will benefit both ecoagricultural initiatives and the individuals involved in them.

Implications for Ecoagricultural Partnership Projects

Building dialogue and collaboration takes time, resources, and trust. Shared and clear ownership is essential to avoid diffusion of responsibilities. Mutual accountability needs to be ensured to avoid misunderstandings and assure people that activities are being run in a fair and transparent manner. Although sharing ownership is important, partnerships also need a "home" within their respective partner organizations' institutional structures, and possibly one specific home for the partnership program itself. The additional workload of managing MSPs is usually not part of existing job descriptions, which may need to be changed as programs become more defined.

Another key lesson is to invest enough time in reflection and metacommunication: the better the relationships are built from the beginning, the firmer the fundament for future developments.

As more and more partnership programs develop, within ecoagriculture and the whole of the sustainable development movement there is a greater need for exchange and learning in terms of process, structures, and tools. Finally, we observe partnership programs being developed mostly on the sheer enthusiasm and passionate dedication of individuals and groups, while external support, particularly for the more open initial developing phases, is hard to get. It would be extremely desirable to develop "seed grant programs" for MSPs, dialogues, and partnerships in ecoagriculture landscapes.

Implications for Individuals Involved in Ecoagriculture Partnerships

The most important lesson for individuals from MSP is likely to consist of a preparedness to learn, and to change. When individuals truly engage in dialogue and collaboration, they will change opinions, behaviors, and networks. The second key lesson is that the complexity and ambiguity of such processes require a lot of coping—with cognitive, emotional, and behavioral challenges. Leaders should consider getting support—an individual coach or mentor, and a facilitator for the group.

Such engagement takes, among other things, *passion* and *patience*. These are among the factors that actors can actually control. Passion and patience together generate persistence and positive drive and complement each other.

There are wider, and deeper, benefits for societies to consider. Ecoagriculture approaches have the potential to contribute to good governance, conflict prevention, and conflict resolution, and thus yield more general societal benefits. Democratic participation, equitable involvement, and transparent mechanisms of influence create ownership and support among stakeholder groups and individual citizens. Successful communication across interest groups and competitors as well as consensus building and joint decision making can increase mutual respect and tolerance and lead societies out of deadlock and conflict on contentious issues.

Considering that these are diverse groups of different stakeholders, partnerships actually provide unique learning opportunities where people have to interact in nontraditional ways. Thus they can breed a culture of consultation without brushing over differences, building consensus without coercion; accountability; shared responsibility; attention to people (rather than documents and institutions); respect for differences; and, ultimately, a culture of unity in diversity. If they culminate in successful, joint delivery of a practical task, they can be considered as an ideal mechanism for overcoming prejudice and stereotypical perception and behavior.

Enayati (2002) provides a comprehensive overview of social psychological research that is relevant to MSPs. Social psychology research has highlighted how our social environment impacts our identities and behaviors, and that we generally underestimate these impacts. Research also confirms that the most effective ways to counter negative stereotyping and prejudicial behavior are not only increased knowledge about and contact with the respective "other" group, but also the positive, practical experience of succeeding in a task jointly undertaken by the different groups.

Many of the foregoing statements may sound like "soft" conclusions and advice. But they are factors we tend to neglect. It is "natural" for experts in ecoagriculture to be excited about the issues and challenges involved in the topic—it is "natural" to focus attention on the issues and challenges rather than on the people and the process. Yet, without an eye on people and process, communication and collaboration is much more likely to fail. The outcome is not only frustration over wasted time and resources, but also a decreased likelihood of successful communication and collaboration in the future. In other words, take your time, your passion, patience, and resources to create a solid and enjoyable learning environment the first time around.

References

Asmal, K. 2000. First World chaos, Third World calm: a multi stakeholder process to "part the waters" in the debate over dams. *Le Monde*, 15 November 2000.

Bawden, R. 1997. The community challenge: the learning response. Invited plenary paper, 29th Annual International Meeting of the Community Development Society 27–30 July, Athens, Georgia.

Buber, M. 1997. *Das dialogische Prinzip*. (The Dialogical Principle). Lambert Schneider, Heidelberg.

Enayati, J. 2002. The research: effective communication and decision-making in diverse groups. In: *Multi-Stakeholder Processes for Governance and Sustainability: Beyond Deadlock and Conflict,* ed. M. Hemmati. Earthscan, London: 73–95.

Hemmati, M. 2004. *Research Agenda for the SEED Initiative: Supporting Entrepreneurs for Environment and Development*. http://www.seedinit.org.

Künkel, P. 2004. Das dialogische Prinzip als Führungsmodell. In: *Organisationsentwicklung,* H 1:64-75 (The Dialogical Principle as a Leadership Model. In: *Organizational Development,* 1:64–75).

Künkel, P. 2005. *Making Use of Dialogue Change Models in Partnership Brokering.* Collective Leadership Institute, Potsdam.

LeResche, D. 2005. The significance of procedural justice for peacemaking processes. Paper presented at the United Nations' Expert Group Meeting entitled, Dialogue in the Social Integration Process: Building Peaceful Social Relations—By, For, and With People, 21–23 November, New York.

Nairobi Declaration 2004. International Ecoagriculture Conference, Nairobi, Kenya. September 27–October 1, 2004. www.ecoagriculturepartners.org/whatis/nairobi declaration.htm.

Sampson, E.E. 1993. *Celebrating the Other: A Dialogic Account of Human Nature.* Harvester Wheatsheaf, New York.

Smith, E.R. and D.M. Mackie. 2000. *Social Psychology*. 2nd ed. Worth, New York.

Tennyson, R. 2003. *The Partnering Toolbook.* The International Business Leaders Forum (IBLF) and the Global Alliance for Improved Nutrition (GAIN), London and Geneva.

Waddell, S. 2005. *Societal Learning and Change: How Governments, Business and Civil Society Are Creating Solutions to Complex Multi-stakeholder Problems*. Greenleaf Publishing, Sheffield.

Wollenberg, E., R. Iwan, G. Limberg, M. Moeliono, S. Rhee, and M. Sudana. 2004. Muddling towards cooperation: a CIFOR case study of shared learning in Malinau District, Indonesia. *Currents* 33:20–24.

Woodhill, J. 2003. Risk society and social learning. In: IAC (ed.) *Linking Participatory Practice and Governance Challenges for a Learning Society: Seminar Proceedings and Notes on IAC's Emerging Work in the Area of Facilitating Multi-Stakeholder Processes and Social Learning*. Annex I, p.B-C. International Agricultural Centre, Wageningen, the Netherlands.

Chapter 20

Restructuring the Supply Chain

Edward Millard

Introduction

As the world's economies have grown, so has the economic pressure on millions of rural families and the environmental pressure on our shared planet. The prices that small-scale farmers receive for the agricultural commodities they produce often barely cover their costs. Prices of major commodities (excluding petroleum) declined by 2000 to 40% of their level in real terms in 1900 (Giovannucci and Koekoek 2003), largely due to more intensive production of fewer varieties, which has created surpluses. Efforts to establish commodity agreements that manage supply for the world market in order to maintain high prices have not generally succeeded. The challenge of developing ecoagricultural landscapes that achieve biodiversity conservation, agricultural production, and rural livelihood goals will only be attained if markets allow them to be financially viable.

Agricultural production has increased more through higher rates of productivity than it has from the conversion of land to agriculture. Nevertheless, agricultural land has expanded at 14 million ha per year since 1980, and this has occurred largely at the expense of forests rich in biodiversity. The area devoted to coffee, cocoa, soy, and oil palm has expanded from under 50 million ha to over 100 million ha between 1970 and 2000 (Clay 2004). A major driver has been the promotion of the agricultural sector by governments hard pressed to meet rural development and foreign exchange earnings targets. Although intensification has brought increased productivity, it often involved loss of natural vegetation and depends on large inputs of chemicals, which exhaust the soil more quickly and cause pollution to critical sources of water on which both people and wildlife depend.

New trends in the international market are stimulating initiatives to address these social and environmental problems. The underlying causes of these new trends are both external and internal to companies that trade and transform agricultural commodities into ingredients and manufactured food products. Can such initiatives ultimately provide market incentives to drive sustainable agricultural production? Can the companies make a profit from adopting buying criteria that improve social and environmental conditions? This chapter analyzes the major initiatives to bring market value to sustainability and suggests which approaches will be necessary to gain support throughout the value chain and enable ecoagriculture systems to attain an important scale of impact.

Market Demand for Products with Social and Environmental Value

This section provides an overview of the market demand for products with social and environmental value, considering both external drivers such as consumer pressure and legislation and internal drivers such as sustainable supply and corporate responsibility.

External Drivers: Consumer Pressure and Legislation

In the past 10 years, consumers living in the parts of the world that consume large amounts of the world's commodities, mostly North America and Europe, are increasingly demanding natural food products that are free from unsafe or unhealthy ingredients, additives, and processes. Concerns for health and diet and for food safety have become major trends influencing the products companies offer. This has contributed strongly to the sales growth of organic foods, which in 2004 reached US$27.8 billion worldwide (IFOAM 2006). The food industry has suffered from a series of safety problems over the past decade, such as the presence of salmonella in eggs and, most notably, the renewed incidence of mad cow disease in Europe, which provoked legislation regarding the international beef trade. These problems also brought home to consumers the link between food safety and the treatment of animals, an issue that the organic movement had already brought to public attention.

Increasing travel and access to international news in the media, as well as some widely read, authoritatively researched publications, consumer guides, and the activities of campaigning organizations such as Rainforest Action Network and Global Exchange have made consumers more aware of the food supply chain. As they have learned about the low prices farmers earn; the poor working conditions of many employees; the effect of heavy agrochemical use on the soil, the waterways, and the people who spray the chemicals; the impact of forest clearing on biodiversity in countries where many of the products they consume originate; as well as the practices that have developed in animal rearing and processing to bring competitively priced goods to the market, they have begun to demand alternatives. They want food that is not only safe and as free as possible from additives but that also respects the natural environment and economic well-being of the farmers who grow it and treats animals humanely. These new consumers "want to know where a product came from, how it is made, who made it, and what will happen when they are done with it" (Ray and Anderson 2000).

Consumer interest in the living standard of producers, health of the environment, and welfare of animals has created a market in which improved environmental and social conditions at the production end of the supply chain and improved product quality of finished products are linked. This requires companies who manufacture and sell consumer products to know what is happening at all stages of the production process, not just at the part of the value chain that they directly manage.

For example, concern over child labor, which had already come to wide public attention in the 1990s around factory employment, became an issue in agriculture when reports spread about the employment of children in conditions equivalent to slavery in West African cocoa plantations. United Kingdom television Channel 4 produced a documentary that was followed in the United States by a series of articles published during the summer of 2001 by two reporters at the *Philadelphia Inquirer* and later spread to the national media. Chocolate manufacturers do not own cocoa plantations, but nevertheless need to respond by demanding changes in their supply chain. In the same way, personal care manufacturers, subject to consumer outcries about animal testing, make demands of the companies that supply them with ingredients, hamburger retailers need to avoid buying beef from cattle grazed on newly cleared rainforest, and paper manufacturers are subject to scrutiny on the trees that were cut down to make pulp. Failure by companies to respond to such scrutiny will affect their brands and reputations.

Consumer movements generally lead to legislative change, because consumers are also voters. Thus a regulatory environment regarding increasing health and safety standards has developed. Companies who sell food products need to have systems that enable them to track their production process from field to table.

Internal Drivers: Sustainable Supply and Corporate Responsibility

It is not just external pressures imposing a wake-up call on companies. Internal forces are also driving them to invest in building a supply chain that incorporates social and environmental responsibility.

First, companies are increasing their understanding of the link between reliable supplies of good-quality materials and a healthy landscape that conserves biodiversity and ecosystem functions. The concept of sustainable production is no longer just an on-farm issue, having to do with what plant varieties and management practices are adopted. Environmentally, it is an issue of the larger landscape that keeps forests standing to harbor biodiversity, provide biological controls of pests and diseases, improve the quality of fresh water, retain nutrients in the soil, and conserve rainfall levels. Socially sustainable production is a factor of the farmers' well-being and access to essential services, such as training, extension, and credit, as well as respect for labor laws. Gaining an understanding of the whole supply chain enables companies to make the most strategic decisions about their investments in it.

Second, good environmental management may reduce costs. Cocoa pests and diseases can destroy up to 30% of the crop, and much more if a disease becomes widespread. Research by the Smithsonian Migratory Bird Center in Panama found that retaining forest cover in cocoa farms provided habitat for the birds and spiders that eat some of the cocoa pests (Smithsonian Institution Migratory Bird Center 2004). Sustaining yields maintains production levels and also ensures farmers do not abandon the crop, which would lead to shortages and higher prices. Chiquita Brands International, which introduced sustainable practices in all of its own 119 banana farms, as well as requiring them in independent supplier farms, claims to have saved US$100 million over 10 years in its production costs partly through technology improvements and partly through reduced use of agrochemicals and better conditions for workers, which improved morale and hence productivity (Rainforest Alliance 2004).

Third, a reputation for positive social and environmental impact is beneficial not only to companies' image but also to their sales performance and employee retention, factors that affect shareholder returns positively. Nowadays, most companies produce corporate social responsibility reports.

Responses in the Marketplace

The trend toward social and environmental responsibility has motivated companies who produce consumer products to address issues throughout their supply chain, to ensure both that their products are in line with consumer concerns and that their supply of materials is secure in the long term. There have been broadly four types of responses: new product development, certification, and individual industry initiative, and company supply chain management.

New Product Development

In the early 1990s there was a fad—it lacked the sustained impact to be called a trend—in favor of food products to "save the rainforest." Several new products appeared with strong conservation messages. Many of these initiatives overreached themselves, promising more than they could deliver in terms of product quality, functionality, and dependability of supply on the one hand and impact on the rainforest on the other. Where they can meet quality and reliability standards, products originating in the rainforest still enter international markets, but with less hype. Companies and consumers value them for their naturalness, as well as for the economic benefits they generate for the local communities who are stewards of the biological resources.

Messages to demonstrate a company's concern for these local benefits appear increasingly both on individual products and also in corporate reports, to demonstrate to customers and shareholders how company operations address global issues of social and environmental concern. For example, the Web site of the largest natural and organic store in the United States, Whole Foods Market, lists a set of core values that span from "Selling the Highest Quality Natural and Organic Products Available" to "Caring About Our Communities and Our Environment" through "Sustainable Agriculture" and "Integrity in All Business Dealings."

As consumers understand more about social and environmental issues, their demand for products with a positive benefit for local communities and the environment increases. Ecotourism, the fastest-growing segment of the international tourism industry, is an example of a strongly growing demand based on consumer understanding of the supply chain. There is a small-scale development within ecotourism now to incorporate visits to farms that practice sustainable management, termed as agroecotourism. An example is tours to coffee farms in Costa Rica.

The new product that has perhaps made the most impact since appearing in the 1990s is shade coffee. Based on research undertaken by the Smithsonian Institution's Migratory Bird Center and others demonstrating that shade coffee farms harbor more biodiversity than farms converted to sun or technified production, a number of coffee products have appeared with notable success in the market. For example, Starbucks Coffee Company introduced Shade Grown Mexico in 1998 and informed its consumers about the relationship between coffee growing systems and conserving biodiversity.

Certification

The essential idea of certification is that an independent organization verifies compliance with a set of standards and awards a seal that the product can display. Certification has grown as companies have realized the need to manage their supply chain and demonstrate the social and environmental value of their products. For example, coffee is certified as organic in 25 countries, as fair trade in 22, and as ecological in 9. Certification provides product differentiation that can give companies competitive advantage. Giovannucci and Koekoek (2003) call it a "global trade passport." There are three main types of certification present in the food market: organic, environmental management, and fair trade.

Organic Certification

The organic certification movement predates the trends that focused attention on social and environmental conditions in tropical countries and had distinct origins. But it lent itself to the cause and there has been a strong expansion over the past decade of products from tropical countries bearing the organic seal. About 40% of organic product sales are in the United States (Clay 2004). However, organic certification has some difficulties as an effective mechanism for demonstrating sustainable agricultural practice:

First, the certified organization incurs a substantial cost in establishing not just the production methods but also the required internal control systems and the costs of inspection, product labeling, and monitoring. As a result, much production that could be certified is not. Another consequence is that certification is often held by a trading company rather than by producers, so that they are not guaranteed the price premium. The organic movement has made concerted efforts to reduce certification costs and increase the availability of certification in tropical countries through training local inspectors, and this is having positive impacts on the access of producers to markets.

Second, the premium that the market is willing to pay for organic certification is variable and inevitably declines as more production becomes available. There is no long-term guarantee that it will meet the costs. Moreover, most food products are traded domestically, and tropical country markets have less willingness and capacity to pay premiums than those in North America and Europe. Many certified products are not able to find a premium market and are sold as conventional.

Third, organic standards do not inherently integrate the conservation of biodiversity in sustainable agriculture. Organic certification agencies are actively looking at ways to increase the importance of biodiversity in their standards (for example, the US National Organic Standards Board approved biodiversity conservation additions into their Organic System Plan Template in 2005).

Fourth, the movement is hampered by governmental control of certification, which manages approval of certification standards and agencies. The legislative authorities in North America and Europe often do not approve standards agreed to in foreign countries. Mechanisms such as equivalence and partnerships between certification bodies are improving market access, but, still, acquiring certification for one country's standards does not necessarily enable a producer to enter another country's market, so that costs may double or triple. Increasingly, producers have to choose between which markets to target because they cannot afford to enter them all.

Certification for Environmental Management

Rainforest Alliance introduced a certification seal to promote sustainable farming in major commodities in tropical countries. This seal offers consumers an alternative to intensive agricultural production with high use of agrochemicals. The certification standards are designed to be flexible and suitable for large or small farmers; for example, they do not require strict adherence to organic methods. Rainforest Alliance works through a Sustainable Agricultural Network to "develop practical standards for responsible agriculture," covering "community well-being, environmental protection, and economic vitality." Its standards address ecosystem and wildlife conservation, fair treatment of workers, community relations, integrated crop management, waste management, conservation of water and soil resources, and planning and monitoring (Dudenhoefer et al. 2004). First established in Latin America, Rainforest Alliance is now extending activities to Africa and Asia.

Rainforest Alliance has been successful in partnering with large food companies. Its collaboration with Chiquita began in 1991 and developed farm management standards for growing bananas, which are a US$5 billion a year industry. The standards were developed collaboratively with farmers, scientists, and industry representatives and comprise ten principles: no deforestation, reduction in pesticide use, protection of wildlife, conservation of soils and water, better pay, environmental education, and housing and safety standards for workers. Chiquita claims the program to be "a stellar example of for-profit and nonprofit organizations working together in good faith and with a common purpose to improve the environment and worker health and safety while at the same time improving bottom-line business performance" (Rainforest Alliance 2004). In 2003, Kraft Foods initiated a partnership with Rainforest Alliance who certify coffee, bananas, citrus, flour, ferns, fruits, and cocoa .

This approach to certification through partnership and collaboration is different from organic certification, in which the certifier determines the standards and has no relationship with the applicant. To be effective in delivering social and environmental benefits depends on a close knowledge of the situation in the producing country and monitoring capacity.

The Smithsonian Migratory Bird Center introduced a certification scheme for coffee called "Bird Friendly," which led to some other coffees entering the US market highlighting the specific relationship between shaded coffee farms and habitat for birds. It also joined forces with the Rainforest Alliance, Conservation International, and Consumer's Choice Council to develop *Conservation Principles of Coffee Production* in 2001.

In September 2004, Rainforest Alliance introduced certified chocolate from Ecuador. The chocolate is sold in up-market restaurants, caterers, and retailers. In its product promotion it emphasizes the relationship between sustainable farming and cocoa quality: "The world's finest tasting gourmet chocolate happens to come from cocoa farmed using traditional shade-grown, small-scale, low-impact techniques." The product's quality derives principally from the native cocoa strain, which yields a high-quality product, supported by traditional knowledge of processing that brings out the full flavor of the bean.

The focus on origin is an established market mechanism for differentiating products. It is always associated with quality, because of the specific characteristics of the locality, such as climate, type of material, and traditional knowledge of growing and processing techniques. The labeling of cocoa origin on chocolate to denote quality is recent. It aims to differentiate products and thereby attract a premium in a market that mostly blends origins to achieve uniform flavor profiles. Divine Chocolate identifies the Ghana origin of its beans in its range of chocolates. The Nature Conservancy has introduced "Conservation Beef," sourced from ranchers directly contributing to biodiversity conservation.

Fair Trade

Fair trade certification began in the Netherlands in 1988 and has spread through Europe and North America. Some 17 national fair trade certification organizations belong to the international umbrella group, Fairtrade Labeling Organizations International (FLO). Each national organization awards its seal to companies that demonstrate compliance with product standards.

The strong growth of fair trade labeling reflects and reinforces consumer interest in social issues at the production end of the supply chain. For example, certified fair trade coffee sales in the United Kingdom at retail prices grew from £15.5 million in 2000 to £65.8 million in 2005 (£1 = about $1.90) (Fairtrade Foundation 2006). Fair trade standards incorporate a minimum price, presently $1.26 per pound of export grade coffee FOB, with an additional 15 cent premium for organically certified coffee.

This growth, at a time when international prices have been very low, has been very important, yielding US$75 million in additional income back to farming communities between 1999 and 2005 (Transfair 2005). Clay (2004) estimates that fair trade affects half a million farmers. Fair trade has become an integral part—about 5%—of the specialty coffee market segment, which is growing at about 15% annually, compared to an overall growth rate in coffee of 1 to 2% (Clay 2004). The specialty market also features origins strongly. A World Bank study of coffees certified under different schemes concluded that specialty coffee is by no means only a niche. It accounts for 17% of global consumption of green beans, 40% of sales, and 50% of profits (Giovannucci and Koekoek 2003).

The Fairtrade Labeling Organization International (FLO) also certifies cocoa, through both drinking products and chocolate bars. In 2004 six countries exported about 1500 tons of fair trade cocoa. Divine Chocolate, based in the United Kingdom and also selling in the United States and elsewhere, is the leading fair trade chocolate company. In September 2004, Equal Exchange introduced into the US market a new range of fair trade chocolate.

The importance of fair trade is that it has greatly increased awareness of low prices and poor labor conditions in producing countries. As a pioneering movement that has contributed to different types of company response in support of sustainable agricultural practices, its value goes well beyond the products it certifies. Nevertheless, as a mechanism for demonstrating good social and environmental practice, fair trade certification has some limitations:

First, although sustainable agricultural practices, including restricted use of agrochemicals, are one of the criteria for certification, there is in practice insufficient monitoring capacity and no technical assistance for farmers to adopt these practices. The US labeling organization Transfair USA describes fair trade certification as the "leading standard for social and environmental auditing of the global supply chain" (Transfair 2005); but few would agree with the environmental side of the claim.

Second, standards are not available for all major commodities. The major product is coffee, whereas others are still under development. Cocoa, bananas, sugar, honey, and nuts also have fair trade certification standards.

Third, the fair trade standard for coffee requires farmers to be part of a cooperative. This excludes owners of larger farms and estates, who may nevertheless treat their workers well and pay them fairly. In its shade coffee program with Starbucks, Conservation International found that cooperatives were not necessarily the most efficient organizations for transferring premiums to producers and that individual farmers selling through private exporters to Starbucks actually earned more. Moreover, not all cooperatives abide by the International Labor Organization's seven principles for cooperatives: voluntary and open membership; democratic member control; member economic participation; autonomy and independence; education, training, and information; cooperation among cooperatives; and concern for community. There have been documented examples of practices that do not favor their members. Fair trade's credibility requires monitoring and auditing of its registered suppliers.

Fourth, some companies may resist paying a fee to label their products, even though they are operating their businesses in a way that meets the criteria. For example, an importer based in Seattle, Washington, USA, the Bainbridge-Ometepe Sister Island Association, pays $1.61 a pound for coffee grown on Ometepe Island in Nicaragua and returns some profits to pay for community-improvement projects. "Ten cents a pound, when we do 14,000 pounds a year, is a lot of money that can go back to Ometepe," said Lee Robinson, the association's treasurer, referring to the licensing fee payable to Transfair (Seattle Times 2004). In fact, Transfair has reduced its premiums to make its program more attractive.

Fifth, the aggressive promotion of some fair trade labeling organizations has sometimes upset companies, which prefer a constructive engagement with partners rather than the threat of campaigns against their operations.

Finally, a perception persists in the market that many fair trade coffees are not high quality. To an extent this derives from a legacy of the movement, which started in the 1960s with a focus on messages to the consumer rather than product quality. It has improved substantially, but the focus on small-scale organizations has prioritized market access for many that have inadequate quality-control mechanisms.

General Challenges for Certification and Sustainable Agriculture

Four further considerations restrict the present capacity of certification approaches as market mechanisms to drive good social and environmental practice.

First, they often conflict with the operating procedures that companies have adopted. Business policies of large retailers make access difficult and expensive for typically small and medium-sized suppliers of organic or fair trade products: a substantial bridge to negotiate. Required quantities are large and supply performance requirements are extensive and rigid. Each new product, which retailers call stock taking units, costs money to put on the shelf.

For coffee roasters or cocoa processors, incorporating new sources of supply can disrupt their business. They invest large sums of money in developing blends that meet consistent flavor profiles required by manufacturers. To change a source of supply represents a big risk of upsetting the consistency of the blend. For manufacturers, investing in strong and stable relationships is usually a more secure basis to ensure the stability and quality of raw material supplies than switching suppliers. They work with their suppliers to develop profiles of the ingredients they need, a substantial investment on both sides. The quantity and quality of supply of agricultural products from tropical countries is subject to a lot of variable factors, such as climate, diseases, availability of technology, and the relative economic attractiveness of other crops for farmers. Food processing companies develop blends that can be largely standardized to allow for these factors.

Second, the proliferation of schemes causes confusion to consumers and may undermine the value of each label by communicating a series of related but uncoordinated messages about the benefits to producers and their environment. There is some rapprochement between fair trade and organic; for example, the International Federation of Organic Agriculture Movements (IFOAM) has incorporated social standards. However, they are not going to come together completely because they represent different ideas and want reasonably to retain their own identities.

Moreover, not all schemes have proper auditing procedures. A study by Consumers International of food products on sale in eight countries found that claims made on product labels "for the most part did not help those consumers who wanted to purchase food products produced in a sustainable way. Indeed, in some cases it complicated choice, and thus prevented changes in market behavior that could positively impact on bringing about sustainable development" (Consumers International 2004).

To address these difficulties, the International Social and Environmental Accreditation and Labelling (ISEAL) Alliance was founded as an association of international standard-setting, certification, and accreditation organizations that focus on social and environmental issues. It supports its members through advocacy, harmonization of approaches to eliminate redundancies, and capacity building to set standards.

Third, the high costs of certification programs can prove to be a barrier for the participation of low-income farmers and communities. Moreover, in ecoagriculture landscapes where farmers seek to diversify products, a proliferation of diverse certification requirements could bear a prohibitive cost.

Fourth, certification schemes are unlikely in and of themselves to reduce expansion by farmers into forested areas; and indeed are criticized for encouraging it by paying price premiums. Hence there is a need for accompanying policy instruments that make forest encroachment unrewarding and for the capacity of authorities to monitor farmer compliance with them.

Individual Industry Initiatives

Faced with a need and interest to apply and demonstrate value to the producing communities, several companies have adopted an individual approach, applying their own standards or working with specialist agencies and nongovernmental organizations (NGOs) to develop them. This is a pragmatic approach to help companies address issues in a way that works for their business operations, rather than presenting a set of standards and pressing for their adoption.

In September 2004, major coffee companies launched The Common Code for the Coffee Community, committing themselves to rewarding good social and environmental practice and allowing independent auditing of the code's operation. The code represents "one of the most sweeping voluntary initiatives undertaken by any industry" (Williamson 2004).

Certification agencies understandably criticize these types of self-defining standards, arguing that voluntary standards will not attain the same rigor as third-party ones, and that only with independent verification can consumers be sure the claims are accurate. A similar tension exists in the timber industry, where the major companies launched their own Sustainable Forestry Initiative, rather than adopt the Forest Stewardship Council standards for sustainable forest management.

The industry's response to the critics is that certification schemes are all very well for niche markets but do not address the needs of the large majority of farmers who will never gain access to them, because of their product quality, or form of organization, or the lack of demand for certified products. Industry codes of conduct are logical responses to acknowledge a need and interest to change buying behavior but through self-regulation and using approaches that are less challenging to company operations.

For the certification agencies, the question is, do they accept that the initiative is being wrested from them, stunting their growth potential and applying what may be a lower standard, but with the benefit that their catalytic role has led to further initiatives that have the capacity to reach mainstream markets, with value for the local people and environment? It may well prove to have a larger positive impact to get a corporate commitment in the supply chain than to get one or two products on the shelf. This is the advantage of a corporate code of conduct approach. For consumers, the question is whether they will be able to cope with even more claims in the market as a guide to their buying decisions, or will the trust generated by certification schemes be dissipated by the bandwagon effect?

A new multicompany initiative is the Sustainable Agriculture Initiative (SAI) Platform, a forum created by the food industry in 2002 to develop more sustainable supply chains. Founded by Groupe Danone, Nestlé, and Unilever, and within two years growing to 18 members, SAI Platform represents the industry's need for a distinct organization that it can manage. Its agenda is "cost-effective precompetitive cooperation to ensure the sustainable supply of high quality agricultural products as basic ingredients for their products . . . the implementation of sustainable practices for mainstream agriculture (not niche markets) on a worldwide scale . . . and ways to increase consumer value and customer confidence as well as overall society's welfare, including environmental quality and farmers' well-being" (Sustainable Agriculture Platform 2004). The Platform has developed principles and practices in a number of commodity groups and is supporting field projects to test them. The Sustainable Food Lab incorporates both industry and nonprofit organizations in a concerted effort to "figure out and demonstrate how to create sustainable food chains for mainstream markets" (Sustainable Food Lab oratory 2004). Concerned by the limitations of niche markets, it has developed what it calls a "U" methodology—taking participants through three stages—observe current reality; retreat and reflect; enact new reality. It is especially concerned with improving the situation of small and medium farmers and fishers and is testing its approach through pilot projects.

Many of the companies involved in these initiatives are members of multistakeholder roundtables, bringing them together with financial institutions, researchers, and other concerned institutions to evaluate the supply chains of a range of commodities and work toward standards of sustainability. The involvement of the financial sector brings increased capacity for economic analysis of sustainable management, as well as the potential to link investment strategies to sustainable practices.

Another voluntary initiative is the Coalition for Environmentally Responsible Economics (CERES). The CERES Principles are a 10-point code of environmental conduct that provides a common basis for companies to report. CERES is a coalition of investment funds, environmental organizations, and public interest groups. Over 70 companies endorse the CERES Principles.

Interestingly, even the fair trade movement itself has undertaken an industry initiative, launching in 2004 a Fair Trade Organization mark that is managed by the International Federation of Alternative Trade (IFAT). Criteria for using it include "fulfilling your monitoring obligations and then being registered as a Fair Trade Organization"; doing so enables the organization "to use the FTO Mark as a special corporate badge of identity on your stationery, your shop front, all your promotional materials, your trade fair stands, in your office and so on" (IFAT 2004). The drivers of this initiative are the lack of FLO standards for many products that the organizations sell and the wish to distinguish themselves as organizations whose whole mission and operations are geared to fair trade, rather than just selling a few fair trade products. These organizations had played a pioneering role, bringing the producer's situation to the awareness of consumers, before certification schemes really got under way.

Supply Chain Management

Another approach to reduce the biodiversity "footprint" of agriculture has been with more comprehensive "greening" of the value chain with individual companies. For example, in Conservation International's corporate partnership with Starbucks, the company provides not only financial resources to support research, training, and extension services but also visits the sites to advise on quality, tests samples, and provides feedback. The company views this as a long-term business investment in its supply chain.

For a conservation NGO, developing corporate partnerships to improve social and environmental impacts in the supply chain is motivated by the potential to reach a large scale. For example, coffee is produced by an estimated 20 million growers around the world, and soy is cultivated on 73 million ha globally. These industries have broad supply chains composed of many individual companies. Rather than aim to work directly with very large numbers of farmers over millions of hectares globally, a conservation NGO can identify points of concentration in an industry that provides a logical entry point. For example, in the soy industry, four companies control 80% of the crushing or oil processing market. These companies are usually not the ones making the direct impact on the ground, but by working with their suppliers to back up their supply chains, they do represent a potentially efficient way to change behavior on a large scale by encouraging improved social and environmental performance. Initiatives for social and environmental responsibility by an industry leader will often spark a similar response by other companies in the industry. The following examples from Conservation International–industry partnerships illustrate diverse strategies.

Coffee Best Practices with Starbucks Coffee Company

Starbucks, the world's largest roaster of specialty coffee, with an extensive retail network, introduced a Preferred Supplier Program in 2001 to foster sustainability throughout its supply chain. The program "provides purchasing preference to coffee suppliers providing green coffee that is grown, processed and traded in an environmentally, socially and environmentally responsible manner" (Starbucks 2004). It evaluates suppliers by a scoring system, with points awarded for their achievement of defined social and environmental criteria. Applicants that achieve Preferred Supplier status earn preferred pricing and contract terms and priority for buying. Applicants who score higher receive a Strategic Supplier status, which carries the additional benefit of a one-year sustainability conversion premium of US$0.05 per pound on all green coffee meeting the program guidelines. In addition, Starbucks also awards a one-year US$0.05 premium for all green coffee shipped by applicants who achieve a certain increase in their score. This program was developed in collaboration with Conservation International, which applied the *Conservation Principles of Coffee Production* in several countries in Latin America where it partnered with coffee producers. Starbucks is also collaborating with Scientific Certification Services to develop

C.A.F.E. (Coffee and Farmer Equity) Practices guidelines as an evolution of the program.

Multiproduct Supply Chain Guidelines with McDonald's

Using guidelines as the basis, McDonald's does not pay premiums for improved environmental performance, but rather uses its buying power and traditionally strong supplier relationships to promote incremental improvements in supplier practices. McDonald's Corporate Social Responsibility Team developed the Socially Responsible Food Supply guidelines, which consist of 10 general principles related to key conservation and sustainability issues—biodiversity, soil, water, airborne emissions, economy, society, animal welfare, waste, pest management, and energy.

The methodology to apply the principles to practices is a scorecard, which suppliers use for self-assessment. McDonald's is identifying tangible, quantifiable indicators for each principle that McDonald's and its suppliers can use to measure performance and set goals for continuous improvement. The self-assessment is designed to provide a flexible format that allows suppliers to understand, monitor, and communicate their social responsibility efforts and environmental performance. It also allows suppliers to record and evaluate their social responsibility management practices. The scorecards are product-specific and consist of several performance indicators in each relevant guideline area. These indicators aim to measure, wherever possible, actual impacts, for example, the amount of water used per ton of product produced, the percentage of total water usage recycled, and the amount of agrochemical runoff and leaching. Biodiversity indicators might include the amount of land set aside, the amount of habitat converted to agricultural uses, the amount of land under organic or integrated pest management systems, or the diversity of species.

Rather than prescribing specific practices, each guideline provides high-level direction by describing desired outcomes. This provides suppliers with the flexibility to use their creativity and technical expertise to identify ways of achieving outcomes that are most efficient and reflect differing geographic conditions. When the program is fully fleshed out, it should provide a practical tool that McDonald's suppliers can use to establish their performance baselines, set targets, measure progress, and report back; and that McDonald's can use to generate meaningful quantifiable data across suppliers and regions for reporting to stakeholders.

McDonald's notes the challenges of implementing the system:

- Developing a system applicable to a wide range of industries
- Accounting fairly for differences due to geography, scale, and suppliers' experience with social responsibility issues

- Developing a cost-effective yet functional approach to data collection and validation
- Establishing appropriate goals and targets that are based upon the latest scientific information and are able to stimulate continuous improvement (McDonald's 2004)

Implementing Forest Set Asides with Bunge

Bunge Ltd. is Brazil's largest soybean processor. Soy is a bulk commodity that, like most of McDonald's products, has no application for niche markets. Yet its production a major cause of deforestation, especially in Brazil. The industry does not have dominant buyers able to dictate conditions to suppliers. Conservation International's approach is to develop incentives to growers that are based not on financial incentives but on the relationship between growers and soy processors. It is working with the processors to develop technical assistance programs that they can use to help growers implement the conservation reserves on their farms that Brazilian law requires. The emphasis is on identifying areas that minimize the opportunity costs of setting aside reserves while maximizing their biological value by placing them strategically. The benefit to the grower is to achieve compliance with the Brazilian Forestry Code in a way that minimizes the cost of retiring productive land and removes concerns about steep fines for noncompliance. The processor gains by improving relationships with growers and thereby also improving its competitive position in securing raw material. Biodiversity is conserved by securing increased areas of land set aside for protection and by introducing a land planning system that prioritizes the land set aside based on conservation value, as opposed to having random disconnected patches of habitat dotting a farm landscape.

Purchasing Guidelines and Certification

The initiatives described above are examples of approaches to sustainable agricultural practices through purchasing guidelines—a general set of conservation principles, indicators, and known best practices that are applied to corporate purchasing policies, product standards, contracts, or other purchasing processes.

The purpose of guidelines is to offer companies a tool for demonstrating preference for products or services based on how they are produced or delivered. By adding a social and environmental dimension to purchasing decisions, over and above standard considerations of price, service, quality, and delivery, they aim to create opportunities for suppliers to distinguish themselves based on environmental performance and thereby create market-based incentives for conservation. Despite some differences in content and structure across the industry programs using guidelines, they have common features:

- Because guidelines do not set minimum standards for performance, they make it easier for suppliers to begin participation, especially those that may be poor environmental performers or those that are not typically engaged in conservation.
- The guidelines approach encourages continuous improvement, with the incentives of stemming competition from other suppliers and meeting clients' requirements to set annual performance targets.
- Guidelines are flexible in their design and application, allowing suppliers to develop the most appropriate practices and procedures for their operations that meet the guidelines.

The purchasing guidelines approach can work in harmony with independent third-party certification. It does not state how verification must be carried out, as long as it is by well-qualified, credible, and independent third parties. Entities that could verify guidelines include existing certification agencies, auditing firms, and NGOs. For example, Starbucks is undertaking training programs for verifiers, so that specialists in each country it sources coffee from can be approved to verify compliance with its C.A.F.E. Practices.

Suppliers participating in certification programs already have the systems and processes operational and verified by third parties that document the social and environmental performance in their supply chain. These same performance and reporting systems could be used to extend into wider guidelines programs, thereby saving time and money, and spreading costs of participation in either program across a broader sales base. Purchasing guidelines can encourage companies to participate in certification programs by providing incentives to begin improving environmental performance and moving toward meeting the minimum requirements of a certification program.

Guidelines are applicable to a wide range of products beyond specialty or niche markets because they are not dependent on premiums or brand recognition. Given the impact on biodiversity from commodities such as soy, it is important to have a means of engaging and influencing producers in these conventional market segments. Because guidelines are developed within existing company purchasing structures and do not require an external management and certification structure, the costs of developing and implementing guidelines programs is relatively low. Because the approach does not depend on price premiums or labels, it is more insulated from changing consumer preferences and gives the approach greater robustness in the face of ever-shifting markets.

Conversely, the guidelines approach has limitations. Because it is not applying a universal set of standards, there is no label to communicate to consumers. As a result, its significance may not be as clear to consumers and it can be harder for companies to tell a story about what they are doing. Strategies for minimizing this problem could include avoiding specific product claims, or, as in the case of McDonald's, to develop quantifiable indicators that provide clear detail on the impact of guidelines implementation.

While having no minimum performance levels makes it easier to engage companies or industries that might not otherwise participate in such initiatives, it also means that making positive social and environmental contributions depends on companies pushing their suppliers to set annual performance targets. Companies as large as Starbucks or McDonald's have powerful leverage to do this and to stimulate competition among suppliers to improve their performance and continue innovation. But not all companies have so much influence over their supply chain.

An example of Conservation International's approach is a new initiative in West Africa, where the US Agency for International Development and the industry are implementing the Sustainable Tree Crops Program (STCP), focusing on cocoa. Conservation International, the Global Environment Facility, and the United Nations Development Program are developing a collaboration with STCP to build an integrated set of social and environmental practices with the intention that industry, government, and farmers will all promote these as the standard for sustainable production.

Conclusion

It is not feasible to protect most areas from existing economic activity, and the most important areas for biodiversity are also often those of high population and major agricultural production. Hence, we need to develop conservation corridors, in which sustainable land use may provide connectivity between isolated forest patches and thereby conserve more biodiversity than would be possible with protection alone. Conservation International's Center for Applied Biodiversity Science undertook a study of major commodities. While warning that "several decades of searching for ways to make exploitation and conservation compatible, including ecofriendly agriculture, sustainable agroforestry and sustainable timber management, have yet to yield successes on a scale sufficient to lay to rest fears about biodiversity loss," it concludes that sustainable agriculture reduces the negative impacts on biodiversity of much conventional agriculture, such as erosion, chemical runoff, and soil degradation on park boundaries. Smithsonian Institution research in Panama found that bird biodiversity in shaded cacao plantations was almost equal to that in the forest; and also that birds do control cocoa pests by eating them.

Incentives for farmers to adopt sustainable agricultural practices cannot come from the market alone. In some countries, such as Brazil, legislation requires farmers to set aside forested land in "legal reserves." Training and technical assistance that result in farmers increasing their crop productivity and diversification can act as strong incentives by resulting in an increase in income. Increasing access to financial services at affordable rates of interest can assist farmers to reduce their costs of production and upgrade their quality and marketing. For financial and extension services to be delivered effectively requires farmers to be organized. Groups are in stronger bargaining positions to negotiate prices for their crop because they manage more production than individual farmers and can operate centers for market information systems, which are increasingly springing up, using Internet technology to keep farmers abreast of market prices. Payment for ecosystem services is an area of strong interest as a mechanism for delivering cash benefits to farmers for adopting sustainable practices and thereby maintaining the health of the ecosystem (see chapter 21).

Ultimately, the market will pay what it can bear and this price will relate strongly to traditional values of quality and reliability of supply. Taking advantage of new consumers willing to pay premium prices makes obvious sense but the potential for growth is limited and, as supply increases in response to the market opportunity, premiums fall. Therefore, initiatives need to look toward mainstream markets and find flexible approaches to respond to the widespread interest among companies to improve their social and environmental performance.

References

Clay, J. 2004. *World Agriculture and the Environment: A Commodity-by-Commodity Guide to Impacts and Practices*. Island Press, Washington, DC.

Conservation International. 2001. *Conservation Principles of Coffee Production*. Conservation International, Rainforest Alliance, and Smithsonian Migratory Bird Center.

Consumers International. 2004. *Green Food Claims: An International Survey of Self-Declared Green Claims on Selected Food Products*.

Dudenhoefer, D., C. Goodstein, and C. Wille. 2004. Rainforest Alliance's sustainable agriculture program: pioneering certification in the American tropics. Presented to the First International Ecoagriculture Conference and Practitioners' Fair, Nairobi, Kenya, September 2004.

Fairtrade Foundation. 2006. www.fairtrade.org.uk.

Giovannucci, D. and F.J. Koekoek. 2003. *The State of Sustainable Coffee: A Study of Twelve Major Markets*. International Institute for Sustainable Development, International Coffee Organization, and the United Nations Conference on Trade and Development.

IFAT (International Federation of Alternative Trade). September 2004. Newsletter.

IFOAM. 2006. *The World of Organic Agriculture: Statistics and Emerging Trends*.

ISEAL Alliance, London. www.isealalliance.org.

McDonald's. 2004. *Worldwide Corporate Responsibility Report*. www.mcdonalds.com.

Rainforest Alliance. 14 September 2004. The Rainforest Alliance unveils its first line of certified sustainable chocolate with a gourmet tasting in New York. Press release. www.rainforest-alliance.org.

Ray, P.H. and S.R. Anderson. 2000. *The Cultural Creatives: How 50 Million People Are Changing the World.* Harmony Books of Random House Inc., New York.

Seattle Times. 21 September 2004. *Big Companies Agree to Set Standard.* Joint press release, 21 September 2004. www.seattletimes.com.

Smithsonian Institution Migratory Bird Center. (2004). Research published on www.nationalzoo.si.edu/ConservationAndScience/MigratoryBirds.

Starbucks Coffee Company. 29 March 2004. CAFÉ Practices Overview. www.scscertified.com/starbucks.

Sustainable Agriculture Platform. 2004. www.saiplatform.org.

Sustainable Food Laboratory. 2004. www.sustainablefoodlab.org.

Transfair. 2005. Transfair USA 2005 Annual Report. Oakland, CA. http://www.transfairusa.org/Content/Downloads/AnnualReport2005.pdf.

Williamson, H. 10 September 2004. Coffee Trade Pact on better standards. *Financial Times.*

Chapter 21

Paying Farmers for Stewardship

Sara J. Scherr, Jeffrey C. Milder, and Mira Inbar

Introduction

Ecoagriculture systems offer farmers and agricultural communities a means to maintain or increase agricultural productivity while helping to conserve biodiversity and ecosystem services. The conservation values resulting from these systems are enjoyed by a variety of groups, ranging from downstream water users to the entire global community. However, the farmers themselves often incur significant costs in implementing these conservation-friendly practices. When farmers are not adequately compensated for the environmental "positive externalities" resulting from ecoagriculture practices, they are less likely to adopt these practices, especially when the costs exceed the benefits enjoyed by the farmer. Accordingly, if ecoagriculture is to be scaled up worldwide, it is critically important to find ways to compensate farmers for the off-site environmental benefits they provide.

Payment for ecosystem services (PES) provides a way of doing just this. PES programs compensate land stewards for providing ecosystem services of value to external beneficiaries, thus helping to align the individual private interests of farmers and other land stewards with the collective interests of the local, regional, and global communities that benefit from ecosystem services. PES transactions are distinguished by two key features: first, they are always *voluntary*, between a willing buyer and a willing seller; and, second, the payment to land stewards is *conditional* upon the provision of the agreed-upon ecosystem services (or actions believed to provide the services).

PES programs have been developed around four main classes of ecosystem services: (1) carbon sequestration, (2) watershed protection, (3) biodiversity conservation, and (4) recreation and landscape beauty. Although all four types of PES can support ecoagriculture, this chapter focuses mainly on biodiversity PES. The chapter first reviews the current state of biodiversity PES in agricultural landscapes, discussing the supply and demand of such services, and drawing on lessons learned from existing programs worldwide. Next, it explores potential benefits and risks to farmers of this approach to promoting conservation in agricultural landscapes. The chapter concludes by identifying obstacles and recommending actions to enable the widespread use of biodiversity PES to support biodiversity conservation and rural livelihoods in agricultural landscapes.

Rationale for Paying Farmers for Biodiversity Conservation Services

Financing and management of natural protected areas has historically been perceived as the responsibility of the public sector and nongovernmental organizations (NGOs). As of 2003, over 102,000 protected areas covered an area of 18.8 million square kilometers, or 11.5% of Earth's terrestrial surface (Chape et al. 2003). However, the last few decades have witnessed severe cutbacks in funding from governments and international public and private donors for the creation and management of protected areas (Jenkins et al. 2003). Increasingly, the donation-driven model for conservation is proving unsustainable because land acquisition for protected areas and compensation for lost resource-based livelihoods are often prohibitively expensive.

Meanwhile, the location of biologically rich areas necessitates that conservation efforts move beyond strictly protected areas. Worldwide, most biological resources are found in populated landscapes, with over a third of the world's land area heavily influenced by cropland or planted pastures, and still more land affected by farm fallows, tree crops, grazing systems, and production forestry (Wood et al. 2000). Almost half of the 17,000 main protected areas worldwide are being heavily used for agriculture. Clearly, biodiversity and ecosystem services cannot be adequately conserved by a relatively small number of strictly protected areas. Instead, conservation is best conceived as part of a landscape or ecosystem management strategy that situates protected areas within a broader matrix of land uses that are compatible with and support biodiversity conservation in situ. Achieving such an outcome will require new, lower-cost mechanisms for promoting biodiversity conservation on private lands.

One possible way to achieve conservation-friendly land uses on private lands is through regulation. This approach, known as the "polluter pays" principle, assumes that ecosystem services are public goods, and that the public's right to these goods trumps the rights of private land stewards to manage their land as they see fit. In reality, though, there has been little political will to mandate, much less enforce, strict regulation on private land management throughout much of the developed world and almost all of the developing world.

Absent regulation, land managers will tend to pursue the most profitable land-use practices, ignoring the economic and noneconomic values of ecosystem services except to the extent that these services benefit them directly. Because conservation-friendly management is often more expensive or less profitable than conventional agricultural management, farmers will tend to overexploit natural resources and undersupply ecosystem services. PES changes the economic equation for farmers by giving them a financial incentive for conservation-friendly management, thus improving the profitability of these practices and encouraging their adoption. Furthermore, experience with similar market-based instruments in other sectors has shown that they can achieve environmental goals at much lower overall cost than regulatory approaches.

A final rationale for paying farmers for biodiversity conservation is to contribute to rural development and poverty reduction. Most obviously, farmers can benefit from an additional income stream that may be less variable than income from agricultural goods. In addition, payments from external beneficiaries can help subsidize the conservation and restoration of ecosystem services that provide important local benefits to farming communities. For example, many low-income farming and pastoral communities are dependent upon forest, freshwater, and aquatic biodiversity for wild foods, medicines, fuels, and farming inputs. Finally, PES programs can improve human capital through associated training and education efforts and through investment in local cooperative institutions (Scherr et al. 2004).

Who Are the Buyers of Ecosystem Services in Agricultural Landscapes?

Five basic types of buyers participate in PES markets and programs, each with distinct motivations:

1. Public sector agencies who seek to secure "public goods" on behalf of their constituencies
2. Private sector companies who are under regulatory obligation to offset environmental impacts and may do so by purchasing ecosystem service credits
3. Private businesses or organizations who seek to secure ecosystem services for their use values or for other business benefits

4. Philanthropic buyers, such as conservation organizations and charitable individuals, who are motivated by the nonuse values of ecosystem services
5. Consumers of ecocertified products who seek to purchase goods produced in ways consistent with their environmental values

This section briefly discusses the scale of demand from each buyer type (based on estimates from the *Ecosystem Marketplace Matrix* [2006]) and provides some examples of biodiversity PES programs in each category.

Public Sector Agencies

Public and quasi-public agencies are the largest buyers of biodiversity conservation services from farmers, with payments totaling at least $3 billion[1] annually, mostly in the United States, Europe, and China. Public sector buyers include international organizations such as the World Bank and Global Environmental Facility; national governments that enact agri-environmental payment schemes; and local governments, which usually engage in PES to provide watershed protection for public water supplies.

The largest public biodiversity PES programs are the agrienvironment payment programs in the United States and Europe, which compensate farmers for providing a variety of conservation-friendly land-use and management practices. Roughly 20% of the farmland in the European Union is under some form of agrienvironment program to reduce the negative impacts of modern agriculture on the environment, at a cost of about $1.5 billion (although much of this land is managed for other ecosystem services, not specifically for biodiversity conservation). In Switzerland, "ecological compensation areas" using farming systems more compatible with native biodiversity have expanded to include more than 120,000 ha (~ 463 sq. mi.) (Biodiversity Monitoring Switzerland 2006). In the United States, seven programs authorized under the 2002 Farm Bill encourage the provision of fish and wildlife habitat on private lands through payments for habitat protection and restoration, or for the presence of wildlife on farms. In 2005, these payments totaled over $4.5 billion.

Outside the United States and Europe, Mexico's public watershed payment program has now incorporated biodiversity benefits (CONAFOR 2007), whereas Costa Rica's national PES program compensates landowners for the conservation and restoration of forests, which may be on or adjacent to farms. The World Bank's BioCarbon fund is one of the largest biodiversity PES programs from quasi-public international organizations, mobilizing $54 million in its first two years of operation (2004–06). This program aims to sequester

[1]All currency figures are provided in US dollar equivalent.

carbon in forests and agroecosystems while promoting biodiversity conservation and poverty alleviation co-benefits.

Private Parties under Regulatory Obligation

Regulation-driven PES results from laws that limit the aggregate level of environmental damage and require parties who exceed their allotted impact to buy compliance credits from other parties (ten Kate et al. 2004). For example, the United States has operated a wetland mitigation program since the early 1980s in which developers seeking to destroy a wetland must buy wetland offsets conserved or developed elsewhere. Such systems, often referred to as "cap-and-trade" programs, have also been successfully established for sulfur dioxide emissions, farm nutrient pollution, and carbon emissions. However, developing such markets for biodiversity is more complicated because it is difficult to establish equivalency units for biodiversity (Agius 2001).

To date, regulation-driven biodiversity PES have been limited to developed countries, namely, the United States, Australia, and France. In the United States, at least $45 million is spent annually on regulatory offsets for biodiversity, including conservation banking. In addition, wetland mitigation banking and tradable development rights programs in the United States often include biodiversity conservation as one of their objectives. In New South Wales, Australia, a salinity control trading scheme has led an irrigators' association to pay landowners to plant trees that combat rising saline water tables while also helping to restore habitat. Recent legislation in Australia allows private landholders who conserve biodiversity values on their land to sell the resulting "credits" to a common pool. The law also creates obligations for land developers and others to purchase those credits (Brand 2002).

Private Parties for Business Reasons

Private companies may purchase biodiversity conservation services to demonstrate corporate environmental responsibility or to secure use values from biodiversity, such as chemical compounds and genetic resources sought by pharmaceutical companies in "bioprospecting" arrangements. In either case, companies invest in biodiversity conservation because of the "business case"—that is, the expected benefit for immediate or long-term profitability. See Box 21.1 for additional motivations of private businesses for purchasing biodiversity services.

Biodiversity payments from private businesses for business reasons are still nascent markets. Still, at least $20 million in voluntary private biodiversity offsets have been documented, half in developing countries. Biodiversity offsets

Box 21.1. The business case for biodiversity service payments

More and more companies are looking at payment for ecosystem services (PES) as a viable mechanism to manage business risks and pave the way for future opportunities. Some of their motivations for engaging in PES are to:

- Comply with existing environmental regulation and policy
- Influence emerging environmental regulation and policy
- Maintain "social license to operate"
- Enhance reputation
- Secure ecosystem services that are critical to ensure the high quality or efficiency of a product that a business is selling
- Embrace strategic opportunities in the new PES markets and business
- Pursue new business opportunities related to core business

Source: Mulder, et al. 2005

are conservation activities intended to compensate for the residual, unavoidable harm to biodiversity caused by development projects (Ecosystem Marketplace 2006).

Philanthropic Buyers

Philanthropic buyers include NGOs, research institutes, foundations, and private individuals who are motivated by protecting nonuse values of biodiversity. A major outlet for philanthropic funds is the purchase of conservation easements, which amounts to about $6 billion annually in the United States, with a large proportion in farm and ranchlands. However, only a small portion of this figure represents the amount paid specifically to conserve biodiversity on agricultural lands.

Philanthropic buyers—especially large conservation NGOs such as The Nature Conservancy—are expected to increase the use of conservation payments and conservation easements as the establishment of new nature reserves becomes more contentious in many regions. Where farmers control land in biodiverse areas, they would be logical beneficiaries of such payments. On the other hand, within the conservation community there remains a heated debate about whether conservation funds should be expended in agricultural settings where native biodiversity may be significantly degraded, or whether investment should focus on lands in a more pristine natural condition. The outcome of this debate will strongly influence the scale of the philanthropic payments to farmers for biodiversity conservation.

Consumers of Ecocertified Products

Ecocertified products constitute a large and rapidly growing market: as of 2006, this market was valued at $26 billion worldwide with a growth rate of 30% annually (Ecosystem Marketplace 2006). Of course, most of the value of ecocertified farm products is for the products themselves, with a relatively small and unspecific premium paid by the consumer for the ecofriendly production practices. Although consumers purchase ecocertified products for a host of reasons (including health, social justice, and environmental concerns), biodiversity conservation is the ecosystem service most closely associated with consumer preference for certified products.

Markets for ecocertified products are very important drivers of biodiversity conservation in agricultural landscapes, but they function differently from direct payments for services and will not be discussed further in this chapter. See Millard (Chapter 20, this volume) for more on evolving ecoproduct markets.

What Types of Biodiversity Conservation Services Can Farmers Provide?

Farmers and agricultural communities can provide biodiversity conservation through a variety of practices. These range from specific plot-level farming practices such as conservation tillage, no till cropping, and organic agriculture to changes in land-use allocations within farms and across entire landscapes to incorporate extensive grazing systems, agroforestry, extractive reserves, and patches or corridors of natural habitat. Although a large and growing literature explores the conservation implications of many such practices (e.g., Buck et al. 2004; Harvey et al. 2005; Schroth et al. 2004; Harvey, Chapter 8, this volume; and Neely and Hatfield, Chapter 7, this volume), this section briefly identifies some practices that may be especially conducive to biodiversity PES. These are divided into three categories: (1) restricting agricultural use, (2) promoting biodiversity-conserving agricultural management, and (3) adopting practices to provide other ecosystem services that incidentally or intentionally also help to conserve biodiversity. See Table 21.1 for a summary of key practices.

Restricting Agricultural Use

Farmers can help conserve biodiversity by maintaining or restoring natural and seminatural habitat patches in the landscape instead of using these areas for agricultural production. This practice is especially important in landscapes where extensive agricultural systems retain a significant amount of biodiversity, but where there are economic pressures to intensify these systems, as is the case

Table 21.1. Farm and landscape management practices that can provide biodiversity conservation services[a]

Ecosystem service provided	*Farm production practices*	*Agricultural landscape management practices*
Restrict agricultural production		
Protect native ecosystems	Protect or restore patches and corridors of natural habitat on the farm, such as wetlands, forests, and prairies	Maintain corridors of natural land among farms and between farms and natural areas; establish protected areas on lands of high conservation value or lands less suitable for agriculture
Conservation-friendly agricultural management		
Improve landscape connectivity for mobile species	Retain or install hedgerows, windbreaks, and live fences; remove impenetrable barriers	Create networks of natural and seminatural areas in and around farms
Protect habitat for native aquatic species	Manage crop and livestock wastes; reduce agrichemical usage	Maintain or establish natural vegetation along stream banks
Protect habitat for native terrestrial species	Protect breeding areas, pure water sources, and wild food sources in and around farm plots; adjust the timing of cultivation activities to avoid interference with species' life cycles; increase the diversity of crop varieties and species on the farm	Create networks of natural and seminatural areas in and around farms; establish community forests, extractive reserves, or other low-intensity multiuse areas
Management for other ecosystem services with cobenefits for biodiversity conservation		
Carbon emission reduction (biodiversity cobeneefit: avoid deforestation)	Reduce the use of burning to clear forests or manage crop residues	Reduce unsustainable slash-and-burn practices
Carbon sequestration in perennial plants (biodiversity cobenefit: improve habitat quality on farms)	Increase the use of perennial crops and tree crops on farms; manage forested areas of farms for conservation and production values	Reforest degraded lands or lands less suitable for agriculture; increase use of agroforestry practices; lengthen fallow periods
Maintain water quality (biodiversity cobenefit: conserve aquatic biodiversity)	Reduce agrochemicals, filter agricultural runoff, soil conservation and runoff management; perennial soil cover	Maintain perennial vegetative filters; use ecofriendly road, path, and settlement construction methods

Table 21.1. (*Continued*)

Ecosystem service provided	*Farm production practices*	*Agricultural landscape management practices*
Salinization reduction (biodiversity cobenefit: reforestation)	Plant appropriate salinity-reducing tree species on farms	Reforest strategic areas of the landscape
Flood control (biodiversity cobenefit: conserve wetlands)	Protect or restore wetlands on farms; retain tree cover; manage soils and ground cover to encourage infiltration of rainwater	Protect or restore wetlands and other riparian areas
Landscape beauty (biodiversity cobenefit: improved habitat)	Establish live fences; plant attractive native species; revegetate land to hide buildings and farm infrastructure	Revegetation in visible areas of the landscape
Recreational access to wild animals for hunting, fishing, and viewing (biodiversity cobenefit: conservation of critical native species)	Restore fishing streams and ponds; maintain salt licks or vegetation attracting wild species	Protect core habitat areas; establish rules for sustainable harvest in natural areas and on communal lands
Pollinator protection (biodiversity co-benefit: conservation of pollinators, the species that feed on them,and their habitats)	Maintain pollinator habitat areas on farm; reduce the use of pesticides	Maintain patches of natural pollinator habitat in the landscape

[a]These practices are divided into the three main headings discussed in the text

in much of Europe (Reidsma et al. 2006) and Central America (Harvey et al. 2005). Restricting agricultural use is also an important strategy at the agricultural frontier in order to protect the world's last large-scale intact forest ecosystems.

Governments and environmental NGOs can pay farmers to restrict agricultural activities by purchasing permanent conservation easements to keep land out of production or by making recurring conservation payments, including conservation concessions (Rice 2003). Because restricting agricultural use by definition involves a tradeoff between agricultural production and conservation, payment to farmers needs to compensate them for the opportunity cost of production, making this a relatively expensive approach to biodiversity PES.

Biodiversity-Conserving Agricultural Management

A lower-cost approach to securing conservation benefits is to pay farmers to manage their land so as to achieve some biodiversity conservation benefits while still allowing for agricultural production. This can be accomplished by switching to more environmentally benign agricultural land uses (such as agroforestry or extensive grazing systems instead of intensive cropping systems), or by adopting agricultural best practices within a given agricultural land use. Where land degradation currently limits both the productivity and the conservation value of agricultural areas, PES can be used to encourage and subsidize restoration, potentially a win–win situation.

A wide variety of payment schemes promote biodiversity-conserving agricultural practices. Payments from both public and nonprofit buyers seek to facilitate wildlife movement across agricultural landscapes by encouraging farmers to establish riparian buffers, create or retain hedgerows and live fences, and establish agroforestry systems.

Management for Other Ecosystem Services, with Biodiversity Conservation Cobenefits

Payments to farmers for other ecosystem services, such as watershed services, carbon sequestration or storage, landscape beauty, and salinity control often provide biodiversity cobenefits, either deliberate or incidental. For example, payments to farmers to sequester carbon by adopting agroforestry practices or reforesting parts of their farm can also enhance the farm's biodiversity value. Payments by hunters or ecotourism companies for access to hunt or view particular species require farmers to maintain viable populations of such species on their farms.

Potential Benefits and Risks to Farmers

Depending on their context, objectives, and design, biodiversity PES programs can involve both benefits and risks to farmers and farming communities.

Potential Benefits

Biodiversity payments can benefit farmers by providing additional sources of income, subsidizing transitions to sustainable production, diversifying farm and forestry portfolios, and providing nonincome livelihood and community social benefits. Direct payments can improve the reliability of income streams, given that other farm income is typically quite variable from season to season or year to year. However, most ecosystem service payments provide only supplemental income to farmers (Scherr et al. 2003); thus they should be considered as providing a catalyst or enabling mechanism to transition to ecoagriculture practices, not as a replacement for farm product–based income. Even a modest level of payment, reliably paid over many years, can provide the increment that makes sustainable resource management viable.

Protecting or restoring ecosystem services for outside buyers can also provide nonincome benefits to farmers, such as improved local water supplies and new forest-based resources, including fuel, medicines, and wild game. Restoration of native vegetation may also help to reduce landslides and control soil erosion and sedimentation. In addition, payments may spur the formalization of resource tenure and the clarification of property rights over ecosystem services. Finally, payments made to community and farmer organizations can be used as a social investment and to build local capacity for enterprise management and development, marketing, and social organization. The PES program in Antioquia, Colombia, provides an example of where PES has provided farmers with a range of nonmonetary benefits in addition to cash payments (see Box 21.2).

Box 21.2. Integrating biodiversity in carbon payments in Antioquia, Colombia

In the Antioquia region of northwest Colombia, intensive land use and violent conflict have caused a deterioration in living conditions among local people. As a result, the nearby watershed has been seriously degraded and much of its hydrological properties and biodiversity lost. In addition, prices for wood processing and demand for local wood products, such as banana boxes and handicrafts, have declined.

A project financed by the International Tropical Timber Organization, Swiss Federal Laboratories for Material Testing and Research (EMPA), and Corporación Autónoma Regional de las Cuencas de los Rios Negro y Nare (CORNARE), seeks to restore the critical biodiversity of this region by paying small-scale farmers for the carbon sequestered by better land management practices in the watershed. By its fortieth year, the project is expected to have offset 750,000 tons of carbon and has already catalyzed a shift to the sustainable extraction of timber and nontimber forest products, connected biological corridors, and trained communities in forest extension, business ventures, and forest ecology. Payments are managed by the San Nicolas Forests Corporation, a coalition of governmental organizations, and benefit 10,000 families in the area. The shift to sustainable agricultural and forest management practices has already restored critical habitat for biodiversity, controlled erosion, and protected the ecological services of the watershed.

Source: Robledo 2003.

Potential Synergies between Agricultural Production and Biodiversity Conservation

A critical question—especially in regions where food security is a concern—is whether managing agricultural systems for ecosystem service provision changes the level of agricultural output and its distribution across space and time. In the long term, and often even in the short term, managing for ecosystem services can increase the production potential of farms by maintaining and enhancing the soil, nutrients, water, and other resources upon which agriculture depends. In other situations, though, managing for ecosystem services requires taking land out of production or reducing the intensity of production, creating a short-term decrease in farm output.

An analysis recently conducted by the PES project Rewarding the Upland Poor for Ecosystem Services (RUPES) identified five types of ecosystem service payments that were especially likely to promote production–environment synergies: maintaining water quality, protecting conservation areas, maintaining biological corridors, restoring tree cover for carbon sequestration, and maintaining landscape beauty for ecotourism (van Noordwijk 2005). This analysis was conducted for landscape settings where farmers tend to be undercapitalized, lack access to external farm inputs, and are often labor constrained. In settings where the opposite is true, such as in many parts of the developed world, the synergy–tradeoff equation will be different. Synergies also tend to be more common in ecologically degraded landscapes, where biodiversity-conserving activities often help restore soil fertility and natural hydrological cycling, thereby benefiting farm productivity or sustainability (Scherr et al. 2007).

Potential Risks

One risk, already discussed, is that biodiversity PES could reduce food production on farms. A related risk is that biodiversity PES programs could cause farmers or rural communities to lose the use or access rights to natural habitats that previously provided them with subsistence or commercial products. Where local people have secure and recognized property rights over natural resources, PES should benefit local people provided that the transaction is truly voluntary and that all users of the resource are represented. But in many landscapes with large remaining areas of natural habitat, local people's rights are customary or poorly defined, so that buyers of biodiversity services may exclude them (intentionally or not) from receiving fair payment. This may lead to a situation where PES becomes a tool for local or external elites to capture the monetary value of important community assets, and even to exclude local people from use of these assets (Smith and Scherr 2003).

Many biodiversity service buyers understandably seek to negotiate long-term deals in order to achieve long-term protection for species or ecosystems.

However, given the dynamic nature of agricultural and land markets, as well as population pressure in many rural areas, farmers correctly perceive limits on future decisions as a significant risk. One solution is to negotiate long-term PES frameworks within which shorter-term contracts can be enacted. For example, Costa Rica's PES program contracts with landowners for five years but will renew their contracts indefinitely.

Barriers to Effective Widespread Use of Biodiversity Payments

Although examples of biodiversity payments have emerged around the world, they have not yet evolved on a scale that makes a globally significant impact on biodiversity conservation and rural livelihoods. Several technical, economic, political, and cultural factors pose barriers to the use of PES at a larger scale. For PES to be globally significant, innovative, and systematic, solutions must be developed to address such barriers.

Technical Constraints

Buyers of biodiversity conservation services will be willing to pay farmers only if they can be reasonably certain that the services are actually being provided. Yet, at present, there is insufficient knowledge about how to measure biodiversity, and lack of consensus on how to develop a currency for valuing biodiversity for PES transactions. Such technical limitations constrain the development of market values for biodiversity.

The challenge in measuring and valuing biodiversity lies in the complex nature of biodiversity itself. Whether examined on the genetic, species, or habitat level, biodiversity is an inherently complex unit to define and quantify. Efforts to quantify biodiversity benefits have typically taken two different approaches. First, when specific conservation benefits can be measured directly and immediately in the field, payments can be made for delivery of these services (for example, where farmers are paid a set amount for every breeding pair of an endangered species found on their land). Second, when the effects of specific land uses and management practices on biodiversity conservation are well understood, the adoption of those uses and practices may be accepted as a proxy and a trigger for payments (for example, farmers may be paid by linear meter for revegetating stream banks for water quality and freshwater biodiversity because this relationship has been widely demonstrated).

To date, a variety of systems have been proposed or implemented for quantifying biodiversity services for the purpose of PES transactions. Table 21.2 summarizes some examples of such approaches.

Table 21.2. Selected examples of approaches to measure biodiversity for payment for ecosystem services (PES)

Metric	*Source or application*	*Description*
Land area	Wetland banking and conservation banking in the United States	Uses a simple hectare by hectare approach (1 ha lost = 1 ha gained) to implement the goals of no net loss of wetlands and no net loss of critical species habitat
Habitat hectares	PES programs in Australia	This approach has identified "benchmark" attributes for Australia's ecosystems and then measures changes in biodiversity against those benchmark attributes to quantify biodiversity lost during development and gained in compensation activities
Environmental benefits index	US Environmental Quality Incentives Program (EQIP)	Uses a standardized index
Land use–based point system	Regional Integrated Silvopastoral Ecosystem Management Project (RISEMP) in Costa Rica, Nicaragua, and Colombia, funded by the World Bank (see Pagiola et al. 2004)	Uses a point system to quantify the biodiversity conservation and carbon sequestration benefits of various crop, pasture, agroforest, and forest land uses; Farmers are paid according to the net change in points attributable to land-use changes on their farm
Landscape equivalency analysis	Bruggeman et al. 2005	Quantitative metric that considers the relative position of habitat patches on the landscape and their role in metapopulation dynamics; Proposed as a way to improve the economic efficiency and conservation benefits of payment schemes that are based solely on the area of habitat created or destroyed
IBSA (biodiversity index for environmental service payments)		Calculates a composite biodiversity value score based on five variables, including conservation importance and species richness relative to the surrounding landscape
Various tools	Business and Biodiversity Offset Program (BBOP)	BBOP, a voluntary collaboration of companies, nongovernmental organizations, and governments, is developing a robust set of tools for quantifying biodiversity in the context of biodiversity offsets (www.forest-trends.org/biodiversityoffsetprogram)

Implementation Constraints

At the implementation level, perhaps the greatest barrier to biodiversity PES is high transactions costs, which can dramatically reduce the proportion of the buyer's price that the seller actually receives. Transactions costs include the cost of providing information about biodiversity benefits to potential buyers; costs of identifying, negotiating, and building capacity of project partners; and costs of ensuring that parties fulfill their obligations (including auditing, certification, and legal costs). Transactions costs tend to increase as the number of individual sellers increases. Thus, especially in the developing world, small farmers and farming communities have been at a serious disadvantage in terms of participating in biodiversity PES, resulting in fewer benefits for them as well as less effective PES programs. This challenge points to the need for intermediary institutions that can coordinate efforts of many small farmers.

A second challenge for biodiversity PES is that biodiversity conservation usually requires efforts that span multiple properties, up to the scale of landscapes or entire ecoregions. Thus a farmer's ability to provide services may depend to a significant degree on how nearby lands are being used and managed. To date, there are few examples of where land is managed and institutionally supported at a landscape scale, but one promising example is the Australian-based pilot processes to establish landscape corridors in the Southern Desert Uplands. This project will pilot the use of payments distributed via an auction format that accounts for the interdependence of bids from neighboring properties. Thus the value of alternative vegetation corridors will depend on strategic cooperation between landholders.

A final implementation constraint is the general lack of accessible information about potential buyers and sellers, business models, prices, and "rules of the game." Typically, information is more available to ecosystem service buyers such as governments and corporations than to farmers, resulting in information asymmetries that can reduce sellers' bargaining power with buyers. At the policy level, farmers in the developing world tend to be poorly represented in establishing the basic "rules of the game," including protections for land and resource rights. Most existing PES programs do not reflect the flexible, locally adapted arrangements required for sustainable and equitable participation by low-income farmers and farming communities.

Cultural Constraints

The PES concept has also encountered cultural resistance from some environmental organizations, indigenous rights groups, and others concerned about the use of market instruments for managing ecosystem services (e.g., Lovera 2005). A common objection is that biodiversity has important noneconomic values and that societies—including landowners and producers—have a basic obligation to conserve biodiversity. The concept of selling species or habitats may also be culturally unacceptable to those who do not accept the "ownership" of nature. For PES to be widely acceptable, it will be critical to frame the approach not as payment for biodiversity itself but as payment for stewardship services that compensate farmers, on behalf of all beneficiaries, for the benefits they provide and the costs they incur in providing those benefits.

Scaling Up Biodiversity PES to Support Ecoagriculture Landscapes

If biodiversity PES is to have a significant impact in agricultural landscapes, the barriers described above must be addressed. In particular, action is needed to mobilize and organize buyers, establish supportive policy frameworks and institutions, engage and support community and farmer organizations, and reduce transactions costs.

Mobilize and Organize Buyers for Biodiversity Services

Markets for ecosystem services cannot exist unless beneficiaries of these services are willing to pay for their provision. Historically, beneficiaries have been hesitant to pay for ecosystem services previously considered free, especially when service providers are unable to exclude beneficiaries from using the services, thus creating a strong incentive to "free ride." Three approaches are likely to be most effective in motivating the private sector to pay for biodiversity conservation services.

First, new regulations can be enacted requiring private actors to minimize or offset their impacts on biodiversity by purchasing credits or engaging in conservation or restoration activities. Second, pressure from a variety of sources can encourage the private sector to take responsibility for conserving biodiversity, again by paying for on-site or off-site conservation and restoration efforts. For example, social advertising, activist movements targeting corporate behavior, and pressure from investors are beginning to influence some firms to avoid investments and activities that harm biodiversity, and to offset impacts that are unavoidable. Pressure from consumers, in the form of purchasing preferences or boycotts, is also motivating corporate social responsibility as well as the proliferation of ecolabeled products that may be produced in a more biodiversity-friendly manner. (See examples in Millard, Chapter 20, this volume.)

A third approach is to integrate biodiversity conservation goals more fully into PES programs for other ecosystem services. For example, contracts for watershed PES aimed at providing hydrological services to downstream users can be designed so as also to protect biodiversity and restore natural habitat (Albrecht and Kandji 2003). Agricultural practices that sequester carbon or prevent greenhouse gas emission—such as planting trees, increasing soil organic matter, adopting agroforestry systems, and refraining from burning forests and crop residues—often have cobenefits for biodiversity, and these benefits can explicitly be encouraged through program design (Watson et al. 2000; Lal 2003; Swingland 2002). However, currently there are few opportunities for farmers to receive payments for carbon sequestration under the Clean Development Mechanism (CDM) of the Kyoto Protocol, the program by which carbon emitters in industrialized countries can offset their emissions by investing in projects in developing countries. At present, forest restoration and regeneration projects are the only land-use changes eligible for generating carbon credits under the CDM, and even these have proven difficult to implement. Therefore, a key policy priority should be to expand the range of carbon-sequestering land-use and land-management practices (with biodiversity cobenefits) eligible for carbon trading in future international greenhouse gas treaties.

Establish a Supportive Policy Framework and Institutions for Biodiversity Markets

Because ecosystem service markets are relatively new, the rules and institutions that will guide and support these markets are still being developed. Both the agricultural and the biodiversity conservation communities need to act quickly and strategically to shape these frameworks to ensure that as ecosystem services markets develop, they are effective, equitable, and operational and are used sensibly to complement other conservation approaches.

Policymakers and public agencies play a vital role in creating the legislative and regulatory frameworks necessary for market tools to operate effectively. This role includes establishing market rules (in the case of public payments or cap-and-trade markets), systems of rights and tenure over ecosystem services and the land that produces them, and mechanisms to enforce contracts and settle ownership disputes. Lessons from other types of market mechanisms can be usefully applied to the design and operation of ecosystem service markets (Bayon 2004).

Farmers' organizations, indigenous groups, rural communities, and their representatives also have an important role to play in shaping future ecosystem service markets. Because new rules may fundamentally change the distribution of rights and responsibilities for essential ecosystem services, it is critical to ensure that rules support the public interest and favor social equity. In addition, international experience suggests that engaging local communities and local governments more fully in PES design and implementation will significantly improve the equity and efficiency of PES programs (e.g., Smith and Scherr 2003).

A challenging issue to be resolved in the design of PES programs relates to the targeting of payments, especially to low-income, rural land stewards. On the one hand, PES could function as a powerful tool for rural development and for advancing several of the Millennium Development Goals by rewarding rural communities that have historically provided good stewardship for ecosystem services of national or international value. Similarly, payments could be targeted to encourage the adoption of ecoagriculture practices or to make them more economically viable. On the other hand, cost-effectiveness considerations may tend to favor payments to land stewards who have historically been bad actors (to encourage improvements in their practices), or for land under a high degree of threat of being converted to less environmentally benign uses. Thus rural communities that have been practicing conservation-friendly land management on a sustained basis may be excluded from receiving payments unless they threaten to switch to more environmentally damaging practices. Design of PES programs must therefore balance the goals of economic efficiency and fairness, using payments to reinforce landowners' stewardship ethic while avoiding perverse incentives that can lead to "environmental blackmail."

Reduce Transactions Costs

As discussed earlier, transactions costs greatly affect the degree to which demand for ecosystem services translates into actual payments at the level of farms and communities. In particular, proactive efforts are needed to reduce the costs associated with obtaining market information, brokering and managing deals between buyers and sellers, and monitoring the provision of ecosystem services.

The availability of market information is growing as the number of PES programs and markets increases and growing numbers of private companies, investors, and traders become involved in PES. Information is critical to reduce uncertainties and risks for market actors due to unfamiliarity with PES, rapid changes in "rules of the game," and difficulties in connecting buyers and sellers. One recent effort to provide such information is The Katoomba Group's Ecosystem Marketplace (http://www.ecosystemmarketplace.com), which offers guidance and case studies on the design and implementation of PES programs while providing a forum for PES innovators around the world to exchange information.

Transactions costs can also be reduced by creating new institutions and financial instruments that package ecosystem services for transaction in the marketplace—for example, by "bundling" biodiversity services provided by large numbers of local producers or by creating investment vehicles that have a diverse portfolio of projects in order to manage risks (Scherr et al. 2004). To convince beneficiaries of biodiversity services to pay for them, better methods of measuring and assessing biodiversity in working landscapes must be developed, along with the institutional capacity to put these methods into practice (see, e.g., Dushku et al., Chapter 14, this volume). Overall, looking to the future, ecosystem service markets will need to be supported by a wide network of knowledge services, exchanges, financial instruments, and advisers, as is now found in other commodity markets.

Conclusions

Recent experience has shown that PES can provide farmers with the additional income needed to protect valuable habitat, restore degraded lands and catalyze a shift to more sustainable land-management practices. Although PES currently exists on a relatively small scale, a growing awareness of the potential conservation and livelihood benefits offered by such schemes is generating new innovations to reduce economic, technical, and political barriers. By learning from past experience and working to address the challenges identified in this paper, practitioners can scale up biodiversity payments to farmers in ecoagriculture systems to yield significant and measurable benefits for biodiversity conservation and rural livelihoods worldwide.

References

Agius, J. 2001. Biodiversity credits: creating missing markets for biodiversity. *Environmental and Planning Law Journal* 18(5):481–504.

Albrecht, A. and S. Kandji. 2003. Carbon sequestration in tropical agroforestry ecosystems. *Agriculture, Ecosystems, and Environment* 99:15–17.

Bayon, R. 2004. *Making Environmental Markets Work: Lessons Learned from Early Experience with Sulfur, Carbon, Wetlands, and Other Related Markets*. Forest Trends, Washington, DC.

Biodiversity Monitoring Switzerland. 2006. *Ecological Compensation Areas (M4)*. http://www.biodiversitymonitoring.ch/english/indikatoren/m4.php.

Brand, D. 2002. Investing in the environmental services of Australian forests. In: *Market Based Mechanisms for Conservation and Development*, ed. S. Pagiola, J. Bishop, and N. Landell-Mills. Earthscan Publications, London.

Bruggeman, D.J., M.L. Jones, F. Lupi, and K.T. Scribner. 2005. Landscape equivalency analysis: methodology for estimating spatially explicit biodiversity credits. *Environmental Management* 36(4):518–534.

Buck, L.E., T.A. Gavin, D.R. Lee, N.T. Uphoff, D.C. Behr, L.E. Drinkwater, W.D. Hively, and F.R. Werner. 2004. Ecoagriculture: a review and assessment of its scientific foundations. Ecoagriculture Partners Discussion Paper No 1. Ecoagriculture Partners and Cornell University, Washington DC. www.ecoagriculturepartners.org/resources/publications.php.

Carreón, S.G. 2007. Commision National Forestal (CONAFOR) Práctica y perspectivas de la compensación ambiental en México: experiencias de la CONAFOR. Presentation given at the Conference for Environmental Compensation for Biodiversity and Development: Towards a Sufficient Compensation. Instituto Nacional de Ecología, 15–16 February 2007, Mexico City, Mexico.

Chape, S., S. Blyth, L. Fish, P. Fox, and M. Spalding (comp.). 2003. 2003 United Nations List of Protected Areas. World Conservation Union (IUCN), Gland, Switzerland, and Cambridge, UK, and United Nations Environment Programme–World Conservation Monitoring Centre (UNEP-WCMC), Cambridge, UK.

CONAFOR. 2007. Presentation: *Environmental Compensation for Biodiversity and Development: Towards a Sufficient Compensation.* Instituto Nacional de Ecologia, Mexico City, Mexico, February 15–16. http://www.forest-trends.org/biodiversityoffset program/Presentations/Mexico%202007/3%10-%20CONAFOR%20Garc%EDa_r .pdf.

Ecosystem Marketplace. 2006. *Banking on Conservation: Species and Wetland Mitigation Banking.* Forest Trends, Washington, DC.

Ecosystem Marketplace. 2006. *Ecosystem Market Matrix.* Vers. 19. Forest Trends, Washington, DC.

Harvey, C., F. Alpizar, M. Chacón, and R. Madrigal. 2005. *Assessing Linkages between Agriculture and Biodiversity in Central America: Historical Overview and Future Perspectives.* The Nature Conservancy (TNC), San José, Costa Rica.

Jenkins, M., S.J. Scherr, and M. Inbar. 2003. Markets for biodiversity services: potential roles and challenges. *Environment* 46(6):32–42.

Lal, R. 2003. *Soil Carbon Sequestration by Agricultural and Forestry Land Uses to Mitigate Climate Change.* Testimony to the Committee on Environment and Public Works, United States Senate.

Lovera, S. 2005. Guest editorial: environmental markets impoverish the poor. The Ecosystem Marketplace. http://ecosystemmarketplace.com.

Mulder, I.K., T. ten Kate, S.J. Scherr. 2005. Private sector demand in markets for ecosystem services: current status of involvement, motivations to become involved, and barriers and opportunities to upscale involvement. Report to UNDP/GEF on Institutionalizing Payments for Ecosystem Services. Forest Trends and Ecoagriculture Partners, Washington, DC.

Pagiola, S., P. Agostini, J. Gobbi, C. de Haan, M. Ibrahim, E. Murgueitio, E. Ramirez, M. Rosales, and J.P. Ruiz. 2004 *Paying for Biodiversity Conservation Services in Agricultural Landscapes.* World Bank Environment Department Paper no. 96. World Bank, Washington, DC.

Reidsma, P., T. Tekelenburg, M. Van den Berg, and R. Alkemade. 2006. Impacts of land-use change on biodiversity: an assessment of agricultural biodiversity in the European Union. *Agriculture, Ecosystems, and Environment* 114:86–102.

Rice, R. 2003. Conservation concessions: concept description. Paper presented at the World Parks Congress, Durban, South Africa, September 2003. Conservation International, Washington, DC.

Robledo, C. 2003. Alternative financing model for sustainable management in the San Nicolas forest. Presentation to Beyond Carbon: Emerging Markets for Ecosystem Services conference, 29–30 October 2003. Ruschlikon, Switzerland.

Scherr, S.J., J.C. Milder, L. Lipper, and M. Zurik. 2007. *Payments for Ecosystem Services: Potential Contributions to Smallholder Agriculture in Developing Countries*. FAO and Ecoagriculture Partners, Rome and Washington, DC. Draft, October.

Scherr, S.J., A. White, and D. Kaimowitz. 2004. *A New Agenda for Forest Conservation and Poverty Reduction: Making Markets Work for Low-Income Communities*. Forest Trends, Center for International Forestry Research (CIFOR) and World Conservation Union (IUCN), Washington, DC.

Scherr, S.J., A. White, and A. Khare. 2004. For Services Rendered: The current status and future potential of markets for the ecosystem services provided by tropical forests. Forest Friends, Washington, DC.

Schroth, G., G.A.B. da Fonseca, C.A. Harvey, C. Gascon, H.L. Vasconcelos, and A.-M.N. Izac (eds.). 2004. *Agroforestry and Biodiversity Conservation in Tropical Landscapes*. Island Press, Washington, DC.

Smith, J. and S.J. Scherr. 2003. Capturing the value of forest carbon for local livelihoods. *World Development* 31(12):2143–2160.

Swingland, I. (ed.). 2002. *Capturing Carbon and Conserving Biodiversity: The Market Approach*. Earthscan, Sterling, VA.

Ten Kate, K., J. Bishop, and R. Bayon. 2004. Biodiversity offsets: views, experience, and the business case. World Conservation Union (IUCN), Gland, Switzerland, and Cambridge, UK, and Insight Investment, London.

Van Noordwijk, M. 2005. RUPES typology of environmental services worthy of reward. World Agroforestry Center, Bogor, Indonesia.

Watson, R., I. Noble, B. Bolin, N.H. Ravindranath, D.J. Verardo, and D.J. Dokken. 2000. *Land Use, Land-Use Change and Forestry*. A Special Report by the Intergovernmental Panel on Climate Change (IPCC), Geneva, Switzerland.

Wood, S., K. Sebastian, and S.J. Scherr. 2000. *Pilot Analysis of Global Ecosystems: Agroecosystems*. Report prepared for the Millennium Assessment of the State of the World's Ecosystems. International Food Policy Research Institute and the World Resources Institute, Washington, DC.

Chapter 22

Policy Implications and Knowledge Gaps

Jeffrey A. McNeely and Sara J. Scherr

Introduction: The State of Ecoagriculture Knowledge and Practice

This book has summarized work devoted to the pursuit of both agricultural development and biodiversity conservation objectives at a landscape scale. Previous experience was poorly documented, and the science is still relatively immature and poorly synthesized across disciplines. Removing major barriers to the widespread development of ecoagriculture landscapes requires answering the following questions:

- How can agricultural production systems contribute to conserving biodiversity while maintaining or increasing productivity? (Part I)
- How can agricultural and natural areas be jointly managed to produce adequate ecosystem services, including wildlife habitat, at a landscape scale? (Part II)
- How can communities, governments, and other stakeholders mobilize and develop the institutions, markets, and policies needed for ecoagriculture landscapes? (Part III)

The preceding chapters have answered these three questions on the basis of the best available current knowledge and highlighted remaining major gaps in knowledge. These gaps provide the basis for future work to spread ecoagriculture more widely around the world.

Agricultural Production in Ecoagriculture Landscapes

Since the 1960s the "improved seed–fertilizer–pesticide–irrigation" paradigm has characterized both industrial agriculture in developed countries and the original Green Revolution in developing countries. This production model involved short-term, plot-level production of a small number of crops, generally in monoculture stands (to increase efficiency in use of external inputs and mechanization and maximize the flow of natural resources to harvestable products). Wild flora and fauna were considered direct competitors for resources or harvested products and were thus eliminated, while water was diverted from wetlands and natural habitats for irrigation. These perspectives have shifted significantly over the past two decades as research has demonstrated the value of agricultural biodiversity in all its forms, including crop and livestock genetic diversity, associated species important for production (pollinators, soil microorganisms, etc.) and wild species who find their home in agricultural landscapes (Gemmill et al., Chapter 9, this volume; Thompson et al., Chapter 3, this volume; Uphoff et al. 2006).

More ecologically benign production systems were retained in many traditional systems that for ecological, cultural, or economic reasons were not effectively incorporated into the industrial model. Such systems sought to build on or interact with natural ecosystems rather than replace them. Different modern approaches have focused on different aspects of ecological synergy, arising from differences in discipline, philosophy, problem focus, or geographic conditions. Agroecology (Altieri 1985), permaculture (Mollison 1990), conservation agriculture (FAO 2001), agroforestry (Garrity 2006), organic agriculture (IFOAM 2000), and sustainable agriculture (Pretty 2005) have focused principally on maintaining the resource base for production through managing nutrient cycles, protecting pollinators and beneficial microorganisms, maintaining healthy soils, and conserving water. They sought to reduce the ecological "footprint" of farmed areas and the damage to wild species from toxic chemicals, soil disturbance, and water pollution. Most focused on farm-scale action, however, rather than coordinating efforts among farmers and others to achieve demonstrable biodiversity benefits at a landscape scale.

To protect wild fauna and flora, ecoagriculture landscapes must provide protection of nesting areas from disturbance, diverse perennial cover for protection from predators, adequate access to clean water throughout the year, territorial access between dispersed population groups to ensure minimum viable populations genetically and demographically, all-season access to food from diverse sources, viable populations of predators and prey, healthy populations of other species with which they are interdependent (such as their pollinators), and biologically active soils. Many of these functions can be supplied by healthy patches and networks of natural habitat, but production areas also play a critical role. To achieve these attributes in production areas, agricultural and conservation innovators are pursuing strategies such as minimizing agricultural pollution of natural habitats, managing conventional cropping systems in ways that enhance habitat quality, and designing farming systems to mimic the structure and function of natural ecosystems. A key challenge for farmers is to do so in ways that also maintain or increase agricultural output, reduce overall production costs, or enhance the market value of their products in order to meet their broader livelihood needs while conserving biodiversity.

Minimizing Agricultural Pollution of Natural Habitats

Reducing agrochemical use and livestock wastes in high-input production systems can greatly benefit wildlife. For example, high-nutrient or toxic runoff into waterways (a problem for both "natural" and synthetic forms of nitrogen) can dramatically reduce aquatic biodiversity. Major advances have been made in methods to reduce and improve the efficiency of fertilizer use through better timing and methods of application.

Agricultural pesticides may also kill nontarget insects and weeds that constitute the food base for insect- and grain-eating species. Integrated pest management systems have effectively used varietal crop mixes, pest monitoring, and management practices to reduce the need for pesticides (Herren et al., Chapter 10, this volume). Cellular and molecular biology have been used to tailor pesticides to affect only specific pests. New ecological and biochemical research techniques are revealing an unexpected sophistication of host–pest relations that could revolutionize agricultural pest control in the future (Wittenberg and Cock 2001)

Pretty's (2005) metareview of farmer experience found gains in both productivity and biodiversity from reduced chemical use in developing countries. For example, the System of Rice Intensification mobilizes biological interactions in plant–soil systems, rather than external inputs, to raise yields significantly while reducing costs (Uphoff et al. 2006). Meanwhile, new, whole-farm planning approaches minimize runoff of agrochemical and livestock waste into aquatic systems by improving storage systems, managing fields to improve infiltration and reduce runoff, and establishing buffer zones to filter pollutants before they enter streams (Coombe 1996).

Managing Production Systems to Enhance Habitat Quality

Farmers and conservationists have modified management of soil, water, fire, and vegetation to transform crop fields into useful habitat for species or to enhance their value as corridors connecting natural habitat areas in the landscape (Clay 2004; McNeely and Scherr 2003). Buck et al. (2006) reviewed 79 studies where investigators quantified biodiversity (usually species richness) associated with 18 specific agricultural practices. The strategy most often correlated with the conservation of wild biodiversity was the maintenance of adjacent hedgerows, windbreaks, or woodlots; 18 studies documented positive correlations with eight taxa. Organic agriculture was correlated with an increase in seven taxa in eight studies (Klein and Sutherland 2003; Kleijn et al. 2006). Shaded tropical crop production (especially coffee and cacao) had higher species richness of three higher taxa according to eight studies (Buck et al., Chapter 2, this volume).

Research has also found that many of these practices provide additional benefits to farmers, such as useful by-products, reduced risk of crop loss during droughts, diversified food and income sources, and reduced vulnerability to environmental risks. For example, following the October 1998 Hurricane Mitch (the worst natural disaster to strike Central America in 200 years), researchers found that farms using agroecological practices suffered 58% less damage in Honduras, 70% less in Nicaragua, and 99% less in Guatemala than those using conventional farming methods (Bunch 2000).

The organic farming industry has only recently begun to develop standards that explicitly address conservation of wild biodiversity. Hole et al. (2005), however, found that a wide range of taxa, including birds, mammals, invertebrates, and cultivated flora, can benefit from organic management through increases in abundance and/or species richness. Management practices such as prohibition or reduced use of chemical pesticides and inorganic fertilizers, protection of noncropped habitats, and preservation of mixed farming are particularly beneficial for farmland wildlife. Though yields from organic systems are still often lower than those in conventional systems, the gap is narrowing, and research is accumulating that shows how agricultural production systems primarily or exclusively dependent on organic inputs can produce superior agronomic and economic results (Uphoff et al. 2006).

Carefully targeted management practices applied to relatively small areas of cropped or noncropped habitats within conventional agriculture may also provide valuable biodiversity benefits (Trewavas 2001). Weibull et al. (2003) found that wild species richness generally increased with landscape heterogeneity on a farm scale, and habitat type had a major effect on species richness for most groups, with most species found in pastures and leys (lands temporarily sown with grass). The level of motivation of the farmer to maintain biodiversity on the farmstead better predicted positive biodiversity outcomes than specific practices.

Approaches are also being developed to integrate both extensive and intensive livestock production systems into ecoagriculture landscapes. Innovations range from modifications in grazing and livestock management that also enhance wild plant and animal populations, to sophisticated rotational grazing and nutrient management systems for intensive dairy and meat production (Neely and Hatfield, Chapter 7, this volume).

Modifying Farming Systems to Mimic Natural Ecosystems

From a wild biodiversity conservation perspective, the ideal agricultural production systems for ecoagriculture landscapes mimic the structure and function of natural ecosystems (Blann 2006; Jackson and Jackson 2002; Lefroy et al. 1999). In humid and subhumid forest ecosystems, farms would mimic forests, with productive tree crops, shade-loving understory crops, and agroforestry mixtures; in grassland ecosystems, production systems would rely more on perennial grains and grasses, and economically useful shrubs and dryland tree species. Annual crops would be cultivated in such systems, but as intercrops or monoculture plots interspersed in mosaics of perennial production and natural habitat areas. Domesticated crop and livestock species diversity would be encouraged at a landscape scale, and intraspecies genetic diversity would be conserved in situ, at least at an ecosystem scale, to ensure system resilience and ecological diversity.

Multistory agroforest systems, tree fallows, and complex home gardens are especially rich in wild biodiversity (Cairns and Garrity 1999; Leakey, Chapter 5, this volume; Schroth and da Mota, Chapter 6, this volume). For example, Siebert (2002) found that canopy height, tree, epiphyte, liana and bird species diversity, vegetation structural complexity, percent ground cover by leaf litter, and soil calcium, nitrate nitrogen, and organic matter levels in topsoils were all significantly greater in shaded than in sun-grown farms, whereas air and soil temperatures, weed diversity, and percent ground cover by weeds were significantly greater in sun farms. Recent research in Central America has identified polyculture combinations and management systems that significantly improve the productivity of coffee, cocoa, banana, timber, and other commercial tree products in these complex systems (e.g., Beer et al. 2000).

New and improved perennial crops can substitute for products now provided by annuals, such as fruits, leafy vegetables, spices, and vegetable oils. Perennial crops can be more resilient and involve less soil and ecosystem disturbance than annual crops, and provide much greater habitat value, especially if grown in mixtures and mosaics (Jackson and Jackson 1999; Leakey, Chapter 5, this volume). Breeding efforts are under way to perennialize annual grains and to mimic ecosystem functions of natural grasslands; in some cases yields are becoming competitive with conventional varieties (DeHaan et al., Chapter 4, this volume). This is a significant research opportunity. Increased demand for livestock products in turn raises demand for animal feed, including for higher-quality pastures, fodder, or inputs for concentrates. Crops for biofuels are poised to become one of the fastest-growing segments of agricultural production, and although short-term investments have favored annual crop sources in the developed world (as a way to absorb subsidy-driven surpluses), grasses, shrubs, and tree sources may be more economic and sustainable options once the technical challenges of processing cellulosic sources are overcome (Ruark et al. 2006).

Major Gaps

The development of agricultural practices and systems that explicitly support wild biodiversity is in its infancy. Buck et al (Chapter 2, this volume) highlight numerous critical knowledge gaps, especially knowledge about the link between diversity and ecosystem function, and the relationships between below-ground and above-ground biodiversity. Methods being used to assess biodiversity impacts are inadequate and generally fail to evaluate the impact on regional or global diversity or to interpret the significance of an individual member of a species found at a particular site. Researchers still find it difficult to link plot-based analysis with landscape-scale impacts (Tomich et al., Chapter 18, this volume).

Even where successful biodiversity and production outcomes are well documented, the underlying biological or ecological mechanisms may be poorly understood. The potential contributions and threats of genetically modified organisms to biodiversity in ecoagriculture landscapes have not been well explored. Little of the existing crop-breeding research in general, much less transgenics, has been considered within an ecoagriculture framework. Rather, most research has focused on addressing problems at the "end-of-pipe" to offset existing problems rather than rethinking the ecological management system, or even considering potential trade-offs of risks and benefits.

Redford and Richter (1999) propose that researchers more systematically assess the impact of different resource management options on specific components of biodiversity (the function, structure, and composition of communities/ecosystems, populations/species, genetic diversity, and option space). Then, where ecoagriculture systems are successful in increasing populations of wild species, new methods for managing them may be needed to minimize conflicts. The major gap is the minuscule level of international and national public investment in research documenting and evaluating existing ecoagriculture production systems, or in pursuing agricultural and conservation research to improve biodiversity-supporting and financially viable production systems.

Biodiversity and Ecosystem Management in Ecoagriculture Landscapes

The ecoagriculture approach encompasses both biodiversity-friendly agricultural production systems and practices, and their management in mosaics with natural areas and other landscape features to meet conservation, livelihood, and production goals. One premise of ecoagriculture is that ecosystem services can come from both production and conservation areas, especially if they are coordinated and managed for that purpose. Improved tools, greater demand for landscape-scale action, and reassessment of long-sustained traditional agro=ecosystems have led to substantial progress over the past two decades in laying out the basic parameters for biodiversity management in ecoagriculture landscapes, if not location-specific guidance. This section highlights some key advances.

New Tools for Landscape Assessment

Despite the importance of agricultural landscapes for biodiversity conservation, only a small fraction of published conservation biology studies has been undertaken in agricultural landscapes (Buck et al., Chapter 2, this volume), so developing a baseline for assessing change is difficult. Most studies of the biodiversity impacts of particular agricultural practices and even the work of biodiversity-oriented groups like the Rainforest Alliance have focused on farm-level indicators. But landscape ecology has begun to provide us with the analytical language and tools to systematically examine the interactions between farmed and unfarmed areas (Forman 1995). The science of "countryside biogeography" has recently begun to work on biodiversity patterns in complex landscape mosaics, which shows how different land-use elements and configurations support different wild species (Daily et al. 2000). Sophisticated landscape modeling and remote sensing tools are becoming available (e.g., Dushku et al., Chapter 14, this volume).

Maintaining Natural Habitats for Terrestrial Species in Agricultural Landscapes

A common goal in ecoagriculture landscapes is to conserve a broad range of terrestrial species native to the area. This includes species that are relatively resilient to habitat fragmentation and agricultural land use, as well as species that are rare or locally or globally threatened, and those that require larger expanses of minimally disturbed habitat. The prospects for achieving this in agricultural landscapes depend on the degree of fragmentation and functional connectivity of natural areas, the habitat quality of those areas, the habitat quality of the productive matrix, and the behavior of farmers.

Efforts to maintain natural habitats in farming areas are longstanding, principally through various types of agricultural set-aside schemes. Based on a meta-analysis of 127 published studies, van Buskirk and Willi (2005) found that land withdrawn from conventional production of crops unequivocally enhances biodiversity in North America and Europe. North American bird species that have suffered population declines reacted most positively to set-aside agricultural land. For many commercial crop monocultures, leaving field margins uncultivated for habitat protection does not reduce total yields if inputs are applied more economically on the rest (Clay 2004).

However, landscape-scale interventions specifically designed to protect habitats for biodiversity (that include but coordinate and go beyond farm- and plot-specific interventions) are much more effective. A recent review of evidence from North America on how much habitat is "enough" in agricultural landscapes (Blann 2006) concluded that habitat needs must be considered within the landscape history and context. Adequate habitat patch size and connectivity must be maintained, but "adequate" must be defined in relation to matrix influence and patch condition (sinks and ecological traps, patch location and configuration, edge effects and boundary zones). Smaller patches of natural habitat may be sufficient if adjacent agricultural patches are ecologically managed. A growing body of research shows that landscape connectivity between large patches of forest can be effectively maintained through retention of tree cover on the farm, such as live fences, windbreaks, and hedges in grazing lands and agricultural fields (Harvey et al. 2004). Thus conservation efforts in agricultural landscapes should focus on protecting (or restoring) large areas of native habitat within the agricultural matrix, retaining elements (such as hedgerows, isolated trees, riparian forests, and other noncropped areas) that enhance landscape connectivity, ensuring heterogeneity at both field and landscape levels, and moderating the intensity of land management (Harvey, Chapter 8, this volume).

Protecting Habitats for Freshwater Aquatic Biodiversity

Protection or establishment of native vegetation buffers along streams, rivers, and riparian systems is critical for biodiversity conservation (Blann 2006). Wetlands should be protected, and the critical function zone of wetlands should be maintained in natural vegetation. The latest guideline in North America is that at least 10% of a watershed and 6% of any subwatershed should be composed of wetlands. Blann (2006) and Molden et al. (Chapter 13, this volume) emphasize the importance of reestablishing hydrological connectivity and natural patterns for aquatic ecosystems. Based on literature review and field experiments, van Noordwijk et al. (Chapter 11, this volume) concluded that watershed functions in agricultural landscapes can be effectively provided through strategic spatial configuration of perennial natural vegetation and planted vegetation, with maintenance of continuous soil cover enhancing infiltration.

Maintaining seasonal flood pulse dynamics in floodplains involves restoring floodplains and protecting them from developments that disconnect rivers through levees and water-level management (Blann 2006). If floodplains must be used for agriculture, ecologists recommend using agroforestry and other approaches compatible with natural cycles rather than monocultures requiring annual plowing and fertilization. Sendzimir and Flachner (Chapter 12, this volume) present an example of river floodplain polyculture in the Tisza River Basin in Hungary that historically exploited flooding as an engine of biodiversity. Natural floodplains, unconstrained by hydroengineering infrastructure, sustained a diversity of habitats and the elevational structure in the landscape. They further maintained hydraulic connections that sustain nursery and migratory functions, stored water during times of drought, and distributed and mixed fallen fruit in novel combinations that stimulated agrobiodiversity and the cultivation of hundreds of varieties of fruits and nuts, as well as fisheries.

Enhancing Agriculture–Natural Habitat Interactions in Landscape Mosaics

Biologically diverse agricultural systems and landscapes can contribute to control of pests and diseases, provide new economic species, and buffer environmental changes and challenges (Jackson et al. 2005; Thompson et al., Chapter 3, this volume). Pollination illustrates potential benefits of natural habitats to production agriculture (Gemmill et al., Chapter 9, this volume). For example, Ricketts (2004) found in Costa Rica that bee species richness, overall visitation rate, and pollen deposition rate were all significantly higher in coffee production sites within approximately 100 m of forest fragments than in sites farther away. Kremen et al. (2002) found similar results for pollinators of watermelon fields near and far from natural woodlands in California.

Major Gaps

The past two decades have revolutionized the potential for landscape-scale assessment and scientific understanding of the ecological functioning of diverse types of agricultural landscapes. A framework for considering key management guidelines and broad parameters is now in place, but empirical or even ecological modeling evidence needed for managing ecoagriculture landscapes (e.g., size and shape of natural areas required to sustain ecological functions, impacts on agricultural productivity of natural vegetation and species) is lacking. Agreed methods do not yet exist for integrated monitoring of livelihood, biodiversity, and agricultural outcomes at a landscape scale, although this challenge is being taken up (Buck et al. 2006). Rigorous understanding of the potential benefits and costs for agriculture of associated wild flora and fauna, and key ecosystem services, is lacking (Jackson et al. 2005).

Molden et al. (Chapter 13, this volume) highlight numerous practices to manage irrigation water in ways that also support biodiversity. But these are not widely implemented because they are not part of the institutions, incentive structures, and education related to irrigation. Little of the new science has been shared with farmers or even with agronomists and other specialized agricultural scientists and technicians. The science is often missing that informs real-life innovations which local people can make to modify ecological impacts of management activities. Technical assistance services for farmers rarely address landscape management issues. Lack of rigorous data and analysis about ecoagriculture impacts and potentials is a key constraint to increased investment in and policy support for ecoagriculture. The complexity of ecoagriculture landscapes and management, multiple objectives, and lack of information on interactions has made it difficult for project or community managers to document outcomes effectively or to compare results across sites.

Institutional Foundations for Ecoagriculture

Achieving ecoagriculture systems requires investment in all key elements—farm production, nature conservation, and associated institutions for collective landscape management. Ecoagriculture landscape innovators often identify their major constraint to be institutional barriers rather than technical or even financial ones (Bumacas et al., Chapter 16, this volume). Key institutional factors include community-level organizations for ecoagriculture action, landscape-scale planning, market incentives to develop financially sustainable ecoagriculture, supportive policies at various levels, and mechanisms to achieve equitable outcomes in ecoagriculture landscapes.

Organization of Communities for Ecoagriculture

A core feature of ecoagriculture landscapes is the role of resident local farming or pastoral communities as key stewards, decision makers, and managers of biodiversity and ecosystem services. Public agencies may operate forests and protected areas, but their viability and sustainability depend on the matrix of private land uses in the landscape.

Farmers and their communities often have strong economic and social rationales for supporting biodiversity conservation: to reduce production costs, raise or stabilize yields; improve product quality; protect their right to farm/herd/harvest wild products in and around protected areas; comply cost-effectively with environmental regulations; conserve biodiversity and ecosystem services critical to their own livelihoods; access product markets that require biodiversity-friendly production systems; earn payments for ecosystem services; or conserve species and landscapes of special cultural, spiritual, or aesthetic significance to them. Economic and social incentives can motivate collective action of local communities. Hundreds of community-based organizations have been documented to mobilize or engage in landscape-scale ecoagriculture initiatives (e.g., Isely and Scherr 2003; McNeely and Scherr 2003; Molnar et al., Chapter 15, this volume; Bumacas et al. Chapter 16, this volume; Rhodes and Scherr 2005). The institutions leading these initiatives are "hybrids" linking conventional farmer cooperatives, rural development committees, and community-based conservation organizations (Buck et al. 2006). In the Philippines, for example, local farmer-based Landcare groups are linked with conservation organizations, municipal governments, and research organizations to revegetate hillsides, conserve biodiversity in populated protected areas, and improve water quality. An important implication of the central role of communities in biodiversity conservation, especially outside protected areas, is that conservation organizations need to embrace and reorient their role explicitly to support local community stewardship in ways that respect and realistically address the central role of agriculture and livelihoods in planning and implementation methodologies (Bumacas et al. Chapter 16, this volume).

Landscape-Scale Planning and Governance

To achieve objectives at the landscape scale requires a process of collective action to support producers and coordinate action among key stakeholders in the landscape, often across sectors with a historical legacy of distrust. Development and adaptation of institutions for engagement, coordination, and governance of ecoagriculture become the critical challenges. Scaling up and sustaining ecoagriculture landscapes that involve multiple stakeholders require a process, and usually an institution, that will enable multistakeholder assessment, planning, implementation and monitoring for adaptive management. Currently, ecoagriculture initiatives take numerous forms, mobilized by community organizations, public agencies, nongovernmental organizations (NGOs), or national/international projects. Methodologies that have been developed to assist the planning and governance process include landscape "visioning" and "scenario-building" processes; participatory landscape modeling; community biodiversity assessments; and guidelines for "adaptive collaborative management" (Buck et al. 2001, Edmunds and Wollenberg 2001). Multistakeholder trust-building processes and negotiation platforms are being adapted to the specific context of agriculture–biodiversity conflict situations (e.g., Hemmati, Chapter 19, this volume). Key elements include creating a "level playing field," engaging strategies for conflict management, defining an appropriate role for "experts," and having neutral facilitation (Jackson et al., Chapter 17, this volume). Diversity of approaches is expected and desirable, but more systematic and comparative evaluation of effectiveness in achieving sustainable processes and outcomes is still lacking.

Developing Product Markets That Support Ecoagriculture

If ecoagriculture systems are to be widely adopted around the world, then incomes (defined to include not only cash but also other livelihood components) for farmers in those systems need to be at least as high, or higher, than in less biodiversity-friendly production systems, and other nonmonetary benefits will be key. Many ecoagriculture systems are, in fact, more profitable or less risky than alternatives. McNeely and Scherr (2003) presented 28 examples that demonstrated clearly positive economic benefits, and another five cases that had a neutral impact on incomes (despite major benefits for wild biodiversity). But in general, product markets for crop, livestock, and wild products do not recognize or value biodiversity or social benefits. Thus communities are organizing themselves regionally to improve market linkages, reduce marketing costs, and connect directly with buyers. Communities need to understand and meet the quality and time demands of interested buyers and to enter into and respect commercial contracts. They also need technical assistance to improve product quality and manage commercial contracts, and to gain access to finance and credit for postharvest product processing and handling facilities/technologies. Development of innovations at all points in the marketing chain can reduce costs for trading, storage, transport, bulking, grading, and other expenses and thus improve returns from marketing products from polycultures and multi-product landscapes.

New market niches are beginning to develop for agricultural products that are certified to be "green." Producers or products are certified by independent third parties to have positive or neutral effects on biodiversity, based on criteria such as reduced agrochemical pollution, protection of natural areas, use of production practices that do not interfere with key natural processes or species life-cycles, and participation in the development of landscape-scale wildlife corridors (Millard 2007, Chapter 20, this volume). A 2005 review by Ecoagriculture Partners found more than 70 such "green" certification systems, ranging from "salmon-friendly" certification of farms protecting critical stream habitats in the northwest United States to "conservation beef" to Rainforest Alliance-certified commodities in Latin America (Dudenhofer et al. 2004; Fukui 2005).

New markets are also developing for products based on sustainable harvest of wild species or the domestication of wild species (such as extracts, spices, medicinals, construction materials, and fruits) (Leakey, Chapter 5, this volume). The use of marketing labels for agricultural products coming from particular geographic regions, originally focused on quality, culture, and taste, is being adopted for products labeled as supporting conservation of high-biodiversity-value landscapes. Demand is growing as the food industry becomes more sensitive to reputation issues around the environment, and advocates promote new institutional procurement policies (Millard, Chapter 20, this volume).

Paying Farmers and Farming Communities for Ecosystem Stewardship

A major potential driver for ecoagriculture landscapes are payments to farmers or herders/ranchers and their communities for conserving biodiversity important to outsiders, and for conserving other ecosystem services using management practices that also conserve biodiversity. Such compensation currently takes various forms, including payments for access to species or habitats (e.g., research permits; hunting, fishing, or gathering permits for wild species; or ecotourism); payments for biodiversity conservation management (e.g., conservation easements, land leases, conservation concessions, or management contracts); tradable rights under "cap-and-trade" regulations (e.g., wetland mitigation credits, tradable development rights, biodiversity offset credits); and support for biodiversity-conserving businesses (e.g., business investments or ecolabeling of "green" products) (Scherr, Milder and Inbar, Chapter 21, this volume).

The size of payments is already considerable, although their effectiveness in achieving biodiversity objectives and in supporting biodiversity-friendly production systems at the landscape scale is quite mixed. The potential future contribution of these new payments and markets to financing ecoagriculture landscapes will depend on the "rules of the game" and institutional capacities currently being developed (Scherr et al 2007).

Promoting Policies That Support Ecoagriculture Landscapes

Ecoagriculture innovators around the world highlight the need for a more supportive policy environment for ecoagriculture, or at least the removal of major policy barriers (Rhodes and Scherr 2005). Core policy needs at local, national, and international scales are (1) compatibility and coordination of agricultural development and biodiversity conservation policies; (2) environmental legislation that embraces the potentials and rights of farming communities as conservators of biodiversity; and (3) removal of public subsidies for agricultural systems and investments that harm biodiversity.

Consumers, policymakers, and investors are beginning to focus on the link between agriculture and conservation, and responding with new demands on the agricultural system, through systems of voluntary certification, industry standards, and government regulation (see, e.g., SCBD 2005, Decisions III/11, V/5, and VI/5). Ecosystem/landscape-scale programs and projects are being initiated by government agencies and NGOs, often in multistakeholder partnerships, and financed through public budgets (e.g., in India and China) and international development loans (e.g., Fernandes 2004). Initiatives to achieve these policy objectives include the Central America Presidents' joint commitment to promote environmentally friendly agriculture, the removal of agricultural subsidies in Australia, and the recommendations from communities to the Millennium Summit. New political coalitions are being formed to promote integrated cross-sectoral policies, bringing in voices and sectors not traditionally involved in either agricultural or conservation policy, such as municipal governments (in the context of political decentralization), urban consumer groups, international financial organizations concerned with screening investments for environmental sustainability, parts of the food industry, public health advocates, and "good governance" movements seeking to reduce wasteful spending on subsidies.

At the international policy level, ecoagriculture strategies are being integrated into the work programs of the relevant international conventions. For example, the Convention on Biological Diversity (CBD) has adopted a new goal of 30% of agricultural areas under wild flora-friendly management by 2010 (CBD 2006), and will focus on agriculture in meetings during 2008, as will the Commission on Sustainable Development. Rules developing under the World Trade Organization will need to be carefully scrutinized to ensure that they do not disadvantage producers in ecoagriculture landscapes.

Some countries, notably Australia, Brazil, and India, have adopted legislation that explicitly recognizes the rights of indigenous and other local communities to manage and conserve forests and natural habitats (Molnar et al. 2004). The CBD and other international bodies are beginning to focus on opportunities for community-led conservation, although many elements of the conservation community are still uncomfortable directly addressing and supporting agricultural development.

Policy changes can enhance the financial viability of ecoagriculture by removing government subsidies and fiscal incentives for biodiversity-harming production systems, in particular subsidies for agrochemical inputs and water, rules for commodity payment support that limit crop rotations, subsidies that favor annual crops over perennials and intensive livestock production systems over grazing systems, and tax incentives for vegetation clearing.

Major Gaps

Although this survey of ecoagriculture innovation reports numerous promising institutional models—at community, landscape, and policy levels—the conditions under which such innovations are most likely to emerge or can be successfully applied are poorly understood. Rural farming communities are largely unrepresented at most international and national environmental policy forums and organizations, and environmental interests are generally absent from farming organizations. Local organizations find it difficult to access the specialized knowledge generated by others and are poorly integrated into ecosystem or watershed planning and policy processes at local, national, or international levels.

New mechanisms exist to reward and finance biodiversity conservation and biodiversity-friendly agriculture, but most are modest in scale and effectiveness. Little research has been done on the structures and institutions in product market chains that facilitate biologically diverse production. Nor has any systematic assessment been done of overall agricultural investment and finance and how it might be shaped to better support biodiversity. Certification processes need to be expanded, streamlined, and designed for landscape-scale impact and to enable low-income people to participate (Molnar et al., Chapter 15, this volume). Still, most demand in developing countries is from domestic markets seeking lowest-cost supply, so it is crucial to focus on reducing costs across the market value chain (not only at the level of the producer). Finance through carbon emission offsets is the greatest unexploited opportunity, but further technical research is needed to lower costs of organizing landscape-scale action and monitoring performance. The trade-offs and synergies among different ecosystem services for different production and conservation strategies need to be more fully understood and addressed.

Effective cross-sectoral political coalitions have seldom arisen to advocate for reconciling conflicting agriculture and environmental policies. Ecoagriculture strategies are not well integrated into public investment plans, including the Poverty Reduction Strategies of low-income countries and donor strategies designed to support the Millennium Development Goals (MDGs). Most international and national policy and legal frameworks separate action on agricultural productivity, ecosystem management, and rural livelihoods, and policymaking institutions reflect this separation. Most policymakers are unfamiliar with the opportunities for ecoagriculture, or of alternative policies and laws that would enable ecoagriculture activities and outcomes. Mainstreaming ecoagriculture will require that strategically important institutions—for policy, research, and markets—modify how they do business so that they can embrace ecoagriculture visions and strategies.

Conclusion: From Community Action to Global Impact

This book reports many examples of apparently successful approaches linking biodiversity conservation with sustainable agriculture. On the other hand, the current knowledge base and institutional arrangements are clearly inadequate to meet the objectives noted above across diverse ecosystems and production systems. To enable ecoagriculture landscape approaches to expand to a globally significant scale will require at least three elements: new knowledge, institutional capacity, and an enabling policy and market environment.

Produce and Share Knowledge for Ecoagriculture

The challenge of shaping agricultural landscapes to meet joint production, conservation, and livelihood goals will require a dramatic scaling up and refocusing of research in national research systems, the centers supported by the Consultative Group for International Agricultural Research, centers of conservation science, national academies of science, and universities. Priorities are to understand the interaction and dynamics of conservation and production areas; to develop production systems (including improved varieties of more diverse domesticated species) that explicitly meet biodiversity objectives and mimic natural ecosystems; and to make more elements of farming systems ecologically sustainable, including industrial processing, packaging, transport, and so forth. Ecoagriculture systems that appear to be successful need to be fully documented, in terms of both landscape-scale outcomes and specific interventions. Mapping of spatial overlays between important agricultural areas (in terms of national product supply or local livelihoods) and developing parameters for biodiversity will be essential.

Build Capacity for Ecoagriculture

Knowledge innovation systems need to be reshaped to provide services to rural resource stewards and to accelerate exchange of practical knowledge among them and across sectors. Rural communities must be acknowledged and define themselves as key stewards of biodiversity conservation. Conservation organizations, public agencies, and others need to reorient their activities to reflect this reality and provide services to community-based organizations as well as to other stakeholder groups. Conservation organizations need to fully embrace farming partners, develop agricultural expertise, and aggressively advocate for sustainable agriculture investment in coordination with conservation strategies. Negotiations between agents with divergent but sometimes overlapping interests will result in new synergies, catalyzing partnerships between the public and private sector, and across disciplines.

Promote Markets and Policies That Support Ecoagriculture

Technical and local organizational opportunities and initiatives for ecoagriculture are unlikely to be successful unless major policy barriers are removed and supportive policies developed. To advocate for this agenda, beneficiaries of ecosystem services provided by agricultural landscapes, new economic actors in the product value chain, and advocates for reinvigorated rural development need to form new political coalitions. In North America, Europe, Japan, Australia, and many developing countries, shifting of government funds from agricultural commodity subsidies to payments for ecosystem services (including carbon emission offsets) in ecoagriculture landscapes could provide initial funding to build institutions and farmer capacity for ecoagriculture. Ecoagriculture offers cost-effective approaches for national investment strategies to achieve the MDGs. Strategic changes in the food industry, institutional procurement, eco-certification of agricultural products, and financial investors' oversight of agricultural investments can be mobilized to shift financial incentives toward eco-agriculture. At the international policy level, opportunities exist to integrate ecoagriculture strategies into the work programs of the international environmental conventions, and to ensure that rules of the World Trade Organization support ecoagriculture landscapes.

The transformation of agricultural production from one of the greatest threats to global biodiversity and ecosystem services to a major contributor to ecosystem integrity is unquestionably a key challenge of the 21st century. Many elements of ecoagriculture landscapes could help to achieve the critical goals of agricultural sustainability, resilience of food systems, and adaptation to climate change as well as rural revitalization. To realize these potentials, the agricultural and conservation research and policy communities will need to reevaluate and coordinate their priorities and strategies.

References

Altieri, M.A. 1985. *Agroecology: The Science of Sustainable Agriculture.* Westview Press, Boulder, CO.

Beer, J., M. Ibrahim, A. Schlonvoigt. 2000. *Timber Production in Tropical Agroforestry Systems of Latin America.* 21st International Union of Forest Research Organizations (IUFRO) World Congress.

Blann, K. 2006. *Habitat in Agricultural Landscapes: How Much Is Enough? A State-of-the-Science Literature Review.* Defenders of Wildlife, West Linn, OR.

Buck, L.E., C.C. Geissler, J. Schelhas, and E. Wolenberg. 2001. *Biological Diversity: Balancing Interests through Adaptive Collaborative Management.* CRC Press, Boca Raton, FL.

Buck, L.E., J.C. Milder, T.A. Gavin, and I. Mukherjee. 2006. *Understanding Ecoagriculture: A Framework for Measuring Landscape Performance.* Ecoagriculture Discussion Paper no. 2. Ecoagriculture Partners, Washington, DC.

Bunch, R. 2000. *Reasons for Resiliency: Toward a Sustainable Recovery after Hurricane Mitch.* World Neighbours, Oklahoma City, OK.

Cairns, M. and D.P. Garrity. 1999. Improving shifting cultivation in Southeast Asia by building on indigenous fallow management strategies. *Agroforesty Systems* 47(1–3): 37–48.

CBD (Convention on Biological Diversity). 2006. *Decision VI/9: Global Strategy for Plant Conservation.*Secretariat for the Convention on Biological Diversity, United Nations Environment Program. http://www.biodiv.org/decisions/default.asp?dec=VI/9.

Clay, J. 2004. *World Agriculture and the Environment: A Commodity-by-Commodity Guide to Impacts and Practices*. Island Press, Washington, DC.

Coombe, R.I. 1996. Watershed protection: a better way. In: *Environmental Enhancement through Agriculture*, ed. W. Lockeretz. Proceedings of a conference held in Boston, 15–17 November 1995, organized by the Tufts University School of Nutrition Science, the American Farmland Trust, and the Henry A. Wallace Institute of Alternative Agriculture. Center for Agriculture, Food, and Environment, Tufts University, Medford, MA: 25–34.

Dudenhoefer, D., C. Goodstein, and C. Wille. 2004. Paper presented at International Ecoagriculture Conference and Practitioners' Fair, Nairobi, Kenya. September 27, 2004–October 1, 2004.

Edmunds, D. and E. Wollenberg. 2001. A strategic approach to multistakeholder negotiations. *Development and Change* 32(2):231–253.

Fernandes, E. 2004. Ecoagriculture Investment: Lessons from the World Bank. Paper presented to the International Ecoagriculture Conference and Practitioners' Fair, Nairobi, Kenya, September 27–October 1, 2004.

Food and Agriculture Organization (FAO). 2001. *Conservation Agriculture: Case Studies in Latin America and Africa*. FAO, Rome.

Forman, R.T. 1995. *Land Mosaic: The Ecology of Landscapes and Regions*. Cambridge University Press, Cambridge.

Fukui, Y. 2005. A global survey of agricultural ecocertification programs. Unpublished Ecoagriculture Partners, Washington, DC.

Garrity, D. 2006. Science-based agroforestry and the Millenium Development Goals. In: *World Agroforestry into the Future*, ed. D.P. Garrity, A. Okono, M. Grayson, and S. Parrott. World Agroforestry Centre (ICRAF), Nairobi, Kenya:3–8.

Harvey C., F. Alpizar, M. Chacon, and R. Madrigal. 2004. *Assessing Linkages between Agriculture and Biodiversity in Central America: Historical Overview and Future Perspectives*. The Nature Conservancy, San Jose, Costa Rica.

Hole, D.J., A.J. Perkins, J.D. Wilson, I.H. Alexander, P.V. Grice, and A.D. Evans. 2005. Does organic farming benefit biodiversity? *Biological Conservation*. 112, 113–130.

IFOAM (International Federation of Organic Agriculture Movements). 2000. *Organic Agriculture and Biodiversity*. IFOAM, Bonn, Germany.

Isely, C. and S.J. Scherr. 2003. *Community-Based Ecoagriculture Initiatives: Findings from the 2002 UNDP Equator Prize Nominations*. Equator Initiative and Ecoagriculture Partners, Washington, DC. http://www.ecoagriculturepartners.org/documents/reports/CIreportdraft12-23%5B1%5D.pdf.

Jackson, D.L. and L.L. Jackson (eds.). 2002. *The Farm as Natural Habitat: Reconnecting Food Systems with Ecosystems*. Island Press, Washington, DC.

Jackson, L., K. Bawa, U. Pascual, C. Perrings (eds.). 2005. *AgroBiodiversity: A New Science Agenda for Biodiversity in Support of Sustainable Agroecosystems*. DIVERSITAS Report no. 4. DIVERSITAS, Davis, CA. www.diversitas-international.org/docs/Inter.%20 Diversitas.pdf.

Jackson, W. and L.L. Jackson. 1999. Developing high seed yielding perennial polycultures as a mimic of mid-grass prairie. In: *Agriculture as a Mimic of Natural Ecosystems*, ed. R. Lefroy, E.C. Lefroy, R.J. Hobbs, M.H. O'Connor, and J.S. Pate. Kluwer Academic Publishers, the Netherlands:1–37.

Kleijn D., R.A. Baquero, Y. Clough, M. Díaz, J. De Esteban, F. Fernández, D. Gabriel, F. Herzog, A. Holzschuh, R. Jöhl, E. Knop, A. Kruess, E.J.P. Marshall, I. Steffan-Dewenter, T. Tscharntke, J. Verhulst, T.M. West, and J.L. Yela. 2006. Mixed biodiversity benefits of agri-environment schemes in five European countries. *Ecology Letters* 9(3):243–254.

Klein, D.J. and W.J. Sutherland. 2003. How effective are European agri-environment schemes in conserving and promoting biodiversity? *Applied Ecology* 40:947–969.

Leakey, R.R.B. and Z. Tchoundjeu. 2001. Diversification of tree crops: domestication of companion crops for poverty reduction and environmental services. *Experimental Agriculture* 37:279–296.

Lefroy, E.C., R.J. Hobbs, M.H. O'Connor, and J.S. Pate, (Eds.) 1999. *Agriculture as a Mimic of Natural Ecosystems*. Current Plant Science and Biotechnology in Agriculture, vol. 37. Kluwer Academic, Dordrecht, the Netherlands.

McNeely, J.A. and S.J. Scherr. 2003. *Ecoagriculture: Strategies for Feeding the World and Conserving Wild Biodiversity*. Island Press, Washington, DC.

Mollison, B. 1990 *Permaculture: A Practical Guide for a Sustainable Future*. Island Press, Washington, DC.

Molnar, A., S.J. Scherr, and A. Khare. 2004. *Who Conserves the World's Forests? A New Assessment of Conservation and Investment Trends.* Forest Trends, Washington, DC.

Pretty, J. 2005. *Sustainability in Agriculture: Recent Progress and Emergent Challenges*. Issues in Environmental Science and Technology, no 21.

Redford, K.H. and B.D. Richter. 1999. Conservation of biodiversity in a world of use. *Conservation Biology* 13(6):1246–1256.

Rhodes, C. and S. Scherr (eds.). 2005. *Developing Ecoagriculture to Improve Livelihoods, Biodiversity Conservation and Sustainable Production at a Landscape Scale: Assessment and Recommendations from the First International Ecoagriculture Conference and Practitioners' Fair, Sept. 25–Oct. 1, 2004*. Ecoagriculture Partners, Washington, DC.

Ricketts, T. 2004. Tropical forest fragments enhance pollinator activity in nearby coffee crops. *Conservation Biology* 18(5):1262–1271.

Ruark, G., S. Josiah, D. Riemenschneider, and T. Volk. 2006. Perennial crops for bio-fuels and conservation. 2006 USDA Agricultural Outlook Forum—Prospering in Rural America. 1–17 February 2006, Arlington, VA. http://www.usda.gov/oce/forum/2006%20Speeches/PDF%20speech%20docs/Ruark2806.pdf.

SCBD (Secretariat of the Convention on Biological Diversity). 2005. *Handbook of the Convention on Biological Diversity*. SCBD, Montreal.

Scherr, S.J., J.C. Milder, L. Lipper, and M. Zurik. 2007 (forthcoming). *Payments for Ecosystem Services: Potential Contributions to Smallholder Agriculture in Developing Countries.* FAO and Ecoagriculture Partners, Rome.

Siebert, S. 2002. From shade- to sun-grown perennial crops in Sulawesi, Indonesia: Implications for biodiversity conservation and soil fertility. Biodiversity and Conservation 11:1889-1902.

Sutherland, W.J. 2004. A blueprint for the countryside. *International Journal of Avian Science (Ibis)* 146:230–238.

Trewavas, A.J. 2001. The population/biodiversity paradox: agriculture efficiency to save wilderness. *Plant Physiology* 125:174–179.

Uphoff, N., A.S. Ball, E. Fernandes, H. Herren, O. Husson, M. Laing, C. Palm, J. Pretty, and P. Sanchez (eds.). 2006. *Biological Approaches to Sustainable Soil Systems*. CRC Press, Boca Raton.

Van Buskirk, J., and Y. Willi. 2005. Meta-analysis of farmland biodiversity within set-aside land: Reply to Kleijn and Baldi. *Conservation Biology*. 19:967–968.

Weibull, A.-C., O. Ostman, and A. Granqvist. 2003. Species richness in agroecosystems: the effect of landscape, habitat and farm management. *Biodiversity and Conservation* 12(7):1335–1355.

Wittenberg, R., and M.J.W. Cock. 2001. Invasive alien species. *How to address one of the greatest threats to biodiversity: A toolkit of best prevention and management practices.* CAB International, Wallingford, Oxon, UK.

About the Authors

Iskandar Abdullaev, Water Management Specialist, International Water Management Institute, Colombo, Sri Lanka

Fahmuddin Agus, Researcher, Indonesian Soil Research Institute, Bogor, Indonesia

Kwesi Atta-Krah, Deputy Director General, Bioversity International, Rome, Italy

Johann Baumgärtner, Professor, Institute of Agricultural Entomology, University of Milan, Milan, Italy

Sandra Brown, Senior Program Officer, Ecosystem Services, Winrock International, Rosslyn, Virginia, USA

Louise E. Buck, Senior Extension Associate, Cornell University and Director of the Ecoagriculture Partners Landscape Measures Initiative, Ithaca, New York, USA

Donato Bumacas, Chief Executive Officer, Kalinga Mission for Indigenous Children and Youth Development, Gapan, Philippines

Delia C. Catacutan, Natural Resource Management Research Officer and Mindanao Programme Coordinator and LandCare Assistant, World Agroforestry Centre, Malaybalay City, Philippines

Gladman Chibememe, Environmenal Activist, Leader of Chibememe Earth Healing Association (CHIEHA), Chiredzi, Zimbabwe

Thomas S. Cox, Senior Research Scientist, The Land Institute, Salina, Kansas, USA

Maria do Socorro Souza da Mota, Forest Engineer and Technical Advisor, Association of the Rural Extractivist Producers of the Left Margin of the Tapajós (APRUSPEBRAS), Santarém, Pará, Brazil

Lee R. DeHaan, Plant Geneticist, The Land Institute, Salina, Kansas, USA

Aaron Dushku, Program Officer, Winrock International, Forest Management Services, Turner Falls, Massachusetts, USA

Connal Eardley, Bee Systematists & SAFRINET Coordinator, Plant Protection Research Institute, Pretoria, South Africa

Zsuzsanna Flachner, Integrated Project Manager, Hungarian Research Institute for Soil Science and Agriculture, and Agrochemistry, Budapest, Hungary

Thomas A. Gavin, Associate Professor, Department of Natural Resources, Cornell University, Ithaca, New York, USA

Barbara Gemmill-Herren, Global Pollination Project Coordinator, Food and Agriculture Organization of the United Nations, Rome, Italy (the views expressed in this publication are those of the author and do not necessarily reflect the views of the Food and Agriculture Organization of the United Nations)

Gianni Gilioli, Assistant Professor, Ecological Entomology, University of Reggio Calabria, Reggio Calabria, Italy

Andrew N. Gillison, Head, Center for Biodiversity Management, Yungaburra, Queensland, Australia

Jerry D. Glover, Agroecologist, The Land Institute, Salina, Kansas, USA

Kurniatun Hairiah, Lecturer, Soil Science Department, Faculty of Agriculture, Brawijaya University, Malang, Indonesia

Celia A. Harvey, Advisor, Climate Change Initiatives, Conservation International, Washington, DC, USA

Richard Hatfield, Senior Program Design Officer, African Wildlife Foundation, Nairobi, Kenya

Minu Hemmati, Advisor in Sustainable Development, Multi-Stakeholder Processes, and Gender Equity, Brussels, Belgium

Hans Herren, Director-General, Millennium Institute, Washington, DC, USA

Toby Hodgkin, Director of the Global Partnership Programme, Bioversity International, Rome, Italy

Coosje Hoogendoorn, Director General, the International Network for Bamboo and Rattan (INBAR), Beijing, China

Bill Howley, Group Vice President, Forestry, Energy & Environmental Services, Winrock International, Arlington, Virginia, USA

Mira Inbar, student, Haas School of Business, University of California-Berkeley, Berkeley, California, USA

William J. Jackson, Director, Global Programs, World Conservation Union-IUCN, Gland, Switzerland

Devra Jarvis, Senior Scientist, Bioversity International, Rome, Italy

Arvind Khare, Director of Finance and Policy, Rights and Resources International, Washington, DC, USA

Wanja Kinuthia, Entomologist and EAFRINET Coordinator, Department of Zoology, Invertebrate Section, National Museums of Kenya, Nairobi, Kenya

Roger R.B. Leakey, Professor, Agroecology, Agroforestry and Novel Crops Unit, James Cook University, Cairns, Australia

David R. Lee, Professor, Applied Economics and Management, College of Agriculture and Life Sciences, Cornell University, Ithaca, New York, USA

Dino Martins, Museum of Comparative Zoology, Harvard University, Cambridge, Massachusetts, USA

Stewart Maginnis, Head of Forest Restoration Program, World Conservation Union-IUCN, Gland, Switzerland

John Mburu, Senior Research Fellow, Centre for Development Research (ZEF), University of Bonn, Bonn, Germany

Jeffrey A. McNeely, Chief Scientist, World Conservation Union-IUCN, Gland, Switzerland

Jeffrey C. Milder, Ecologist and Land Use Planner, Ph.D. Student, Cornell University, Research Associate, Ecoagriculture Partners, Ithaca, New York, USA

Edward Millard, Senior Manager of Sustainable Landscapes, Sustainable Agriculture Division, Rainforest Alliance, London, UK

David Molden, Principal Researcher, International Water Management Institute, Colombo, Sri Lanka

Augusta Molnar, Director, Communities and Markets Program, Rights and Resources Initiative, Washington, DC, USA

Constance L. Neely, International Education Director, Holistic Management International, Albuquerque, New Mexico

Stefano Padulosi, Senior Scientist, Bioversity International, Rome, Italy

Cheryl A. Palm, Senior Research Scientist, The Earth Institute at Columbia University, New York, USA

Dr. Timothy Pearson, Program Officer, Winrock International, Arlington, Virginia, USA

Ranjitha Puskur, Scientist, International Institute of Water Management, Hyderabad, India

Claire Rhodes, Program Manager, Community Knowledge-Sharing, Ecoagriculture Partners, London, UK

Sara J. Scherr, President, Ecoagriculture Partners, Washington, DC, USA

Götz Schroth, Advisor, Land Use Strategies, Conservation International, Washington, DC, USA

Jan Sendzimir, Researcher, International Institute for Applied Systems Analysis, Laxenburg, Austria

Sandeep Sengupta, Project Officer, Forest Conservation Programme, World Conservation Union-IUCN, Gland, Switzerland

David Shoch, Carbon Markets Specialist, The Nature Conservancy, Charlottesville, Virginia, USA

Rebecca Tharme, focal point, RAMSAR, co-leader, Working Group on Water Resource Management, lead, Agriculture cross-cutting group, International Water Management Institute, Colombo, Sri Lanka

Thomas P. Tomich, Director, Human and Community Development, University of California–Davis, Davis, California, USA

Judith Thompson, Science Writer, Bioversity International, Rome, Italy

Dagmar Timmer, Sustainability Consultant and Founder of Resourceful Solutions Consulting, Vancouver, British Columbia, Canada

Norman T. Uphoff, Professor, Department of Rural Sociology, Cornell University, Ithaca, New York, USA

Meine van Noordwijk, Regional Coordinator, Southeast Asia Regional Research Programme, World Agroforestry Center, Bogor, Indonesia

David L. Van Tassel, Plant Biologist, Land Institute, Salina, Kansas, USA

Sandra J. Velarde, Programme Officer, Alternatives to Slash and Burn Programme, World Agroforestry Centre, Nairobi, Kenya

Bruno Verbist, Research Associate, Division of Soil and Water Management, Department of Land Management and Economics, Faculty of Bioscience Engineering, Katholieke Universiteit, Lueven, Netherlands

Index

Island Press Board of Directors